Perspectives for Agroecosystem Management

Perspectives for Agroecosystem Management

Balancing Environmental and Socio-Economic Demands

Edited by:

P. Schröder

Department Microbe-Plant Interactions
GSF-National Research Center for
Environment and Health
Neuherberg, Germany

J. Pfadenhauer

Vegetation Ecology, Center for Food and Sciences
Weihenstephan, Technical University of Munich
Munich, Germany

J.C. Munch

Institute of Soil Ecology, GSF-National
Research Center for Environment and Health
Neuherberg, Germany

ELSEVIER

Amsterdam • Boston • Heidelberg • London • New York • Oxford • Paris
San Diego • San Francisco • Singapore • Sydney • Tokyo

Elsevier
Radarweg 29, PO Box 211, 1000 AE Amsterdam, The Netherlands
Linacre House, Jordan Hill, Oxford OX2 8DP, UK

First edition 2008

Library of Congress Cataloging-in-Publication Data
A catalog record for this book is available from the Library of Congress

British Library Cataloguing in Publication Data
A catalogue record for this book is available from the British Library

ISBN: 978-0-444-51905-4

For information on all Elsevier publications
visit our website at books.elsevier.com

Printed and bound in the United Kingdom
Transferred to Digital Printing, 2011

Working together to grow
libraries in developing countries

www.elsevier.com | www.bookaid.org | www.sabre.org

ELSEVIER   BOOK AID
            International   Sabre Foundation

Contents

List of Contributors

Albrecht, H., Chair of Vegetation Ecology, Center of Life and Food Sciences Weihenstephan, Technical University of Munich, Freising

Anderlik-Wesinger, G., Chair of Vegetation Ecology, Center of Life and Food Sciences Weihenstephan, Technical University of Munich, Freising

Auernhammer, H., Technical University of Munich, Department of Life Science Engineering, Chair of Agricultural Systems Engineering, Freising-Weihenstephan

Barthel, J., Chair of Vegetation Ecology, Center of Life and Food Sciences Weihenstephan, Technical University of Munich, Freising

Berkenkamp, A., Institute of Soil Ecology, GSF – National Research Center for Environment and Health, Neuherberg

Demmel, M., Bavarian State Research Center for Agriculture, Institute for Agricultural Engineering, Farm Buildings and Environmental Technology, Freising

Ebertseder, T., University of Applied Sciences Weihenstephan (FH), Freising

Gattinger, A., Institute of Soil Ecology, GSF – National Research Center for Environment and Health, Neuherberg

Gayler, S., Institute of Soil Ecology, GSF – National Research Center for Environment and Health, Neuherberg

Gerl, G., Institute of Soil Ecology, GSF – National Research Center for Environment and Health, Neuherberg

Gutser, R., Chair of Plant Nutrition, Center of Life and Food Sciences Weihenstephan, Technical University of Munich, Freising

Hartmann, H.P., Institute of Soil Ecology, GSF – National Research Center for Environment and Health, Neuherberg

Heuwinkel, H., Technical University of Munich, Department of Plant Sciences, Chair of Plant Nutrition, Freising-Weihenstephan

Huber, B., Institute of Soil Ecology, GSF – National Research Center for Environment and Health, Neuherberg

Kainz, M., Chair for Organic Farming, Center of Life and Food Sciences Weihenstephan, Technical University of Munich, Freising

Kühn, N., Chair of Vegetation Ecology, Center of Life and Food Sciences Weihenstephan, Technical University of Munich, Freising

Küstermann, B., Chair for Organic Farming, Center of Life and Food Sciences Weihenstephan, Technical University of Munich, Freising

Lang, A., Institute of Environmental Geosciences, Department of Environmental Geosciences, University of Basel, Switerland

Loos, C., Institute of Soil Ecology, GSF – National Research Center for Environment and Health, Neuherberg

Maidl, F.-X., Technical University of Munich, Department of Plant Sciences, Chair of Organic Farming, Freising-Weihenstephan

Mattheis, A., Chair of Vegetation Ecology, Center of Life and Food Sciences Weihenstephan, Technical University of Munich, Freising

Mayer, F., Chair of Grassland Science, Center of Life and Food Sciences Weihenstephan, Technical University of Munich, Freising

Munch, J.C., Institute of Soil Ecology, GSF – National Research Center for Environment and Health, Neuherberg

Noack, P., Geo-konzept gmbh, Gut Wittenfeld, Adelschlag

Palojärvi, A., MTT Agrifood Research Finland, Environmental Research/ Soils and Environment, Jokioinen

Pfadenhauer, J., Chair of Vegetation Ecology, Center of Life and Food Sciences Weihenstephan, Technical University of Munich, Freising

Priesack, E., Institute of Soil Ecology, GSF – National Research Center for Environment and Health, Neuherberg

Reents, H.J., Chair for Organic Farming, Center of Life and Food Sciences Weihenstephan, Technical University of Munich, Freising

Rothmund, M., Technical University of Munich, Department of Life Science Engineering, Chair of Agricultural Systems Engineering, Freising-Weihenstephan

Ruser, R., Institute of Plant Nutrition (330), University of Hohenheim, Germany

Schloter, M., Institute of Soil Ecology, GSF – National Research Center for Environment and Health, Neuherberg

Schmid, H., Chair for Organic Farming, Center of Life and Food Sciences Weihenstephan, Technical University of Munich, Freising

Schmidhalter, U., Technical University of Munich, Department of Plant Sciences, Chair of Plant Nutrition, Freising-Weihenstephan

Schröder, P., Institute of Soil Ecology, GSF – National Research Center for Environment and Health, Neuherberg

Sehy, U., Institute of Soil Ecology, GSF – National Research Center for Environment and Health, Neuherberg

Sommer, M., Institute for Soil Landscape Research, Leibniz-Centre for Agricultural Landscape Research (ZALF), Müncheberg

Sprenger, B., Chair of Vegetation Ecology, Center of Life and Food Sciences Weihenstephan, Technical University of Munich, Freising

Weber, A., Chair of Plant Nutrition, Center of Life and Food Sciences Weihenstephan, Technical University of Munich, Freising

Wehrhan, M., Institute for Soil Landscape Research, Leibniz-Centre for Agricultural Landscape Research (ZALF), Müncheberg

Weinfurtner, K., Fraunhofer Institute for Molecular Biology and Applied Ecology – IME, Division Applied Ecology, Schmallenberg

Weller, U., Soil Physics, Helmholtz Centre for Environmental Research, Halle (Saale)

Zipprich, M., GSF – National Research Center for Environment and Health, Institute of Biomathematics and Biometry, Neuherberg

Preface

During the last century, agricultural production in Western Europe developed from little scale farming oriented on regional characteristics and site specificities into high technology and maximum yield level based systems. Farms that had formerly employed sophisticated crop rotations coupled with animal production moved into two-crop-systems or monocultures and transfers of protein rich feeds for animal production. The application of synthetic chemicals, fertilizers and pesticides, and the use of modern equipment led to high production on more productive soils and to food security. High power machines enabled deeper tillage and improved soil management, as well as economical use of resources, but also less employment in rural areas. In view of nutritional value and quality, food reached highest standards.

However, the intensively used landscape showed serious limits of a maximum yield and maximum income agriculture. Chemical contamination of water, air and soil, soil erosion by water and wind, and a drastic decrease of biodiversity in uniform production systems led to rapid deterioration of the environmental quality of agroecosystems. The trophic state of rivers increased from nitrogen and phosphorus inputs. Residues of nutrients and pesticides, often applied in high doses, were leached into other natural ecosystems, into aquatic systems and finally to the ground water aquifer and the marine ecosystems. Climatic relevant trace gazes were emitted to the atmosphere in amounts proportional to the high doses of nitrogen fertilizers. Numerous plant and animal species disappeared after the change from the traditional low tech system to the high tech land use in Central Europe. And furthermore, contamination of food with pesticide residues and nitrate lead to doubts about the healthy value of this food production.

It was in the seventies of the 20th century when the first thoughts arouse to minimize these negative effects on the environmental quality in Germany. Impact of chemicals was quantified and legislation started to regulate the application of pesticides and of organic residues like slurries. However, the economic and political constraints for farmers, the needs of societal development for more food security were decisive. A first active answer on the management scale was the establishment of organic farming by some farmers during the last decades and, more extended in some regions, an adoption of integrated farming which minimizes chemical and technical impact as far as possible in a sustainable way. Yet, the real effect of land use change on species richness, on soil erosion, groundwater quality and trace gas emission was hardly investigated. New knowledge was necessary toward the understanding of ecosystems, including all scales from the arable field to the entire landscape. The challenge was to understand the functioning of agricultural ecosystems, farms, landscapes, to understand these biological systems under human steering, to understand the whole impacts of cultivation and interrelationships with adjacent ecosystems in order to establish a sustainable use of agricultural soils. Designing landscapes with high food quality and minimization of impacts at all scales would realize high yield to assure food security for the future world population. At the same time, such new management systems had to support the economic survival of farms and farmers.

In 1989, scientists of the GSF – National Research Center for Environment and Health (Friedrich Beese, now at University of Göttingen) and of the Technical University of Munich (Gerhard Fischbeck, Jörg Pfadenhauer, Udo Schwertmann, among others) created an interdisciplinary research network for the investigation of ecosystem processes in a real agricultural landscape, FAM (= Forschungsverbund Agrarökosysteme München; Munich Research Association for Agricultural Ecosystems). The aim of FAM was to work out instruments and to develop strategies for a sustainable agricultural land use. Three central hypotheses were: If land is cultivated according to the principles of sustainability and ecological compatibility,

1. adjoining systems will not be exposed to excessive quantities of carbon, nitrogen and phosphorus compounds and foreign organic substances (xenobiotics),
2. the diversity of plants, animals and microorganisms as well as of their communities will be greater, and rarer species will also be able to establish sustainable populations,
3. the economical use of the resources required for the production of food will be increased, and the quality of the products will be maintained at a high level.

FAM comprised a variety of scientific disciplines that had rarely been in cooperation before, so agronomy, plant and animal production, agro-techniques, soil science, population biology of plants and animals, socio-economics as well as mathematics, geography, modelling and more. In a concerted action, the German Federal Ministry of Sciences and Education and the Bavarian Ministry of Science, Research and Arts assured the financial support of this research network.

A farm, large enough and suitable for the FAM was rented from the Scheyern Benedictine Monastery about 40 km northwest of Munich in the district of Pfaffenhofen/Ilm. The Monastery is situated in the Tertiary Hill Landscape (part of the South German Molasse Basin) and was installed as the Scheyern Experimental Station. This farm had formerly been managed according to the local rules of maximum yields, with an effective application of synthetic chemicals and being exposed to hilly typical erosion and leaching losses. It was now adapted and optimized toward environmental sound agriculture. Two farming systems were established, an organic and an integrated one. Organic Farming applies the principle of closed cycles of nutrients and resources within the farm. Avoiding external energy supply (in the form of mineral fertilizers and pesticides) is postulated to improve and maintain the quality of biotic and abiotic resources. Integrated Crop Cultivation aims at avoiding erosion and soil consolidation by crop rotation with cover crops, application of chemicals according to the pest threshold principle, use of wide tyres, and a reduction of soil tillage. Studying these farming systems should allow understanding transformation processes and the development of soil fertility as well as matter fluxes, so of water and solutes at scales from the field to ponds, on the farm and to the catchment area, from the soil surface to the ground water. Besides, it should clarify interrelationships leading to losses of floral and faunal biodiversity on one hand and the needs for their reestablishment on the other hand. The entire area was redesigned in order to minimize soil erosion and to establish more diversity in the landscape, so by fallows, hedges and more structural units in the landscape, mostly oriented toward environmental functions (retention areas for eroded material, habitats for flora and fauna and more).

A dense 50×50 m sampling grid was established to allow monitoring of changes. New techniques of positioning and remote sensing were adapted to agricultural scale for establishment of yield maps and analysis of yield parameters. GPS-positioning of harvesters occurred worldwide for the first time in Scheyern. Remote sensing equipment, partially mounted on tractors, was adapted for soil and stand analysis, for terrain analysis and topography assessment. The basis of precision agriculture was developed. At the same time, lab scale methodologies especially molecular tools and approaches as well as stable isotope probing were employed to analyse environmental samples and processes, especially with regard to their biological functions.

The project demonstrated the possibility to maintain a highly productive agriculture together with lower environmental impacts on the basis of knowledge and technologies; it showed also options for further amelioration of organic and integrated farming practice.

Results were discussed with farmers and led to new management procedures on the farm, some being adopted also by farmers in this region. Impacts of the application of the new technical systems were analysed by several scientific domains of the project.

We are pleased to thank to the multitude of persons involved in this long-term research, to thank to the authors of the chapters of this book, of this latest effort to aggregate a multitude of results, knowledge and conclusions from the project FAM. The results of our joint project were published in details in 500 scientific papers, 60 PhD-Theses and presented on numerous international scientific meetings of all scientific research domains. Annual reports were distributed to decision makers and stakeholders.

An important product of the Scheyern project is the worldwide best-documented agricultural area terrain on the farm and landscape level as well as the corresponding farming systems. This documentation will provide the basis for newly needed environmental research considering the actual developments in land use toward renewable resources.

Striving to present in broad and aggregated form the multiple results and overall conclusions of the Scheyern project, we hope that the present book may be instructive in view on appropriate use of agricultural ecosystems. It may also deliver basic knowledge to further perspectives and future developments of agroecosystem management.

The Editors

Munich, in January 2007

Part I

Approaching Sustainable Agriculture

Chapter 1.1

Outline of the Scheyern Project

P. Schröder, B. Huber, H.J. Reents, J.C. Munch and
J. Pfadenhauer

1.1.1 Background

Agricultural practices have undergone intense changes over the last decades which are evidenced by significant advances in technology and mechanisation, specialisation of agricultural undertakings, discontinuation of labour-intensive farming branches, increased field size, drainage of wetlands and the removal of hedges and boundary strips. In addition, more and more yield-raising production means (fertiliser, pesticides and fuel) have been used, tillage treatment has intensified and the import of animal fodder and fertilisers has increased. Further, an increase in livestock per hectare of land has resulted in higher amounts of manure per hectare, thus polluting the groundwater. Parallel to this development, there is a decrease in available work force, and farmers and their families are increasingly employed in areas outside of agriculture. Moreover, agricultural buildings and farmyard sites are utilised for other purposes and agricultural road networks are enlarged. The urbanisation of former farming villages, as well as the sealing of surfaces, has increased; flowering

Perspectives for Agroecosystem Management
Edited by P. Schröder, J. Pfadenhauer and J.C. Munch

meadows are now hardly found, whereas 'standard' green areas are expanding. In addition, agricultural research and advisory services have been for a long time principally oriented at enhancing production and lowering costs and expenditure of human labour. The negative consequences of these developments are known: increased erosion, occurrence of fertilisers and pesticides in groundwater, soil compaction from heavy machinery, depletion of fauna and flora due to large crop fields and destroyed habitats, and an inappropriate mass livestock husbandry. Animal production concentrates in certain regions, with the consequence of an uncoupling of the production of feed and the generation of high amounts of waste and pollution of soils and groundwater. The fact that most of the livestock is heavily dependent on medication causes further threats to end users and the production basis.

The insight that agricultural productivity and sustainability must be unified is slowly catching on. Agriculture has to nourish the population and provide plant and animal resources for secondary industry processing. Farming assures the economic existence of individual farms and the rural community by utilising land for effective agricultural production. This provokes conflicts with other land use demands such as industrial, transportation and housing development, the production of drinking water or preservation of natural resources for recreation and wildlife refuges. In the national and global exchange systems, Germany is among the world-leading net import nations of food, in spite of having fertile soils, favourable climate and surplus agricultural production in some areas. Only these food transfers from abroad allow agricultural production below yield optima, a high rate of meat consumption and the loss by impermeabilisation of 900 ha of arable land every week (which could nourish approximately 2000 people; Bundesamt für Bauwesen und Raumordnung, 2002).

It is logical that all forms of agriculture cause changes in the balances and fluxes of the pre-existing ecosystem, thereby limiting self-regulatory ecosystem (resiliency) functions. The intensive agriculture of the past, with its strong reduction of landscape structures and vast decoupling of energy and matter cycles, has caused stress and degradation of the production basis which has already led to a loss of these resiliency functions in many regions; massive influence has been also exerted on neighbouring compartments. This has resulted in the well-known problems of pesticide loads, high phosphate loads to surface waters via over-fertilised soils or erosion as such. To overcome the economic, social and political inadequacies leading to ecological degradation, the demand for sustainable agricultural management needs to be transposed into knowledge-based practical instructions and political regulations on a regional scale. Thus, applied research for a sustainable and ecologically compatible land use aimed at sufficient food production is ever so important.

Leading up to the 1980s, most agroecosystem research had been concerned with yield effects and yield optimisation on a field scale and with matter dynamics in small watersheds. This approach gave short- to mid-term results from implemented land use systems. Therefore, as the *FAM Munich Research Network on Agroecosystems (Forschungsverbund Agrarökosysteme München)* was founded in 1989, the discussion regarding ways to minimise environmental damage caused by agricultural production lacked a fundamental understanding of long-term ecosystem processes and their interaction with economic and technical compulsions. A favoured solution at that time – from an ecological perspective – was the reduction of agricultural intensity by either setting aside land or implementing organic farming or extensive agriculture. At the same time, and becoming increasingly important, integrated farming systems have been suggested as a means towards sustainable and resource-conserving agriculture.

Understanding ecosystems as functional units that contain organisms interacting with each other and the abiotic environment, agricultural systems have to be regarded and classified as specific ecosystems.

Ecosystems might be analysed and classified according to their dominating processes, specific structures or functions (Figure 1). Contrary to natural systems, human activities interfere in agrosystems with almost all structural elements and processes in order to enhance productivity, secure yields and foster selected species – all to supply food and energy to a food web outside of the agricultural ecosystem.

Research concept

In 1989, a group of scientists from the Munich area gathered to tackle the burning problems of intensive agriculture and to propose novel solutions for agroecosystem management. They formed the core

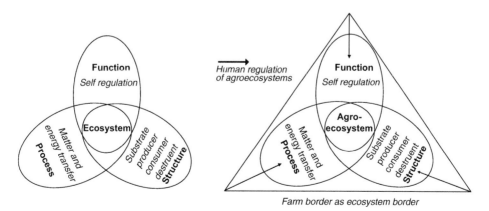

Figure 1 Ecosystem and agroecosystem – Human influence.

group of the *FAM Munich Research Network on Agroecosystems*. The overall aim was to establish guidelines for an ecologically compatible, site-adequate and sustainable management of rural landscapes, while maintaining high productivity levels. To acquire such guidelines, an extensive research concept was elaborated, which comprised the following points:

- Methods should be developed which (1) are sensitive enough to record and quantify early-stage, area-specific changes in the system; (2) are precise enough to allow for long-term assessment of developments; (3) can be carried over to other systems; and (4) also permit economical evaluations of occurring changes to the ecosystem and production sites.
- Changes should be documented, assessed, evaluated and prognosticated.
- Methods for the restoration of damaged ecosystems should be developed.
- Environmentally sound and sustainable land use strategies should be developed and transferred to agricultural practices.
- Models should be developed to aid the future development of agroecological activities.
- Changes should be recorded, assessed, forecast and evaluated.

This agroecosystem-oriented, interdisciplinary research aimed to record, assess, forecast and evaluate the time course of the environmental and socio-economic impacts of management-induced changes on different structural and scale levels (Table 1).

Table 1 Experimental levels studied by the FAM Munich Research Network on Agroecosystems.

Level	Description
Landscape	Tertiary Hills, district of Pfaffenhofen/Ilm (400 km^2), southern Bavaria, Germany
Landscape section	Research Station Scheyern (150 ha)
Farm	Organic farming (68 ha), integrated farming (46 ha)
Field	Between 2 and 7 ha on the Research Station Scheyern
Plot	16–600 m^2
Model ecosystem	Exposure chambers, monolith lysimeter, pure-air green house, model ponds (0.1–10 m^2)
Laboratory systems	Cultures of organisms, soil aggregates, minirhizotrons, microcosms (0.01–0.1 m^2)

The 150 ha former cloister estate Scheyern was selected as the area under investigation. The research station is situated in the Tertiary Hill slopes, a typical Bavarian agrarian landscape, demonstrating all problems associated with intensive agricultural use. Following a 2-year inventory phase, the research station was reorganised and subdivided into two farming systems: organic farming and integrated farming. To understand changes at field and farm levels, of the water dislocation to ponds and groundwater, and in the development of biodiversity, a total run time of 15 years was defined as the project duration. The farming systems should be self-reliant, economical sustainable and competitive and act as a normal commercial farm in the region.

FAM hypotheses

The FAM defined three central hypotheses (Table 2). These hypotheses called for a holistic approach, including information on farming, energy and matter fluxes in the agroecosystem as well as on adjoining ecosystems. Instruments sensitive enough to detect and quantify early-stage, site-specific system changes with sufficient accuracy to forecast future development and spatial distribution needed to be developed. Furthermore, it became increasingly important to set up instruments which could be transferred to other systems, as well as allow for an economical and ecological evaluation of encountered changes.

Table 2 Three central hypotheses of the FAM Munich Research Network on Agroecosystems.

If land is cultivated according to the principles of sustainability and ecological compatibility …	Hypothesis
… adjoining systems will not be exposed to excessive quantities of carbon, nitrogen compounds and xenobiotics.	I. Land use can conserve or re-establish ecosystem control functions.
… the diversity of plant, animal and microbial organisms and their communities will increase; rare species will be able to establish sustainable populations.	II. Land use can conserve or re-establish habitat functions.
… the economical efficiency of resources required for food production will be increased, and the quality of products will be maintained at a high level.	III. Land use can conserve or re-establish the economical and ecological productivity.

1.1.2 Research station and project phases

FAM study area

Research Station Scheyern (Figure 2) was leased for a time period of 15 years, divided into several phases: inventory phase (1990–1992), landscape redesign (fall/winter, 1992/1993), project phase (1993–1998) and future project phase (1999–2005).

The FAM project was carried out at Research Station Scheyern, a 150 ha cloister estate located 40 km north of Munich in southern Bavaria. Scheyern is situated at an altitude of 445–498 m above sea level in the Tertiary Hills, a landscape demonstrating typical problems associated with intensive agricultural use, such as erosion, soil compaction, groundwater contamination, impoverishment of flora and fauna and having only few existing hedges and fallow strips.

Climate and soil characteristics

Average annual precipitation in the Scheyern area is 803 mm, and the mean annual temperature is 7.4°C. Research Station Scheyern represents the boundary between the loess–loam clay ridge and the loess–loam sand ridge tertiary landscape (clay contents varying from 90 to 450 g kg^{-1}), whereby the tertiary hilltops and eroded slopes are of sand–gravel–clay composition. However, approximately 85% of the research station is covered

Figure 2 Aerial view of Research Station Scheyern. (For colour version of this figure, please see page 421 in colour plate section.)

by a thin (<2 m) loess–loam or by loess deposits, and the area is divided by three main valleys, two of which bear a line of ponds.

Landscape history

The Tertiary Hills, located between the pleistocene moraines of the alpine glaciers and the Danube River, have coarse- and fine-grained deposits originating from the upper sweet water molasses and are largely covered by thin quaternary loess layers. The landscape is characterised by asymmetrical valleys shaped by uneven loess deposition, solifluction and erosion. The natural habitat is primarily composed of woodruff, oak and beech forests. Agricultural activities in this area taking place since the younger Stone Age and, in particular, the intensive agriculture of the last decades have resulted in a marked reduction of forests, hedges, lynchets and buffer strips. The Bavarian Tertiary Hills represent one third of Bavaria's agricultural region and demonstrate the typical pressures of intensive agriculture: compaction, erosion and overdressing, low nutrient efficiency and subsequent pollution of surface and groundwater, as well as impoverishment of flora and fauna.

1.1.3 Two farming systems

The organic farming system

The goals of the organic farm were to establish closed nutrient and resource cycles, on a 68.5 ha area having predominantly low-sorption soils, following the principles of organic farming according to *Naturland* and *Bioland*, which are members of the German Association for Organic Farming (AGÖL).

A seven-field crop rotation was set up on 31.5 ha arable land with (1) lucerne-clover-grass-meadow (fixing N over 1.5 years, harvested as forage); (2) seed potatoes with mustard intercropping; (3) winter wheat with cover crop; (4) sunflowers (oil) with undersowing of lucerne-clover-grass-meadow; (5) lucerne-clover-grass-meadow as forage; (6) winter wheat with white clover undersowing; and (7) winter rye with undersowing of lucerne-clover-grass-meadow. The organic farm also had a 95-head cattle herd for meat production, 25 ha of grassland and 3.5 ha set-aside land meant for succession.

Ecological aims were achieved by (1) banning mineral fertiliser and pesticides; (2) minimising external energy and matter supplies, damages to fauna and flora and matter exports to surface and ground water; and (3) optimising N cycling in the crop rotation. Tillage intensity was reduced to a level in which weed control and soil conservation efforts were balanced; ploughs were only used when necessary. Wide tyres and use of combinations reduced both frequency and impact of vehicular traffic. Mulching was implemented; cover crops and underseeds with diverse species conserved the soil, controlled weeds and enhanced habitat diversity. An important factor in organic farming was the selection of crop varieties having a broad resistance against pests and which were competitive to weeds. Especially in the case of potatoes, resistance against *Phytophtora infestans* and Colorado beetles represented a decisive factor. Manure was applied at appropriate soil and weather conditions and immediately incorporated to minimise NH_3 volatilisation losses.

Integrated farming system

Integrated farming system was established on 46 ha having well-buffered soils. The arable land (30 ha) was cultivated with a four-field crop rotation with cover crops: (1) winter wheat; (2) potatoes; (3) winter wheat; and (4) maize. The remaining area was used as grassland (1.8 ha) and fallow land (8.8 ha). Potatoes and wheat were grown as cash crops; maize was used to feed 45 fattening bulls living in neighbouring farms. The slurry was brought back to the fields.

To achieve sustainability, integrated farming reduced tillage frequency and intensity; moldboard ploughs were not used. Harrowing and chiselling were favoured methods in Scheyern. Wide tyres and combinations minimised soil compaction and erosion. Cover crops, wheat and maize stubble mulch protected the soil surface and enhanced soil faunal and microbial activity. Further tillage was avoided by direct and no-till planting in mulch of the previous crop or cover crop. Leaching was reduced by using crop varieties with high ability to compete for nutrients, water and light. Crop varieties exhibiting adequate resistance to most important diseases were preferred. Fertilising strategies were optimised to the plants' needs, and pesticides were used not for prevention (pest threshold principle) but only when necessary.

Inventory phase (1990–1992)

The FAM project began with an inventory phase. During the 2-year preliminary period, the arable land (Figure 3) was uniformly cultivated with winter wheat (1991) and spring barley (1992), according to the principles of conventional farming. The aim of this phase was to record differences between the various sites at the research station and to create uniform starting conditions for subsequent project phases. A grid of 600 measuring points (50 m $\times$ 50 m) was established throughout the experimental sites, allowing an inventory of various conditions and parameters. Information obtained at these grid points was used to create a database (http://www.gsf.de/FAM/adis.html) used by the FAM research groups and for geoinformation systems (GIS). In addition, modern mathematical methods such as geostatistics and pedotransfer functions facilitated the generation of detailed maps of the entire research station.

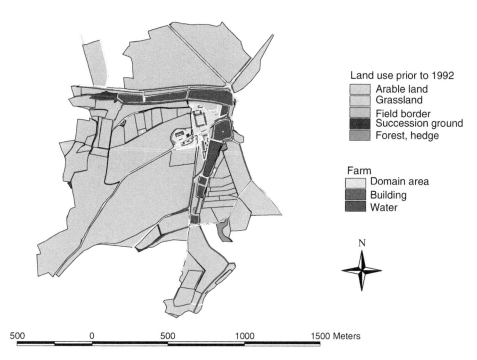

Figure 3 Research Station Scheyern before land use redesign in 1992 (from Schröder et al., 2002). (For colour version of this figure, please see page 421 in colour plate section.)

Landscape redesign (fall/winter, 1992/1993)

Following the inventory phase, in fall 1992, partitioning of arable and grassland was redefined, considering aspects of nature and resource conservation (Figure 4). Plot size was reduced, and forest edges, hedges, field boundaries and fallow grounds were created. As countermeasure against erosion, buffer zones were created along brooks and ponds and on slopes, and grassland was established in the river valley. The landscape was redesigned to (1) avoid fertiliser and pesticide input into water bodies; (2) minimise erosion, water losses, soil compaction and fertiliser and pesticide overlapping on headlands; (3) enable more site-specific farming; (4) reduce effort and expenses required for agricultural purposes; (5) enhance the net income and aesthetic value of the farms and the biodiversity (fauna and flora); and (6) improve the recreational function of the landscape.

Furthermore, the central part of the research station (farmland) was divided along the main watershed into two farms: organic and integrated, each striving for ecological and economical sustainability (Figure 5). At the northern part of the research station, a 39 ha plot was further subdivided into experimental plots to conduct detailed studies on management-induced changes (Figure 6): A: arable land, W: grassland, F: set-aside land, Ö: organic

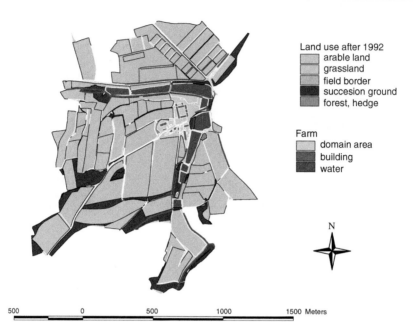

Figure 4 Research Station Scheyern after land use redesign in 1992 (from
Schröder et al., 2002).(For colour version of this figure, please see
page 422 in colour plate section.)

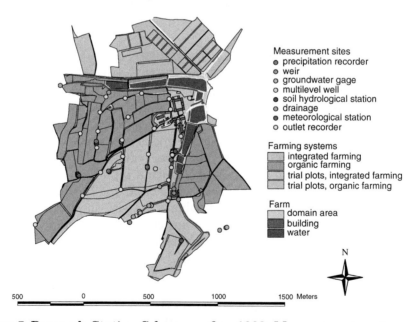

Figure 5 Research Station Scheyern after 1992. Management systems and
permanent monitoring equipment are shown (from Schröder et al.,
2002). (For colour version of this figure, please see page 422 in
colour plate section.)

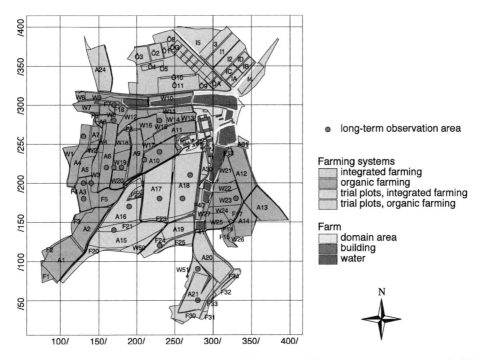

Figure 6 Research Station Scheyern after 1992. Management systems, field
numbers, grid system and long-term observation areas are shown
(from Schröder et al., 2002). (For colour version of this figure,
please see page 423 in colour plate section.)

farming, trial plots, I: integrated farming, trial plots. The grid co-ordinates
(given as: xxx/yyy) allowed to identify each point at the research station.

The field size was reduced to cut erosive slope lengths and facilitate
site-specific farming. Fallow strips (between 3 and 15 m wide) were re-
introduced between fields (field margins) and along adjoining ecosystems
(surface water and woods), to increase wildlife refuges and buffer capacity.
Hedges were planted with indigenous species or implemented as dead
wood pilings. Dams were erected to stop run-off at field borders. Steep
arable land was transferred into grassland or set aside for succession (permanent fallows). Grassland seeds were enriched with meadow species, and
grazing intensity was reduced at very wet or dry sites. A review of the project
phase between 1990 and 1992 is provided in Tenhunen et al. (2000).

Project phase 1993–1998

After landscape redesign, different managing approaches of the organic
farm and the integrated farm began, and trial plots in the northern part of
Research Station Scheyern were set up for detailed studies (Figures 5 and 6).
Permanent monitoring was possible by broad-measurement equipment;
processes were analysed causally. About 30 subprojects dealt with several

topics, e.g. soil properties and erosion, water and matter fluxes, crop plants, security of yield, plant nutrition, xenobiotics, diversity of flora and fauna, economic aspects, social acceptance, modelling and forecasting. In this phase, the main focus was to evaluate the economical and ecological changes induced by the different management systems. To reach this aim, indicators were sought and validated. In short, our investigations led to the following results, concerning the three main hypotheses:

Hypothesis 1: Land use can conserve or re-establish ecosystem control functions

The adapted management helped to reduce the transfer of C and N compounds as well as xenobiotics to adjacent systems. Through conservation tillage (direct drilling, mulching, reduced axle loads) coupled with the new plot design, soil loss was reduced and the soil pool of organic N in microbial biomass was increased, in spite of reduced N fertilisation. Emissions of N_2O from arable land remained high. Nitrate and phosphate transfers to surface and ground water were significantly reduced. In the studied aquatic ecosystem, high nitrate and phosphate pools facilitated a high biomass macrophyte production. This biomass production buffered 50% of phosphate and 30% of nitrate input.

Another stress from former land use is atrazine, which is still detected in some ground water and deposition samples, even though application has been banned for several years. Of the applied pesticides, 0.1–0.2% of the dose were detected in surface waters after 1–2 weeks following application. With sufficient rain, a rapid movement in the soil was observed. Conservation tillage improved infiltration capacities. As a side effect, the amount of interflow (60–180 cm below ground), which is a typical phenomenon in the Tertiary Hills, and matter transport via interflow increased.

Adapting and improving the geophysical procedures of ground-penetrating radar and electromagnetic induction enabled the mapping of mean water and clay contents in a high spatial resolution ($5 \, m \times 5 \, m$). These data formed the basis for the validation of spatially extrapolating models and precision farming.

Hypothesis 2: Land use can conserve or re-establish habitat functions

Changing the production system – according to the principles of a sustainable and ecological compatible agriculture – helped increasing biological diversity of plant, animal and microbial communities. For the soil fauna, clear reactions to the changed management were indicated by an altered dominance structure of collembola and an increase in earthworm biomass. The reduced pest control efficiency in organic farming by the banning of chemical pesticides led to an increase in the abundance and species diversity of weeds. An increase in abundance was also observed in the integrated farm resulting from reduced tillage (increase in the surface diaspore pool and increased germination). In the integrated farm, increased competition by weeds was successfully encountered with an adaptation of herbicide applications. This in turn led to a significant decrease in rare and endangered species. An increased biodiversity was observed for several faunal groups.

Hypothesis 3: Land use can conserve or re-establish the economical and ecological productivity

It was verified that sustainable and environmentally friendly land use requires sparing employment of natural resources. Especially, the resources soil and water were significantly saved by the changed land use methods. At the same time, the input of energy per ton of wheat was diminished, although weed control efforts increased dramatically. Therefore, both production systems were shown to be economically sustainable.

Energy inputs in the integrated farm were highly influenced by mineral fertilisation. N-fertiliser efficiency was increased by improving application strategies. Yield mapping with the aid of GPS revealed a high spatial variability. Further improvements in N efficiency with respect to these differences in yield potentials would be possible by utilising precision farming.

Project phase 1999–2003

In the project phase 1999–2003, management of the organic and integrated farms continued as before. Monitoring of processes and conditions, recording of landscape redesign-induced changes and causal analysis of processes were also further carried out. The integrated farming system was optimised by precision farming. The establishment of indicators for sustainable land use was an essential requirement for the evaluation methods. Last, models were to be further developed so as to enable forecasting of processes up to the landscape level.

The *FAM Munich Research Network on Agroecosystems* was a co-operation between the Center of Life and Food Sciences, formerly known as the Agricultural Faculty of the Technical University of Munich in Freising-Weihenstephan, and the GSF – National Research Center for Environment and Health in Munich-Neuherberg. Approximately 30 scientific groups collaborated on this project. The German Federal Ministry for Education and Research provided funding for the FAM project, as a centre of ecosystem research. The Bavarian State Ministry for Research and the Arts contributed the overhead costs and for the agrarian management of Research Station Scheyern. The GSF – National Research Center for Environment and Health in Munich-Neuherberg and the Technical University of Munich in Freising-Weihenstephan participated with their own financial resources. FAM was part of the international ecological research program 'Man and the Biosphere' (MaB) of the UNESCO (United Nations Educational, Scientific and Cultural Organization). Thanks to U. Weller (UFZ, Leipzig) for conceptualising the maps of Research Station Scheyern.

References

Bundesamt für Bauwesen und Raumordnung, 2002. In Kampe D, Ed., Positionspapier – Handlungsschwerpunkte von Raumordnung und zur langfristig vorbeugenden Hochwasservorsorge, Bonn.

FAM database: http://www.gsf.de/FAM/adis.html.

Schröder P, Huber B, Olazábal U, Kämmerer A, Munch JC, 2002. Land use and sustainability: FAM Research Network on Agroecosystems. Geoderma, 105, 155–166.

Tenhunen J, Lenz RJM, Hantschel R, 2000. Ecosystem Approaches to Landscape Management in Central Europe. Part III: Investigations in an Agricultural Catchment in the Tertiary Hills of Southern Germany. Springer-Verlag, Berlin, Heidelberg, New York. Ecological Studies, vol. 147, 173–269.

Chapter 1.2

Sustainable Land Use by Organic and Integrated Farming Systems

H.J. Reents, B. Küstermann and M. Kainz

1.2.1 Organic farming system in Scheyern

Principles

The design of organic farming systems and their cultivation measures are based on the knowledge of the structural elements and functional relationships in natural ecosystems. Existing relationships are supported and strengthened under the ecological approach with the target to make food production efficient. The design of the farming system in Scheyern mirrors various principles of natural ecosystems. Food cycles and mass fluxes were considered in crop rotation patterns and livestock keeping. The input of materials from industrial processes was avoided or kept on a very low level. Inputs, vital for growth, had to be provided by biological processes (e.g. N_2 fixation, mobilisation of minerals). Means for the control of pathogens as well had to be of natural origin. Diversity schemes in crop production were designed to contribute to the conservation of regulatory cycles, especially with a view to pathogens and weeds. However, the

Perspectives for Agroecosystem Management
Edited by P. Schröder, J. Pfadenhauer and J.C. Munch

temporal and spatial dominance of some species (cash crops) may seem contradictory to this objective, but it is necessary for food production. A special criterion is the recognition of an intrinsic value of animals and thus the organisation of livestock keeping in a way that would safeguard their requirements. The mentioned principles of ecological farming were implemented by agronomic measures (Auerswald et al., 2000). Design and utilisation of agro-ecosystems were adapted to the tolerance and productivity of the given sites (natural, human and economic resources). Part of this adaptation was the establishment of crop rotations with an essential share of legumes, which brings atmospheric nitrogen into the system. The resulting N and C accumulation should help maintain or even raise the humus content in the soil. Above this, crop rotations combined with tillage were able to control weed infestation and plant health. To keep soil compression as low as possible, it was controlled by crop rotation and appropriate machinery (e.g. wheel load). Only farming systems with livestock and own forage production are able to fully implement the principle of nutrient cycling and exploit it with high flexibility. Stock management and feed composition were organised with attention to the natural behaviour of the animals and their feed preferences. Diseases and pests had to be handled exclusively by measures of system management or by substances issued from natural or biological processes. Purposeful farmscaping was to supplement the ecological measures on the operated area (Altenweger et al., 1998; Furchtsam et al., 1995, 1996; Gerl et al., 1999; Gerl and Kainz, 1997, 1998a, 1998b; Hofmann, 2005; Kainz et al., 2001, 2002; Reents et al., 1999; Weller et al., 2000; Weller and Kainz, 1999).

Structural design of the organic farm in Scheyern

The total area of the farm Scheyern was subdivided between the two farming systems with regard to the water catchment borders. Organic farming was executed on the more sandy and non-arable sites (grassland), as typical for the region (too steep or wet and close to waters). To follow the principles of site-specific farming with nutrient cycling, the grassland was used as pasture. Mother cow keeping with 'baby beef' production was set up including grazing in summer and in-house feeding in the winter months. Keeping dairy cattle, which would have fit the farm structure much better, was not possible for organisational reasons. To implement the principle of nutrient cycling, livestock was of central importance. During the grazing period, a short cycle developed with the chain pasture–animal–dung–pasture. Nutrient export due to increasing stock numbers was relatively small; however, the behaviour of the animals led to a redistribution of matter within the grazed area. In arable farming, ruminant keeping requires forage legumes in the crop rotation as precondition for N and C

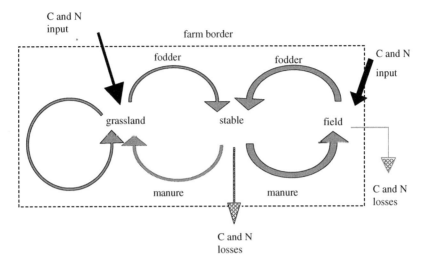

Figure 1 C and N fluxes in the organic farm in Scheyern.

supply to the field. The wider mass flux on farm level included the production of forage on fields and grassland and its transport to the stable. Via the excreta output in the stable period in winter, when feed preserves were given, nutrients and organic matter were brought back to the cropping areas (Figure 1).

Even when publications about organic farming often refer to this mass flux on farm level as closed mass cycle, additional matter is supplied by the plants (N_2 by legumes, CO_2 fixation, mineral mobilisation by plant roots), or losses occur in husbandry (weight gain of animals for the market, N and C loss from stable dung) and the soil (i.e. nitrate leaching, N_2 by denitrification). Therefore, management measures had to be oriented on a long-term extension of the mass cycle without increasing losses, thus supporting and even improving the productivity of the farm.

Crop rotation pattern

After the decision had been taken in favour of a farming system with livestock keeping, the essential element for implementing ecological principles by management measures was the design of the crop rotation. The crop rotation pattern had to meet numerous demands which can be summarised as follows: It was necessary to provide the feed basis for livestock. The input of farmyard manure and the cultivation of legumes were the basis for corresponding yields of cash crops. The succession of root crops, cereals and green manure crops led to an alternation of humus degradation and accumulation. This alternation favoured soil conservation and plant diversity. Last but not least, the call for the provision of a broad variety of foodstuffs for the population was to be satisfied. In view of the outlined criteria, the following rotation pattern for the main crops was designed and implemented in the years 1992–1994: *Lucerne-clover-grass mix–potato–wheat–sunflower–lupin–wheat–rye*. This crop rotation pre-supposed an optimal N_2 fixation on both legume fields, together with the grown catch crops. The lupin fields, however, had to be abandoned because of increasing development of anthracnosis, which lowered the productivity and favoured strong weed growth with additional volunteer sunflowers. So this crop rotation field was substituted in 1995 also with lucerne-clover-grass. Table 1 shows the final crop rotation (*lucerne-clover-grass mix–potato–wheat–sunflower–lucerne-clover-grass mix–wheat–rye*) with an evaluation of the ecological and economic target values.

The decisive criterion for evaluating the sustainability of arable farm-ing is the nitrogen cycle (Figure 2) of the crop rotation in combination with livestock keeping. The main source of nitrogen supply was N_2 fixation in the two rotation fields under lucerne-clover-grass. The fixed nitrogen was transferred to the stable via the N accumulated in the soil on the one hand and, to the larger extent, via the forage on the other; from there as solid and liquid manure to the subsequent crops in the rotation. The nitrogen fixed in catch crops was transferred to the subsequent crops in the rota-tion mainly via the humus pool in the soil. The nitrogen gain from the atmosphere was counter-effected by the export of N in cereals, potatoes and animal products sold on the market. In addition to this, losses occurred in the stable or during the spreading of organic fertiliser, mainly in gaseous state as NH_3 and N_2, or in the field in form of nitrate after intensive tillage. Both loss potentials can be reduced by improved man-agement measures (Rühling et al., 2005).

Tillage

The objective of sustainable and soil-conserving management requires, in addition to crop rotation, soil preparation in such a way that the best preconditions are provided for both plant growth and soil conservation (Figure 3).

The establishment of lucerne-clover-grass sown under rye or sunflow-ers allowed to omit one basic tillage operation to each crop. The purpose was to achieve a better biological structure of the soil. Earthworms had more time to create biopores and found an increased food supply in form of crop residues at the surface.

For potatoes, wheat after clover-grass and winter rye, ploughing was carried out to a depth of 26–28 cm. Thus, the root layer was loosened and provided a maximum root–soil contact. Simultaneously, organic matter was incorporated into the soil (lucerne-clover-grass or stable manure), and

Table 1 Crop rotation of the organic farm beginning in 1994/1995 and the agronomic, ecological and economic target values.

Year	Main crop	Intercrop	Target values
1	Lucerne-clover-grass		Forage production, N_2 fixation, weed control, humus formation, subsoil developing, soil conservation
2	Potatoes	Sowing of mustard into the potato crop, early August till harvest in September	High sales revenue, reduction of deep-rooted weeds, nitrate uptake, soil and groundwater protection, easier harvesting
3	Winter wheat	Legume mixture	Sales revenue, straw for bedding, weed control, N_2 fixation, humus formation, soil conservation
4	Sunflowers	Undersown lucerne-clover-grass	Sales revenue, K uptake from the subsoil, feedstuff (pressed cake), soil conservation, early yield of forage
5	Lucerne-clover-grass		Forage, N_2 fixation, humus formation, weed control, subsoil opening, soil conservation
6	Winter wheat		Sales revenue, straw for bedding
7	Winter rye	Undersown lucerne-clover-grass	Sales revenue, straw for bedding, weed control, nutrient decomposition, early and increased productivity of forage crops, soil conservation

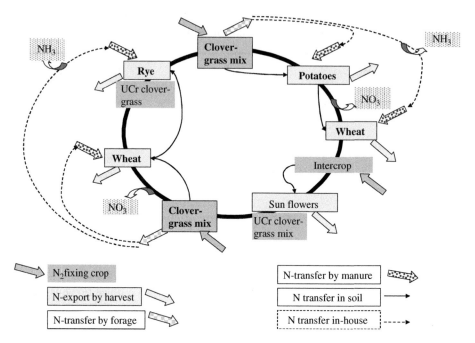

Figure 2 Essential N fluxes within the crop rotation und at interfaces to livestock keeping in organic farming in Scheyern.

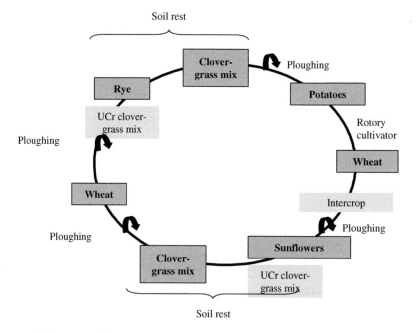

Figure 3 Tillage in the crop rotation.

competitive weed growth was reduced. To avoid an incorporation of not picked up potatoes, which would occur as volunteers in the successive crop, soil preparation between potatoes and wheat was made with the rotovator without turning the soil up.

Crop-related cultivation technologies – Lucerne-clover-grass

Main targets of lucerne-clover-grass cultivation were forage yield and N_2 fixation; secondary targets were weed regulation, humus formation, erosion control, and subsoil opening. The seed mixture of the three main components lucerne, clover and grass allowed to make use of their specific needs of growth and of the different ecological responses. Under the heterogeneous site conditions in Scheyern, this helped stabilise crop yields; with 17.6%, the variation coefficient was lower than in the other crops of the organic crop rotation. In drought periods, for example, particularly lucerne on sandy sites was still able to develop enough biomass, whereas for the other crops (except for some grass species), water supply was not sufficient. In wet periods and on wet sites, however, clover and most grasses grew better than lucerne. The lucerne-clover-grass mix was used as overwintering crop, usually sown under rye in spring. Undersowing was generally successful. Beginning in 1998, the forage mixture was already sown together with rye in autumn. This improved plant establishment, mainly of the grass component. In some years, however, rye tended to lodge, and the undersown crop developed excessively. This was favourable for lucerne-clover-grass, but outgrowth occurred in rye. After the harvest of rye cover, one or two cuts of the forage were taken, in the main harvest year even three. When the mixture was sown under sunflowers, the forage population developed weaker than under rye due to more shading and the later harvest of the sunflowers. No problems were recorded with the sunflowers. The forage crop could not be used in the sowing year because of the sunflower stalks; they had to be mulched in autumn or winter. Then the undersown crop used to grow better than in case of resowing in spring, when especially lucerne reached maturity rather late. When three cuts per year were taken on average, 79–131 dt DM ha^{-1} were harvested (mean level: 99 dt DM ha^{-1}) (Table 2). With underseed, forage yields of 16 (minimum) to 53 (maximum) dt DM ha^{-1} or 30 dt DM ha^{-1} on average were achieved after the cereals had cleared the field, which demonstrates the importance of this cropping technology for the forage output.

The lucerne-clover-grass mix was ploughed under as close as possible before the successive crop, that is, immediately before wheat sowing in autumn or before potato planting in winter or spring. For the purpose of site-specific tillage, operations on heavy soils with the tendency to lump formation were postponed to the winter (January), and on light soils with a tendency to leaching to spring.

Table 2 Yields (dt ha^{-1}) and coefficient of variation (CV) of the main crops
in the organic farming system in Scheyern.

	Lucerne-clover-grass main crop	Lucerne-clover-grass undersown	Potatoes	Winter wheat	Spring wheat	Rye	Sunflower
Average	98.9	30.3	238.8	37.3	31.5	37.7	26.9
Median	101.3	27.9	205.0	35.2	33.4	36.9	26.0
Maximum	131.0	53.0	342.0	58.9	35.6	65.8	33.0
Minimum	78.8	16.5	137.1	21.3	25.0	28.5	19.8
CV (%)	15.9	36.1	28.6	27.0	14.5	26.7	18.1

Potatoes

Main targets of potato cultivation were maximum yields and high revenues from the sale of seed potatoes; secondary target was controlling deep-rooted weeds. Potato varieties were selected according to the demand in the market, supported by own experience gathered in running projects on potato growing. The nutrient supply was guaranteed by the pre-crop lucerne-clover-grass, supplemented by about 300 dt ha^{-1} stable manure. The seed potatoes were pre-germinated and planted as early as possible, mostly not before the end of April. Weeds were regulated mechanically in two to four operations with a roller hoe. Depending on prognosis and infestation, Cu preparations were applied against late blight (*Phytophthora infestans*) and Bt (*Bacillus thuringiensis*) against the Colorado beetle. Since the production target were seed potatoes, the crops were routinely checked for virus attacks; seriously infested plants were eliminated. Till the year 2001, the foliage used to be cut at the end of July and beginning of August, to restrict the size of the seed tubers and to avoid *P. infestans* spores to be washed in. In addition to this, till 2001 mustard was inter-sown in order to absorb nitrate from the soil, to cover the ground and thus protect it from erosion, and also to suppress photophilic problem weeds such as common chickweed (*Stellaria media*), to crumble the soil and to produce green biomass. Harvesting took place not before early September to give the tubers enough time for developing sufficient firmness of the skin. After the machine harvest, the seed tubers were stored in pallet boxes in a computer-controlled storehouse and later, in the winter months, graded and bagged. Throughout the years, the yields varied between 167 and 342 dt DM ha^{-1} (maximum); the mean yield was 239 dt DM ha^{-1} (Table 2). Thus, they ranked on the level of the potato yields obtained in organic farms in the years 2000–2002 However, it must be remembered that here seed potatoes were propagated, which requires that plant growth be stopped at a defined size of the tubers. The annual yield differences were caused mainly by the weather conditions. In wet and cool years, yields tended to be below average, and in warm and dry years above average.

Wheat

Main targets of wheat cultivation were high yields and sales revenues for seed propagation; secondary target was straw for animal bedding. Wheat varieties were selected according to demand, experience gathered in own experiments, rating in other tests and the 'list of varieties'. Wheat occupied two positions in the crop rotation, after potatoes (root crop) and after annual lucerne-clover-grass. In some years, winter wheat had to be substituted by spring wheat due to inappropriate soil conditions at the moment of sowing (wetness). After potatoes, the field was worked with the cultivator (non-turning soil preparation, to let remaining potatoes freeze in the winter) and then wheat was sown. The lucerne-clover-grass mix as pre-crop was ploughed under just before sowing. Depending on sowing time and soil conditions, 350–420 germinable grains were sown per square metre. Weed control was performed with a spring tine weeder. One operation was scheduled in autumn and one or two in spring; the exact processing depended on the conditions in the given year. Intersowing white clover together with the last combing operation as done in the first years was given up in favour of an intensive mechanical control of couch grass after harvest. The yields of winter wheat varied between a minimum of 21.3 dt ha^{-1} in 1994 and a maximum of 58.9 dt ha^{-1} in 1996; averaging the years, a yield of 37.3 dt ha^{-1} was achieved (Table 2). Spring wheat, which was grown five times on different fields throughout the years, reached 31.5 dt ha^{-1} on average. Thus, yields corresponded to the level cited in the 'agricultural reports of the federal government'. Since in organic farming, crop rotation is the decisive element for controlling the yielding capacity, it was especially interesting to compare wheat yields after different pre-crops. It turned out that the mean yields throughout the years were by 10% increased after lucerne-clover-grass compared with potatoes as pre-crop (Table 3); also was the variance slightly lower (CV). Thus, in single cases, lucerne-clover-grass can be regarded as the more favourable pre-crop, but its position in the crop rotation must finally be determined on the basis of the overall performance of the system.

The production target 'higher-than-average yields' was reached in some years only; usually, the mean level of organic farming was reached. Seed material, however, brought higher prices than wheat as cash crop for consumers' nutrition.

Table 3 Wheat yields (dt ha^{-1}) and coefficient of variation (CV) after the pre-crops potatoes and lucerne-clover-grass in the years 1994–2004.

maximum	Pre-crop potatoes	Pre-crop lucerne-clover-grass
average	36.1	40.0
median	33.9	37.7
maximum	57.9	61.9
minimum	18.1	23.9
CV (%)	32.9	28.5

Rye

Main targets of rye cultivation were high yields and sales revenues for seed propagation till 2001; secondary targets were straw as bedding material, cover crop for undersown lucerne-clover-grass, weed suppression, nutrient decomposition, and soil conservation. Varieties were selected according to demand and rating in comparative variety trials. The cultivation technology was elaborated step by step under the guidance of the research association for agro-ecosystems, the FAM. At the beginning, sowing was performed in early October, as usual in the region. The forage crop was undersown in spring. After some years of project running, sowing was moved up to the first decade in September with simultaneous intersowing of the undercrop. This method favoured the pre-winter development of rye and a reliable growth of the undercrop. There was, however, a higher risk of lodging. In 2004, yields were extraordinarily high (65.8 dt ha^{-1}). In other years they ranked between 40.8 and 28.8 dt ha^{-1}; the mean yield was 35 dt ha^{-1} and thus only slightly lower than in case of winter wheat (Table 2).

Sunflowers

Main targets of sunflower cultivation were oil yield for human nutrition and thus high sales revenues; secondary targets were flowering in the crop rotation, an aesthetic landscape effect and K uptake from the subsoil. Sunflower varieties were selected according to rating in the 'list of varieties'. The seed rate was 7.5–8 grains m^{-2}, sown by mid-April. In the beginning, the inter-row distance was 37.5 cm (uniform seed spacing), later 50 cm and finally 75 cm, to provide longer penetration of light and, resulting from this, a better development of the undercrop. Weed management shifted from two harrowing and two hoeing operations to harrowing before the emergence of the sunflowers with simultaneous intersowing of the undercrop. At the moment of emergence, field borders were treated with burnt lime to prevent slugs from immigrating. Dock plants were removed by hand. Grain yields varied between 19.8 and 33.0 dt ha^{-1}, the mean was 26.9 dt ha^{-1} (Table 2).

Livestock and grassland management

The concept of organic farming in Scheyern, i.e. matter cycling, N supply and site-specific use of fields, could only be successful in combination with cattle husbandry. For breeding, a herd of Simmenthal cows was established. A German Angus bull was purchased for crossing to get calves with higher meat yield and meat quality. The calves were sold as baby beef with about 450 kg live weight. In the first years, as typical for the region, the animals were kept on pasture in summer and in-house in winter, where mainly silage from the own forage fields was fed. The housing system was an open stable with deep bedding for resting and a non-roofed paddock with feeding table. In the course of the FAM experiment, cattle keeping was adjusted to the possibilities offered by the system, which means increased forage output thanks to changes in the crop rotation (lupins replaced by a second lucerne-clover-grass field) allowed to raise stock numbers and to intensify feeding. In 1996, the animals were divided into two herds: (1) the herd of mother cows with their calves of 1/2 to 3/4 years, which in the vegetation period were kept exclusively on grazing and in winter in the stable, and (2) the herd of weaning calves which in summer were sent out to pasture for hours and in winter housed in a separate stable with deep bedding, paddock and outside feeding until they had gained the final fattening weight of 600 kg. On January 1, 1999, the herd of breeding cows included 29 cows, 1 heifer and 17 calves; the mean age of the cows was 7.8 years with a span from 3 to 13 years. By the end of 2000, 32 mother cows and 26 calves were kept. The herd of weaners for fattening comprised 45 animals.

Feeding and grassland productivity

In summer, the herd of breeding cows was kept on rotational grazing on 28 different meadows. Plant cuts not needed for grazing were preserved for winter feed in form of silage and hay. The grassland yields were determined on the basis of the forage cuts taken for ensiling and hay making on the one hand and animal performance (subsistence need and meat gain) on the other. With 68 and 74 dt DM ha^{-1}, respectively, averaging the years (Table 4), yields ranked on the normal level of the site; the recorded deviations resulted mainly from varying precipitation. When we consider the yield level, it must be remembered that the grassland got no fertiliser, which means that plant regrowth was mainly stimulated by the share of legumes and on pasture by the return of nutrients in excreta and urine from the grazing animals.

Table 4 Mean dry matter yields (dt DM ha^{-1}) and coefficient of variation (CV) of grassland in the organic farming system.

	Pasture	Meadow
Average	62.2	64.9
Median	61.8	67.5
Maximum	71.7	84.7
Minimum	48.2	47.1
CV (%)	9.0	20.2

1.2.2 Integrated farming system in Scheyern

Principles

Integrated farming comprises cropping methods and other agricultural production techniques which fulfil both ecological and economic demands. Suitable methods of agronomy and crop production are to be harmonised in compliance with site specifics. The farmer has to adjust his management measures concerning variety selection, crop rotation, cultivation technology, plant nutrition and plant protection to the natural environment. This also includes an optimal soil conservation, for example, by environment-friendly management systems and purposeful fertilisation and pest control. At the same time, it must be excluded that groundwater and surface waters as well as adjacent biotopes become polluted by matter input; typical landscape elements are to be safeguarded. The selection of pesticides depends on the degree of ecological tolerance and their application on economic considerations (http://www.bauernhof.net, 2006).

Structural design of the integrated farming system in Scheyern

The integrated farm was established as arable farming system with cattle fattening on a very limited grassland area. Farms of this type with all their typical problems such as handling of liquid manure or the cultivation of maize are frequent in the region. Integrated Farm Scheyern was established in view to reduce negative impacts on the environment and to find possibilities for the optimisation of the operational result with the goal of high yields. There was no cattle fattening within the farm boundaries; it was simulated by corresponding mass fluxes. Cattle fattening was assumed on the basis of feeding maize silage, farm-produced cereals and the external purchase of protein-supplemented feed. The animals were supposed to stand on slatten floors as common. This allowed storing the collected liquid manure up to 6 months before spreading it on the fields in spring or summer. The roughly 25% maize in the crop rotation corresponded to 45 bulls per year and to a calculated output of liquid manure of about $18\,\mathrm{m^3\,ha^{-1}\,a^{-1}}$ arable land. Crop rotation in such a system has to satisfy the following demands: The forage output for cattle fattening should be sufficient and of high energy content. Good profit should be gained from other cash crops. Since mainly the row crops maize and potatoes were able to fulfil these two criteria, cultivation had to be oriented on soil and environment protection to the highest possible extent. Resulting from this, the following crop rotation was established (Table 5).

The principles of the farm management design and the objective of the crop rotation were to be implemented by the following agronomic and cultural measures. For reducing erosion in winter wheat cropping, attention was paid to an optimal sowing time and to sufficient pre-winter development of the plants. Non-turning tillage should leave the straw on the ground as long as possible to reduce water flow-off. Prior to the row crops maize and potatoes with their high erosion potential, a cover crop had to provide a protective layer on the field in the juvenile phase of the crops. Fertilisation as well had to be adapted to the site-specific and agronomic conditions. Thanks to the preceding management measures in the former

Table 5 Crop rotation of the integrated farm and related agronomic, ecological and economic target values.

	Main crop	Intercrop	Target values
1	Maize		Basic feed supply for cattle fattening, energy for the feed ration
2	Wheat	Summer ridges for potatoes with Intersowing of mustard	Sales revenue via yield and quality, erosion control in the potato crop of the following year = dam stability
3	Potatoes	Intersowing of mustard in summer after clearing the potato foliage	High sales revenues, soil conservation, nitrate uptake
4	Wheat	Mustard as undercrop	Sales revenue via yield and quality, erosion control in winter and spring, after maize sowing

years and a good storage capacity of the soils, P and K were sufficiently available. Liming, however, was necessary. N fertilisation had to be coordinated with the system of reduced tillage, this involved partially higher N doses than in a system with ploughing. The continuously declining S immissions made sulphur supplies to wheat necessary. The application of fungicides had been planned in such a way that the given yield potential could be guaranteed. Spraying rates were reduced by combining them with liquid fertilisers. Residual weed plants were tolerated if problematic species (cleavers, creeping thistle) were thoroughly controlled. The technological objective was a largely soil-conserving passage on the ground, executed by the combination of different machine operations with wheels as broad as possible and low pressure of the tyres (Altenweger et al., 1998; Furchtsam et al., 1995, 1996; Gerl et al., 1999; Gerl and Kainz, 1997, 1998a, 1998b; Hofmann, 2005; Kainz et al., 2001, 2002; Reents et al., 1999; Weller et al., 2000; Weller and Kainz, 1999).

Crop-related cultivation technologies – Wheat

Main targets of wheat cultivation were high yields and revenues from seed propagation. Varieties were selected according to demand and rated according to 'list if varieties'. Tillage for wheat sowing was guided by the idea of soil conservation with pre-crop–related decision of the single measures (Table 6). The basic implements were cultivator, rotary harrow and seed drill.

Desirable was a sowing date (first decade of October) that allowed the plants in most years to reach EC 23 in autumn. Thus, the soil surface

Table 6 Cultivation of winter wheat after different pre-crops.

	After potatoes	After maize
Tillage	Directly before wheat sowing, in case of early potatoes directly after harvest	Directly before wheat sowing
Criteria of tillage	Loosening wheel tracks without turning the soil, maximum depth 10–12 cm, levelling	Loosening wheel tracks without turning the soil, levelling
Ground coverage	Intersowing mustard into the ridges before harvest left plant residues on the ground surface (maximum 20% ground coverage).	Mais stubble mulching, remaining stalks should be incorporated (Fusarium re-infection), ground coverage by root residues

became rather resistant against erosion, and overwintering was more successful with a prolonged tillering phase and better ear differentiation. In years, when the weather conditions did not allow sowing in autumn, spring wheat was grown. The seeding rate varied depending on the date between 250 and 320 (400) germinable kernels per square metre, this quantity was increased by 10% when poor emergence was expected. Besides the current demand on the market, the following criteria were important for variety selection: (1) broad resistance spectrum, especially to *Fusarium* and ear diseases and (2) yield components: single-ear type, rather long and stable. From 1993 to 1997, *Atlantis* and then *Petrus* were grown. The exceeding of threshold values was rather likely for *Septoria tritici* and *Drechslera tritici-repentis*. The seed material was dressed with a preparation that acted also against dwarf bunt. Liquid cattle manure, an all-nutrient fertiliser produced by the farm system, had to be integrated into the fertilisation scheme despite its difficult handling. Particular attention was paid to spreading the liquid manure at lowest ammonia loss and towards the highest possible use efficiency of the crop. According to the management scheme, $22 \, m^3 \, ha^{-1}$ had to be applied to wheat. The spreading required a ground with good trafficability in order to carry the corresponding axle loads. Frosty days were avoided to evade damages of the wheat plants by cauterisation. In the late stage of tillering, nitrogen was to be provided via liquid manure. After spreading, cool and moist weather was desirable to keep NH_3 emissions low. Liquid manuring involved the risk of covering the weed plants up, which might reduce the efficiency of herbicide applications within a space of up to 3 weeks. By the time, the following approach turned out to be optimal: The first N dosis was given as mineral fertiliser by about mid-March, the second as ammonium-urea solution together with herbicides (reduced application rate), and roughly 1 week later liquid manure, but not later than to EC 30, distributed by a 15 m drag hose spreader. The ammonium N contained in the liquid manure was used by the crop to 100%; the total nitrogen appeared in the balance sheet to 100% after the deduction of losses (model calculations). Mineral fertiliser to wheat was given mainly in form of nitrogen (Table 7). Plant protection measures were sometimes combined with the application of low Mg or Mn doses, depending on the weather situation. The dosage of N fertiliser was determined in view of the expected removal. In the juvenile stage (to EC 30), fertilisation was carried out very carefully to avoid excess inputs which might lead to uncontrollable plant growth. As from EC 31, dosage was oriented on the site- and weather-specific yield optimum. Uptakes of 120–$200 \, kg \, N \, ha^{-1}$ by the plants required a supply of 130–$190 \, kg \, N \, ha^{-1}$.

Active agents were used which could control the existing weed flora and, combined with AU fertilisation, tolerated reduced application rates. On some subfields, cleavers (*Galium aparine*), creeping thistle (*Cirsium*

Table 7 Mean N rates to winter wheat on the example 1999, (kg N ha^{-1}).

	Stage	Fertiliser type	kg N per hectare
First N	Early vegetation	AU solution	40
Second N	Tillering	Liquid manure	33
Third N	EC 31/32	AU solution	50
Fourth N	EC 39	AU solution	50
Total			173

arvense) and, beginning in 2000, bindweed (*Convolvulus arvensis*) in their later growth stages were fought, depending on the need. To a certain degree, grass weeds (wind grass) and other weeds were generally tolerated. The cultivated crops were comparatively stable, medium high with a long distance between flag leaf and spike. Nevertheless, lodging resistance was stabilised with CCC, and side shooting was supported. As common, 0.5 l ha^{-1} CCC was applied to EC 21–25 and 0.3 l ha^{-1} to EC 30–31. The second dose of 0.3 l ha^{-1} CCC was applied irrespective of temperature. It should retard the main shoot and induce a uniform development of side shoots. The grain moisture content at harvest was always below 15.5%. Between 1993 and 2004, wheat yielded 69.6 dt ha^{-1} on average with a range from 40 to 88 dt ha^{-1} (Table 8). When we ignore the low yield in 2003, which was caused by the special long summer drought, the yield trend throughout the test period was rising. Besides enriched experience and knowledge on farm management, a certain adjustment of the site to non-ploughing soil-conserving tillage may also have attributed to this trend. In the integrated system as well, wheat was sown after two different pre-crops, potatoes and maize. A 10-year comparison showed that the mean wheat yields after potatoes exceeded those after maize by 10%; the yield stability as well was increased (Table 8). Extreme levels in the crop rotation were recorded after maize. The reasons for the superiority of potatoes as precrop might be the good aeration and mixing of the soil during the tuber harvest despite rather soil-conserving seedbed preparation. Deficiencies of maize are the increased production of high-C root and plant residues, less soil aeration by and after harvest and a higher pressure of fungi on the subsequent wheat population.

Straw was regarded as valuable humus resource (>50% of the humus supply). It was chopped (<8 cm) by the combine in the field and left there. As cover material or after shallow incorporation, it helped reduce erosion and offered food for earthworms, thus contributing to a favourable soil structure. Despite these measures, straw manuring could not compensate the humus balance. Soil preparation before mustard sowing as pre-crop

Table 8 Wheat yields in the period 1994–2004 (dt ha^{-1}) and coefficient of variation (CV), differentiated for the pre-crops potatoes and maize.

	Pre-crop potatoes	Pre-crop maize
Average	76.0	68.4
Median	75.9	65.9
Maximum	89.2	97.0
Minimum	61.0	51.3
CV (%)	12.2	19.5

was performed with a cultivator by not later than August 20, to guarantee sufficient plant development before the winter.

Maize

Main targets of maize cultivation were high mass and energy yield for bull feeding. In some years, silage maize was substituted by grain maize for experimental reasons. Then, the main target was maximum grain yields. Secondary target was soil and groundwater conservation. Varieties were selected according to the 'list of varieties'.

Maize as row crop in the Tertiary Hills with slopy fields and partially high percentage of silt in the soils involved a high risk of erosion. Above this, necessary herbicide applications in the early growth stages represented a serious danger of xenobiotics input into surface water and groundwater. Therefore, 'soil and groundwater conserving cultivation' as secondary target was a special criterion for the cultivation technology. Soil preparation for maize sowing already took place in the summer of the preceding year and included a shallow incorporation of straw and one passage of the cultivator (see the subsection 'Wheat'). After this, the mustard stand was established whose main purpose was to protect the soil: root development in autumn, ground coverage and protection against erosion under maize in spring. Maize was sown into the (frost-killed) intercrop without additional soil preparation using a slotted precision drill (for example, from F.A. Becker, Kleine). Soil preparation and sowing technique were aimed at reaching >50% soil coverage. The frosted mustard plants formed the upper layer, below which residues of the wheat straw from the preceding year were present. The target could not be reached in all years because the mustard plants had not always reached the required length in autumn. Sowing was performed at a soil temperature of >8°C in a depth of 5 cm, in moist soil rather shallow, in dry soil 4–5 cm deep. The seed material had been dressed with Thiram and Mesurol (against frit fly and phytophagic pests). In view of the defined targets, variety selection required special attention of certain growth characters influencing yield formation (total DM, MJ NEL ha^{-1}), stable resistance and share of cobs. The fact that the crop rotation included two wheat links made resistance to *Fusarium* a further important criterion in the variety selection. Our approach was based on the assumption that maize removes from the soil about 220–250 kg N and 50–60 kg P. In some years, about 30 m^3 manure ha^{-1} were given to the pre-crop before maize (this is equal to 90 kg total N). In the following spring, about 120 kg N was given as mineral fertiliser. There were also years when no liquid manure was given to the pre-crop at all. In such cases, liquid manure was spread in maize up to a plant height of 60 cm using a high-clearance tractor. Under the normally growing mustard pre-crop, shatter grain and unsown weeds emerged, which made it necessary to apply a total herbicide in spring. The general goal was to keep the populations free of weeds from the 3-leaf to the 10-leaf stage. Mechanical measures were not used. Control measures against fungus attack and insects were not necessary. Harvesting took place with self-driven choppers which blow the chopped material to accompanying trailers. The resulting soil compaction had to be considered in soil preparation for wheat. Initially, it was planned to grow silage maize for bull fattening. Because of the fact that these bulls were not kept in the regarded farm, in some years different reasons led to the change in the production target in favour of grain maize (experiments, production technique). Grain maize yielded from 70 to 111 dt ha^{-1} with 87 dt ha^{-1} on average. The yields of silage maize varied over the regarded period from 737 to 327 dt ha^{-1} with a mean yield of 493 dt ha^{-1} (Table 9). It becomes evident that the yields of silage maize varied much more than the grain yields; nevertheless, the former were clearly rising throughout the regarded period, which, similar to the situation in cereal growing, can be explained by the adaptation of the site to improved agronomic aspects.

Potatoes

Main targets of potato cultivation were high sales revenues from high yields and seed production. Secondary target was soil and groundwater conservation. Varieties were selected according to demand.

The local conditions for potato growing are difficult in Scheyern because of the very high erodibility on the slopy areas. On some fields, clay levels are far too high for potatoes. Therefore, the guiding principle in all activities was consequent soil conservation. The risk of potato cropping in slopy regions is erosion. Planting potatoes parallel to the slope may reduce erosion at the beginning. In case of strong rainfall, however, ridges may break and favour gully erosion. Therefore, the cultivation technology was aimed at building the ridges as stable as possible and thus to clearly reduce the risk of erosion. To reach the above-described target, ridging was

begun already in the summer of the preceding year. After a phase of optimisation, the set task was implemented on the basis of the following
scheme. Tillage and ridging were performed with a ridge-forming cultivator developed in Scheyern which loosens the soil to a depth of 22 cm in the
first operation; then large-size discs throw up the dams which are shaped
by special dam-shaping blades. In a second step, mustard seeds were distributed over the ridges by a sowing equipment attached on the ridging
cultivator. The cultivator tines moving over the ridge were turned upward,
and the space between the dams was loosened a second time. Initially, the
green mustard plants favoured the protection of the dams in autumn and
early winter; they supported aggregate formation by the roots and then
the uptake of nitrate for the purpose of no groundwater contamination.
After mustard had been killed by frost, the plant residues served as protection against erosion. Risks of this approach were partially non-uniform
emergence of the mustard seedlings and germination of shatter grain.
Under favourable weather conditions in autumn, crops might reach a
height of 130 cm; the plants were killed by freezing not later than in
December. The seed potatoes were planted with a fully automatic potato
planter with a special tandem control which had been adjusted to this purpose. Because of the preformed ridges, higher ground clearance was necessary. Before the drop tube, a disc coulter had to be attached to cut off
plant residues and to open the dam. Other than on a usual potato planter,
the covering discs had to have a larger diameter because of the pre-shaped
ridges. It was important to not destroy the dam bottom which had formed
a rather stable connection with the remaining soil on the surface; thus, the
risk of erosion was lowered, particularly, the danger of dam ruptures (Gerl
and Kainz, 1998c, 1999; Kainz et al., 1999; Seuser et al., 1999). After planting the tubers, the ridges were interspersed with coarse, but high-porous
lumps. When the potatoes broke through, ridging was performed with a
roller hoe. At that moment, the residues of the cover crop mustard had
already dried and began to rot; therefore, the cover was carefully
chopped and incorporated into the soil, which decreased the remaining
mustard biomass on the surface to 10–20%. Much higher coverage was
left by the overwintering cover crop bird rape because it developed much
more and heavily lignified biomass which as well was degraded not
before April. Herbicide treatment with a soil-active means took place
immediately after ridging, when the soil was covered to less than 20% by
fine-structured material. Higher coverage rates partially shielded the
soil off and prevented a reliable effect of soil-applied preparations. In
such cases, foliar treatment post emergence was used (for example,
Metribuzin and Rimsulfuron). The decisive fungal pests in potato growing were *P. infestans* and *Alternaria solani*. The control of *P. infestans*
was executed according to the program SYMPHYT. As first step sys-
temic pesticides were applied (for example, Metalaxyl-M), followed by

Table 9 Yields of the main corps (dt ha^{-1}) of the integrated farming system.

	Potatoes	Winter wheat	Spring wheat	Grain maize	Silage maize
Average	368.1	69.6	53.0	86.8	493.3
Max	500.0	88.2	70.7	111.6	737.0
Min	276.5	39.8	36.6	70.0	327.5
CV (%)	19.7	18.7	28.7	16.2	27.1

contact herbicides (for example, Fluazinam). Attention was paid to a reliable effectiveness against A. *solani* in July and August because without control this disease had caused considerable damage (addition of maneb). The leaves of food potatoes were mulched by mid-September. For maturing, the tubers remained in the soil for another 3 weeks. For seed production, the crop was desiccated, when the starch content had reached at least 11% and first signs of maturity became visible, accompanied by a corresponding size of the seed tubers. Desiccation was executed as early as possible to exclude virus infections of the tubers. The yield range of 276–500 dt ha^{-1} in the years 1993–2004 (Table 9), goes back do to the weather and site conditions on the one hand (different fields) and to the different production targets (here not allocated) on the other.

1.2.3 Discussion

The established farming systems in Scheyern are built up by their own principles. The interaction of the different farm sections – plant production with the differentiated crop rotation, grassland and livestock – is of utmost importance for the success of the organic farming system. Cycles of matter and nutrients could be arranged in a way resulting in a sufficient N supply to the arable crops and good yields from livestock and fields. Inside the integrated farming system, the nitrogen supply can be assured by import of mineral fertiliser, so nitrogen fixation by legumes was not necessary and the feeding of the livestock could built upon maize. For this reason the methods of soil conservation – non-turning tillage, direct seeding – became more important.

The achievement of objectives in the two farming systems can be followed by different parameters. The food for livestock in the organic farming system was provided by grassland and the fodder legumes as a part of the crop rotation. The yield level of these legumes and the grassland were on the regional level over the investigation period of 10 years and did not show any trend. After the change from lupin to a second field of lucerne-clover-grass, the fodder volume increased and the livestock could be increased also. For this reason a higher amount of manure was available for the crops. The potatoes benefited from this in a specific way. The averaged yield over 10 years of 238 dt ha^{-1} exceeded significantly the yield of organic farms in Germany. In the same period we achieved a trend of increasing yields of 7.3 dt ha^{-1} a^{-1} and reached largely the goals for this crop. The yields of the two grain crops wheat and rye were similar to yield level and trends in organic farming in Germany; for wheat we got a rate of $+0.7$ dt ha^{-1} year^{-1} in the investigated period. For rye the rate was lower if the outstanding yield of 65 dt ha^{-1} in 2004 was not taken into consideration. For the sunflowers the goals could not be reached completely. The averaged yield of 27 dt ha^{-1} was high level for organic farming but there was a negative trend of -1.0 dt ha^{-1} a^{-1}. One of the reasons was the immigration of the slug pest *Arion lusitanicus*, which damaged plants in the early growing stage in some years. But most important was that plant space was changed from 37 cm $\times$ 37 cm to at last 75 cm $\times$ 18 cm, what was favourable for the management.

Within the integrated farming system, the potatoes, winter wheat and corn yielded on a regional typically level or in some years also a little higher. During the 10-year period, all crops showed a positive trend of the yields. The silage maize had a significant higher yield of 493 dt ha^{-1} than the region and an increasing rate of 24.3 dt ha^{-1} year^{-1}. The main reasons for this success have been the cultivation techniques and the advancements in plant breeding. It is remarkable that these results could be achieved at a very high level of soil conservation. The different measures like not-turning tillage, direct seeding and cover plants reduced the soil erosion significantly (Fiener and Auerswald, 2006; Huber et al., 2005). The leaching of nitrate and pesticides could be decreased, too.

The results of the two farming systems at the research station in Scheyern have shown that the goals verbalised at the beginning of the project could be achieved. The yields reached the region typical level or even better, and at the same time, the environmental achievement was improved significantly.

Acknowledgements

The scientific activities of the *FAM Munich Research Network on Agroecosystems* were financially supported by the German Federal Ministry of Education and Research (BMBF 0339370). Overhead costs of Research Station Scheyern are funded by the Bavarian State Ministry for Science, Research and the Arts.

References

Altenweger A, Preitsameter B, Neumeier K, Kainz M, 1998. Zielkonforme Bewirtschaftung der Flächen der Versuchsstation und Betreuung der Versuchsanlagen. FAM-Bericht 22, 295–300.

Auerswald K, Albrecht H, Kainz M, Pfadenhauer J, 2000. Principles of sustainable landuse systems developed and evaluated by the Munich Research Alliance on Agroecosystems (FAM). Petermanns Geographische Mitteilungen 144, 16–25.

Fiener P, Auerswald K, 2006. Rotation effects of potato, maize and winter wheat on water erosion from cultivated land. Advances in GeoEcology 38, 273–280.

Furchtsam F, Kuch G, Kainz M, 1995. Erfassen der Stoff- und Energieflüsse im Zuge der Landbewirtschaftung. FAM-Bericht 5, 487–494.

Furchtsam F, Kuch G, Kainz M, 1996. Erfassen der Stoff- und Energieflüsse im Zuge der Landbewirtschaftung, Erfassen von Parametern für die Schlagkartei. FAM-Bericht 9, 293–298.

Gerl G, Festner T, Gutser R, 1999. Langfristige Erfassung nutzungsbedingter Veränderungen der Bodeneigenschaften und des Pflanzenwachstums sowie der Stoffflüsse auf Schlag- und Betriebsebene zur Ableitung von Indikatoren auf die Auswirkung der Nachhaltigkeit der Bewirtschaftungsmassnahmen. FAM-Bericht 32, 51–60.

Gerl G, Kainz M, 1997. Erfassen der Stoff- und Energieflüsse im Zuge der Landbewirtschaf-tung. FAM-Bericht 13, 271–274.

Gerl G, Kainz M, 1998a. Erfassen von Parametern für die Schlagkartei sowie Stoffflüssen im Zuge der Landbewirtschaftung – Kumulierter Energieaufwand für die Erzeugung von Kartoffeln; ein Vergleich zwischen ökologischem und integriertem Landbau. FAM-Bericht 22, 289–294.

Gerl G, Kainz M, 1998b. Zielkonforme Bewirtschaftung der Flächen der Versuchsstation und Betreuung der Versuchsanlagen Optimierung einer standortangepassten, erosionsmindernden Kartoffelproduktion. FAM-Bericht 22, 301–306.

Gerl G, Kainz M, 1998c. Kartoffeln in Sommerdämme pflanzen? Top Agrar 12, 54–57.

Gerl G, Kainz M, 1999. Erosionsschutz im Kartoffelanbau – Senfeinsaat nach der Krautbeseiti-gung. Kartoffelbau 50/7, 270–272.

Hofmann M, 2005. Modellierung der Stickstoff- und Kohlenstoffflüsse des Versuchsgutes Scheyern. Diplomarbeit Weihenstephan.

Huber B, Winterhalter M, Mallén G, Hartmann HP, Gerl G, Auerswald K, Priesack E, Seiler KP, 2005. Wasserflüsse und wassergetragene Stoffflüsse in Agrarökosystemen. In: Osinski E, Meyer-Aurich A, Huber B, Rühling I, Gerl G, Schröder P (Eds.), Landwirtschaft und Umwelt – ein Spannungsfeld. Ergebnisse des Forschungsverbunds Agrarökosysteme München (FAM). Oekom-Verlag München, 57–98.

Kainz M, Gerl G, Habermeyer J, Schieder A, 1999. Mulchpflanzverfahren als Erosionsschutz-maßnahme. Kartoffelbau 50(8), 298–301.

Kainz M, Göhring, Kimmelmann S, 2001. Zielkonforme Bewirtschaftung der Flächen der Versuchsstation. FAM-Bericht 48, 209–217.

Kainz M, Kimmelmann S, Hackelsperger F, 2002. Zielkonforme Bewirtschaftung der Flächen der Versuchsstation. FAM-Bericht 53, 215–221.

Reents HJ, Kainz M, Weller H, 1999. Zielkonforme Bewirtschaftung von Flächen der Versuchs-station und Betreuung der Versuchsanlage. FAM-Bericht 32, 61–66.

Rühling I, Ruser R, Kölbl A, Priesack E, Gutser R, 2005. Kohlenstoff und Stickstoff in Agrarökosystemen. In: Osinski E, Meyer-Aurich A, Huber B, Rühling I, Gerl G, Schröder P (Eds.), Landwirtschaft und Umwelt – ein Spannungsfeld. Ergebnisse des Forschungsverbunds Agrarökosysteme München (FAM). Oekom-Verlag München, 99–154.

Seuser K, Gerl G, Kainz M, 1999. Mulchpflanzverfahren bei Kartoffeln – Erosion wirkungsvoll verhindern. Sonderausgabe Bodenschutz heute: Erosion vermeiden, Erträge sichern, Fördergemeinschaft Integrierter Pflanzenbau e.V.

Weller H, Auernhammer H, Pfadenhauer J, Schmidhalter U, 2000. Zielkonforme Bewirtschaf-tung der Flächen der Versuchsstation. FAM-Bericht 39, 177–186.

Weller H, Kainz M, 1999. Zielkonforme Bewirtschaftung der Flächen der Versuchsstation. FAM-Bericht 32, 173–178.

http://www.bauernhof.net/lexikon/kpl.htm. 1.9.2006

Part II

Management of Heterogeneous Systems

Part II

Chapter 2.1

Effects of the Management System on N-, C-, P- and K-fluxes from FAM Soils

R. Ruser, G. Gerl, M. Kainz, T. Ebertseder, H.J. Reents,
H. Schmid, J.C. Munch and R. Gutser

2.1.1 Introduction

The *Forschungsverbund Agrarökosysteme München* (FAM) pursued the purpose to develop concepts for a sustainable and environmental sound land management which also account for soil conservation. Therefore, the reduction of soil erosion was defined as an essential purpose of the tasks of the FAM in this undulating area in which the top soils show distinctive sealing characteristics, as for example high silt contents (Sinowski, 1995). During a strong rain event in August 1992, extremely high amounts of soil material were lost by erosion ($9 \, t \, ha^{-1} \, a^{-1}$), resulting in the implementation of erosion-diminishing measures in both of the management systems (integrated and organic). The influence of these measures on the dynamics of single nutrients is to be evaluated. The understanding and the quantitative classification of the nutrient fluxes in agroecosystems is one of the most important prerequisites to achieve and document an efficient, sustainable and environmentally sound land use. Nutrient balances (difference between input and output) of properly managed agricultural farms must be well adjusted for most nutrients or slightly positive for N to compensate for the unavoidable nutrient losses to the environment. It is the general purpose of agricultural management to keep these losses on an absolute minimum to avoid emissions about the aerial or water path or an unwanted strong accumulation in soils (Vos and van der Putten, 2000).

During the 12 years of the FAM two soil inventories were conducted with a temporal distance of 10 years. So, there was the chance to investigate not only short-term but also longer-term effects of the land use on the nutrient fluxes. The balance of the nutrient fluxes for this period permitted the aggregation of single annual data spanning more than two crop rotations on the integrated FAM farm or about more than one complete crop rotation on the organic FAM farm. The significance is raised due to the aggregation of the recorded data because the influence of single annual outliers is reduced accordingly. This was shown by trace gas investigations at several locations (Kilian et al., 1998; Kaiser and Ruser, 2000). An important task of the FAM was to assess the measured nutrient fluxes within both management systems by means of independent balances and modelling procedures, to compare the results critically with each other and to weigh the single model outputs concerning their applicability and the compulsions linked with it in the application mutually. After a short description of the methodical attempts the results are shown for the examined nutrients and are discussed.

2.1.2 Materials and methods

Soil inventories in 1991 and 2001

In 1991, two years after an initial uniformly management of all arable soils at the FAM research station, soil samples from the A_p horizon were taken using a 50×50 m grid (Weinfurtner, 2002, 2007). At each grid point 3–5 undisturbed soil cores were collected. Additionally, the soil from 4 to 6 soil augers was homogenised and a representative aliquot was transported to the laboratory. The mean thickness of the A_p at all sampling points provided for the integrated farming system was 23.8 cm, the corresponding value for the future organic management system was 22.4 cm.

The bulk density of the soil was determined after drying the undisturbed soil cores at 106°C. Soil texture and skeleton were detected gravimetrically after sieving and sedimentation analysis. The disturbed soil samples were sieved <2 mm. C_{org}- and N_t-contents were measured using a CN analyser equipped with a thermal conductivity detector. P- and

K-concentrations were determined in CAL-extracts, the pH was determined in 10^{-2} M $CaCl_2$ solution. The volumetric data and the nutrient concentrations allowed for the calculation of the nutrient contents per area.

To characterise temporal changes of the nutrient contents of the soils due to different management practices, the same sampling design and the same analytical procedures were repeated in 2001 (Weinfurtner, 2002, 2007; Gutser et al., 2003). The sampling protocol 2001 also considered the new A_p horizon (10 cm depth) which developed as a result of the shallow soil management on the integrated farm sites. In August 1992 heavy rainfall induced a strong soil erosion event. The mean soil loss of the topsoil of all arable fields was 1.9 cm for the integrated farm, and 0.5 cm for the organic farm. Since the sampling depths in both inventory years were identical, it can be concluded, that the samples also contained low amounts of soil material from the B horizons for the sampling in 2001.

Nutrient input/output balances

During the complete duration of the FAM experiment a balance was available which was capable of weighing harvest products or organic fertilisers on field or on single lot scale. Simultaneously, moisture and nutrient contents (N, P, K, C) of the harvest products were determined. This procedure enabled as far as possible for the exact quantitative and qualitative calculation of nutrient flux data which were the basis for nutrient balancing and modelling approaches (Kainz et al., 2003; Ebertseder et al., 2003). The nutrient balances were determined on different levels that are briefly described in the following:

(i) The base for the N-, P- and K-nutrient flux calculations in context with the *farm-gate balance* approach were input and output data of the whole agricultural farm. These data are largely available from simple receipts collected by the farm manager. Bach and Frede (2005) summarised the following input data for N balances: N-fertiliser (mineral and secondary fertiliser as well as purchased organic fertiliser), externally produced feeding stuff and seeds, new animals entering the farm, the atmospheric N deposition and N_2 fixation of legumes, where the latter can only be estimated or modelled by the farmer. Output data were the nutrient removals via market products from plant and animal production and probably further exports, as e.g. the transfer of organic fertilisers to other farms in order to maintain acceptable N loads from animal husbandry or from biogas reactors. The farm-gate balance contains the whole N surplus: it is composed of the N surplus of the stable and of the field balance of a farm. The balance is easy to calculate, more reliable and less afflicted with errors as compared to balances on the field level, where internal nutrient flows remain unconsidered between the different levels (Breitschuh et al., 2004; VDLUFA, 2006). This black box approach does not allow for the assessment of single production systems within one farm. Despite the adherence of tolerable N surpluses on the farm level, high positive N balances may occur on the level of single fields (Bach and Frede, 2005). Therefore, to achieve a reliable optimisation of nutrient fluxes on the farm a further balance on field level has to be calculated.

(ii) The *stable balance* as a measure to assess the animal husbandry system takes account for the nutrient import via the acquisition of feeding stuff and external animal as well as the input of intra-farm produced feedings. These data are compared with nutrient losses from farm exports as for instance animal market products, organic fertilisers and gaseous NH_3 losses in the stable and during the storage of animal excrements.

(iii) Only *field balances* on different scales (total farm area or single field) allow for the differentiated assessment of single production methods. They represent a very good consulting tool and facilitate the introduction into fertilisation optimisation (Bach and Frede, 2005; Embert, 2004). Under common practical conditions the amount of organic fertiliser has to be roughly calculated and the estimation of the yield of grassland or forage crops remains difficult. This particularly affects the reliability of the N balance from farms with animal husbandry (Gutser, 2006). For both FAM farms predominantly measured data were available for the nutrient balances. We calculated the nutrient balances from input and output data on the level of the total managed farm area, of the total arable area or of the total grassland area. The balances of the single fields were aggregated up to the total farm level. The nutrient input data comprises values of mineral and organic fertiliser, nutrient import by seeds, N depositions (NO_x, NH_3) and N_2 fixation by legumes (Heuwinkel et al., 2002; Hülsbergen et al., 2000). Nutrient fluxes in the form of intra-farm removal of harvest products (straw and hay) and the outflow of market products were used as output. On both sides (input and output side), changes in stocking rates (animals, feed, organic fertiliser etc.) were taken into consideration.

The nutrient balances calculated with the different balances approaches comprise the potential for changes of the total nutrient stocks in soils, and for losses via atmospheric or aqueous paths. It depends on the method of nutrient balancing whether NH_3 losses in the stable during the storage of organic fertiliser or after the application in the field are included in the calculations, where the farm balance is the only method which accounts for the total NH_3 losses (VDLUFA, 2006).

The above mentioned balance approaches were not sufficient to describe the C-fluxes based on the comparison of output and input data because the main input data, as e.g. C exudation and temporally high resolution results on the regeneration and the mortality of roots were not measured. Therefore, we used a C-balancing approach which only needs input parameters that are easy to determine (Leithold et al., 1997; Körschens and Schulz, 1999). This method compares the crop-specific humus loss with the humus reproduction from organic substances. It was prepared by an experts group of the VDLUFA (2004). It is already applied in the context with the cross compliance commitments for the payout of EU funds to German farmers and the concurrently claimed maintenance of soil organic matter.

Modelling approach with REPRO and questionings on representative farms of the administrative district Pfaffenhofen

The balance oriented model REPRO was applied for the calculation of the C and N fluxes. This model was developed in the Institute of Agronomy and Crop Science of the University Halle. Several ecological and economic assessment methods are integrated and more than 200 indicators can be evaluated optionally (Diepenbrock et al., 1997-1998; Hülsbergen, 2003). Due to the special data structure of the FAM the model REPRO – originally developed for the East German dry region – it could also be adapted to the Tertiary Hilly region in South Bavaria and evaluated accordingly (Hülsbergen et al., 2002). An advantage of the model is the input parameters which can simply be determined. It is constructed modularly and links material fluxes at farm level, so that output data of a module flow directly as input data into the following module. Furthermore, the modular construction allows adapting the calculation of the respective purpose (Hülsbergen et al., 2000).

To arrange the scales of the nutrient fluxes of both FAM farms to a regional level, questionings were carried out to the material fluxes on farms of the administrative district Pfaffenhofen as an area reference in 2002 (Gutser et al., 2003). Complete data sets were sampled on 16 integrated managed farms and on eight organically managed farms in the surrounding region of the FAM. These material fluxes were modelled with REPRO. Rühling et al. (2005) showed that both farms show typical farm structures for the region Pfaffenhofen, even if the field area of the organic farm with 32 ha lies clearly under the regional average of 82 ha. Another essential difference existed in the animal stocking density of the integrated FAM farm. This was much lower with 0.6 AU ha^{-1} (where AU is animal units) compared to the average of 1.9 AU ha^{-1} on the regional level. The organic farm corresponded with 1.1 AU ha^{-1} possibly to the average of the organically managed cattle-holding farms of the region. REPRO classified the animal stocking density of the integrated FAM farm as low and those of the organic farm as optimum (Rühling et al., 2005). The stocking rates of the regional integrated farms were consistently classified 'high'. Another calculated indicator was the crop type diversity according to Hülsbergen (2003). It was computed as 'high' for both FAM farms and consequently as 'enriching' for the agricultural landscape and as 'stabilising' for agroecosystems.

2.1.3 Results

Influence of the management system on the C-fluxes

The practices of reduced soil tillage to reduce soil erosion resulted in the development of a new A_p horizon with a thickness of 10 cm. The addition of organic material (crop residues, green manure and organic fertilisers) in this zone increased the C_{org} concentration and the C stocks (1250 kg C ha^{-1}) as compared to an A_p horizon of the same thickness prior to the installation of the FAM. Simultaneously a reduction of the total C stock was observed in the depth from 10 to 23.8 cm (-2450 kg C ha^{-1}), and this loss could not be compensated by the increase of the stock in the upper 10 cm (Tables 1 and 2). Over the period of 10 years, the C_{org} stocks of the soils of the integrated farm were reduced by 1200 kg C ha^{-1}. The erosion induced loss of 1.9 cm soil thickness may be one reason for the not nearer quantified uncertainties concerning the C_{org} losses in the soils of the integrated farm. Furthermore, a high and intensive turnover of C in the upper layer of the A_p horizon as a reason for the changes of the C_{org} stocks will be discussed later.

As compared to the sampling in 1991 the C_{org}-stock of the topsoils of the organic farm in 2001 were increased by 1800 kg ha^{-1} (Table 2). This

Table 1 C_{org}-, N_t-, P_{CAL}-, K_{CAL}-stocks and C-to-N ratios of the topsoils of the integrated managed FAM fields[a].

	1991			2001			Δ
	0–10.0 (kg ha^{-1})	10.0–23.8 (kg ha^{-1})	Σ 0–23.8 (kg ha^{-1})	0–10.0 (kg ha^{-1})	10.0–23.8 (kg ha^{-1})	Σ 0–23.8 (kg ha^{-1})	
C_{org}	18 700	23 500	42 200	19 950	21 050	41 000	−1200
N_t	1 910	2 450	4 360	2 160	2 420	4 580	+220
C/N ratio[b]	9.7	9.7		9.2	8.7		
P_{CAL}	125	173	298	102	144	246	−52
K_{CAL}	274	362	636	243	253	496	−140

[a] The initial thickness of the A_p horizon was 23.8 cm. As a result of reduced soil management, a new A_p horizon developed with a thickness of 10.0 cm.
[b] Dimensionless.

humus accumulating effect could not be explained solely by the C amount of surface applied plant residues (straw or green manures) and organic fertilisers (liquid or solid manure, compost) (Figure 1).

The changes in the C stocks of the arable fields can be estimated by a humus balance. It takes into account the crop-specific humus decomposition of soil organic C which is compared with crop and management specific C supply from roots, aboveground plant residues and organic fertilisers (Figure 2).

Table 2 C_{org}-, N_t-, P_{CAL}-, K_{CAL}-stocks and C-to-N ratios of the topsoils of the organically managed FAM fields.

	1991	2001	Δ
	025.8 cm (kg ha^{-1})	025.8 cm (kg ha^{-1})	0–25.8 cm (kg ha^{-1})
C_{org}	44 200	46 000	1800
N_t	4 475	5 040	565
C/N ratio[a]	9.9	9.1	
P_{CAL}	291	253	−38
K_{CAL}	681	571	−110

[a] Dimensionless.

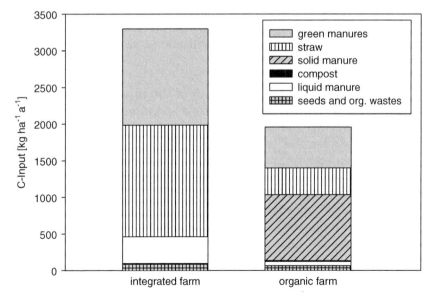

Figure 1 Input of organic C including surface applied plant residues and organic fertilisers for arable fields of the integrated and organically managed FAM farm (mean values for the period 1993–2001).

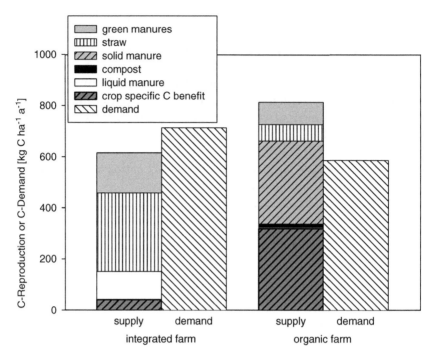

Figure 2 Crop specific humus demand (mean value over a whole crop rotation) and humus reproduction via aboveground plant residues
and organic fertilisers on integrated and organically managed
arable FAM sites (covering the period 1993–2001). The balance
was calculated according to Leithold et al., 1997.

The annual input on the arable sites of the integrated managed farm
was 3300 kg C ha^{-1} a^{-1} whereas the corresponding value for the arable
sites of the organically managed farm was approximately 1/3 lower (2100
kg ha^{-1} a^{-1}). However, the humus balance calculated according to Leithold
et al. (1997) revealed a humus accumulation only for the organically managed farm (+215 kg C ha^{-1} a^{-1}), whereas the high amounts of C input did
not adequately account for the C demand of the crop rotation resulting in
negative C-values of approximately −100 kg C ha^{-1} a^{-1}.

This result emphasises the importance of the crop rotation for the
humus budget of soils. One main reason for the humus enrichment of the
organically managed soils was the high portion of clover–grass mixtures
which accounted for approximately 35% of the total arable area. Schmid
et al. (1997) determined the root biomass of all crop types cultivated at the
FAM research station immediately after harvest (Figure 3). Obviously the
high amounts of C input by clover–grass mixtures favoured humus
enrichment in soils.

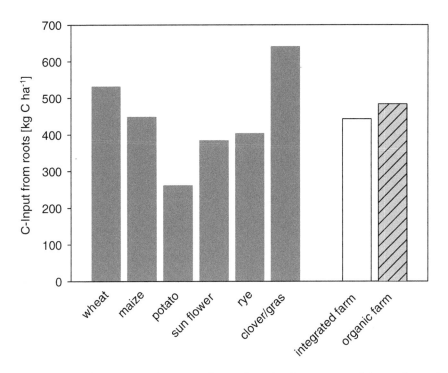

Figure 3 Annual net root C input from the single crops cultivated on the FAM farms and mean C input calculated for the two different management systems at the FAM research station (Schmid et al., 1998; Gutser et al., 1998).

Beside the net amount of C input the crop-specific humus demand was also influenced by the whole rhizodeposition of the crops. Due to difficulties in quantification of these gross amounts of C input, they have often been underestimated. Schmid et al. (1997) demonstrated on a wheat cropped soil that the gross C input was about 3–4-fold higher than the net C root biomass at the time of milk ripeness (Table 3). For wheat and barley, rhizodepositions between 200 and 500 kg C ha^{-1} a^{-1} were reported (van Noordwijk et al., 1994; Swinnen et al., 1995). The data from Schmid et al. (1997) do not account for passively or actively exuded C compounds from roots. Recently, it was shown that these exudates significantly contribute to stable humus fractions in FAM soils (Marx et al., 2006).

Additionally, the increasing effect of clover–grass mixtures on the humus content of soils was also a result of the extended soil dormancy which reduced humus consumption as compared to soils with intensive soil management measures (plough, hoe etc.).

A root system of nutrient deficient plants may be more distinctive as compared to the root system of plants without deficiency. Many investigations on

Table 3 Mean shoot and root production (gross production) as affected by soil texture and farming system, percentage of gross root biomass to the total biomass of winter wheat at milk ripeness in 1996, and range of this ratio during the sampling period 1995–1997 (Schmid et al., 1998).

Soil texture	Farming system	Shoot weight (g DM m^{-2})	Root weight (g DM m^{-2})	Root/Total biomass	
				%[a]	Min–max[b]
Loamy	Integrated	1197	388	20	15–24
Sandy	Integrated	1009	382	24	15–28
Loamy	Organic	1318	315	19	18–22
Sandy	Organic	797	325	31	25–44

[a] In 1996.
[b] 1995–1997.

that topic were summarised by Marschner (1990) for N-deficiency and by Steingrobe et al. (2001a) for P-deficiency. The results from Steingrobe et al. (2001a) were supported by the investigations of Schmid et al. (1997) concentrating on the effect of soil texture and management system on root growth at FAM sites (Table 3). They could show that the portion of root bound C to the total plant C was significantly higher at sandy study sites with low nutrient availability on both farms when compared with the C distribution within plants at loamy sites with a high nutrient storage capacity. The highest portions of root C to the total plant C were found on sandy sites of the organic farm. They clearly exceeded the portion of all other investigated sites with a mean proportion of 31% and a maximum of 42%. One of the main problems concerning the demand oriented nutrient supply in organic farming systems is the optimum temporal synchronisation of the N-availability from crop residues, legumes and organic fertilisers (manure and composts). Temporal N-deficiency with an increased translocation of assimilated C into the root might also be one reason for the higher humus contents of the organically managed soils at the FAM research station.

As compared to the time before FAM started, the application of machinery with more traction and the resulting deeper ploughing depth may also have contributed to the increase of the C contents on the organic farm (Rühling et al., 2005). Ancient soil material from the B horizons with additional free sorption sites was now included into the new A$_p$ horizon.

The conversion from conventional to a reduced soil management system did not affect the C$_{org}$ contents during an investigation period between 5 and 12 years (Angers et al., 1997). They reported a higher C storage in 30 to 60 cm depths in conventional managed soils, whereas reduced soil management increased the C$_{org}$ contents in the topsoil.

However, the total amounts of stored C did not statistically differ between the two soil management systems. Angers et al. (1997) concluded that on a regional scale, recently established systems with reduced soil management have a low potential for further C storage in soils as long as the amount of plant residues can only slightly be affected by soil management measures. The magnitude of the annual C_{org}-increase or -loss within the two farming systems was in good accordance with results from investigations on the effect of crop rotation or of the soil management system on the humus contents after soil management changed within a similar investigation period between 5 and 12 years (Pimentel et al., 2005; Halvorson et al., 2002; Jarecki et al., 2005; Angers et al., 1997).

Figure 4 compares the humus balances (REPRO and VDLUFA) for both FAM farms with the results of the soil inventories. Generally, the results of the REPRO model and the soil inventories were in good agreement. On the integrated farm the annual C losses from arable fields ranged from -100 to -120 kg C ha^{-1} a^{-1}. On the organic farm the corresponding values were between $+180$ and $+215$ kg C ha^{-1} a^{-1}.

The estimation procedure for the humus dynamics in soils used by the VDLUFA (2004) was predominately developed on the basis of data from conventionally and integrated managed farms. It is not well suited for the

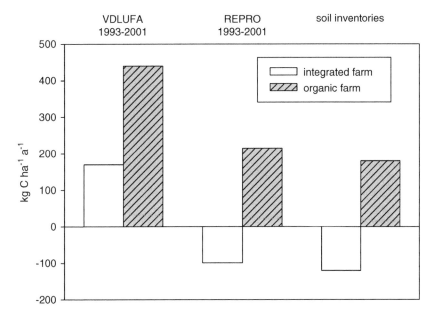

Figure 4 Humus C balance for the arable fields of the two FAM farms calculated according to the method of the VDLUFA (2004), the HE-method (Leithold et al., 1997), and as the result of the two soil inventories in 1991 and 2001.

calculation of humus balances from organically managed farms. When compared to the measured C_{org} data from the soil inventories, the VDLUFA methodology (which is used in a modified form for the EU cross compliance regulation) consistently overestimated the humus enrichment in arable soils. This was especially evident for the organically managed soils. The main reason for this phenomenon was the underestimation of crop-specific humus decomposition rates (Table 4).

According to the classification of the VDLUFA (2004) the humus balance of the integrated farm resulting from the soil inventories was classified as 'low' (-120 kg C ha^{-1} a^{-1}). This can be tolerated in C-enriched soils in the medium term. The humus balance of $+180$ kg C ha^{-1} a^{-1} calculated with the results from the soil inventories for the organically managed sites was grouped in the category, high and was also tolerable in the medium term, especially when soils are low in C.

Using REPRO the humus supply of the arable soils of the investigated farms in the district Pfaffenhofen was evaluated as 'high' ($\approx$43% above

Table 4 Main elements of the humus balance for arable fields of the FAM farms calculated according to the method of the VDLUFA (2004) and REPRO (Leithold et al., 1997) in kg C ha^{-1}a^{-1} (Mean value 1993–2001).

Balance	Balance method			
elements	VDLUFA FAM farm		REPRO FAM farm	
	Integrated	Organic	Integrated	Organic
Humus demand	−464	−270	−713	−586
C benefit[a]	59	237	41	319
Straw application	307	66	307	64
Green manure	140	86	157	87
Addition of organic fertilisers	130	324	110	342
Solid manure		313		325
Liquid manure	130	4	110	6
Others		7		12
Humus balance	173	443	−99	215
Evaluation of the supply[b]	D	E	B	D

[a] Depending on cultivated crops or on crop rotation (esp. legumes, catch crops, fallow ground etc.).
[b] Evaluation classes according to VDLUFA (2004): A = very low (< -200 kg C ha^{-1}a^{-1}); B = low (-200 to -76 kg C ha^{-1}a^{-1}); C = optimal (-75 to $+100$ kg C ha^{-1}a^{-1}); D = high ($+101$ to $+300$ kg C ha^{-1}a^{-1}); E = very high ($> +300$ kg C ha^{-1}a^{-1}).

optimal value), this result was also computed for the N supply. The organically managed farm did not differ from this pattern whereas the supply on integrated FAM fields was only about 80% of the optimum range (90–110% of a well-adjusted supply). Values above the optimum supply of C and N in soils represent a risk potential for undesirable C and N losses to the environment.

N fluxes at the FAM sites – influence of the management system on the N fluxes

Calculated from the results of the soil inventories, the N_t stocks of both operating systems increased within the 10 years investigation period. The increase was 57 kg N $ha^{-1} a^{-1}$ in the organically managed soils and 22 kg N $ha^{-1} a^{-1}$ in the integrated managed ones (Tables 1 and 2). It is supposed that also changes within the organic N fractions have arisen. Al-Kaisi et al. (2005) and Liang et al. (2004) reported an increase of the N_{org} stocks after the rearrangement from conventional on minimum soil tillage, and this inevitable had to result in an enhanced N mineralisation. Obviously the conditions for N mineralisation were degraded by the diminished soil tillage accordingly, so that N mineralisation rates were similar to the conventional treatment.

The C-to-N ratios were decreased by the increase of the N_t stocks in the soils of both FAM farms. The originally C-to-N ratio of 9.7 decreased to 9.2 (integrated farm, 0–10 cm), to 8.7 (integrated farm, 10–23.8 cm) and to 9.1 (organic farm). On the one hand, in the organic system the decrease of the C-to-N ratio might be due to the entry of organic matter rich in N (legumes and rotted stable dung with narrow C-to-N ratios), a generally high N supply by legumes (98 kg N $ha^{-1} a^{-1}$ from N_2 fixation) and, on the other hand, on the removal of harvest products with a wide C-to-N ratio (grain straw). Obviously, an intensive turnover of organic matter resulting in a strong C depletion and a relative N enrichment has taken place in the uppermost 10 cm of the soils of the integrated farm. The intensive turnover of the organic matter coupled with a strong O_2 consumption could be a possible reason for the high N_2O emissions at the study site Scheycrn (Ruser et al., 2007).

Figure 5 shows the N surpluses of the two FAM farms on the farm-gate level and on the field level for the investigation period 1995–2001 (during this seven-year-period the crop rotation of the organically managed farm with seven different fields was completely realised). The N surpluses of the farm-gate balance (farm with the total agricultural managed area as reference level) includes the supply via deposition, N_2 fixation by legumes, and the total NH_3 losses in the stable during excrement storage and after application on the fields. The surpluses on the integrated farm completely account for NH_3 losses from stable and liquid manure storage,

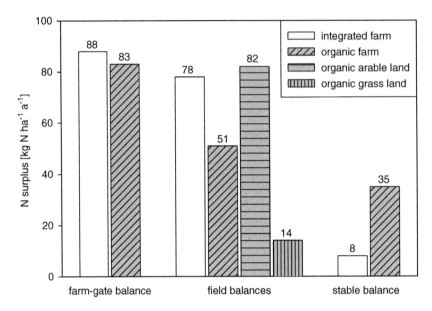

Figure 5 N surpluses of the farm-gate balance, of the field balance (for the organically managed farm this balance is also split into arable and grassland balance), and stable balances for the two FAM farms for the period 1995–2001.

because the animal husbandry of this farm was only simulated: the liquid manure was imported in equivalent amounts according to default animal densities from other farms in the neighbourhood of the FAM research station. Therefore, the N surpluses of the integrated FAM farm (farm-gate balance) consider only the NH_3 losses during field application where the emission from stable and liquid manure storage have to be taken into account in the balances for the farms which provided the liquid manure. The N surpluses of the arable and grassland sites also account for atmospheric N deposition, N_2 fixation by legumes and NH_3 losses after solid or liquid manure application.

The N surplus of the stable balance was estimated by the aid of modified default data on NH_3 losses as affected by the animal housing system (solid and liquid manure) provided by the KTBL. Up to now mean N_2 fixation rates of 69 kg N ha^{-1} a^{-1} were determined. The application of innovative isotope techniques in situ and the exact quantitative capture of the legume proportions in legume grass mixtures by NIRS allowed for the revision of this value to now 98 kg N ha^{-1} a^{-1} (Heuwinkel et al., 2002).

The N surpluses of the farm-gate balance and of the field balance of both management systems barely differed. Taking the atmospheric deposition and the symbiotic N_2 fixation into account, the mean N surplus for the arable sites was 78 kg N (integrated farm) and 82 kg N ha^{-1} a^{-1} (organic farm). The total N surpluses on the farm-gate level were 88 and 83 kg N ha^{-1} a^{-1}, respectively.

The N surplus on the extensively used grassland sites of the organic farm was very low (14 kg N ha^{-1} a^{-1}), because the animal excrements have been applied mainly to the arable soils.

The N balances on farm-gate or field level varied temporally. This is particularly true for the organic farm as affected by the animal density, by the portion of legumes within the crop rotation, and by internal nutrient cycling (consideration of leased areas). The calculation of balances with REPRO for the years 1998-2000 (a period with slightly modified management) showed a N surplus of 93 (integrated farm) and of 96 kg N ha^{-1} a^{-1} (organic farm) on the farm-gate level, of 81 (arable fields integrated) and of 36 kg N ha^{-1} a^{-1} (total agricultural area, organic) on the field level, where the N surplus of the organically managed arable fields was 56 kg N ha^{-1} a^{-1}. The corresponding value for the organically managed grassland sites was 11 kg N ha^{-1} a^{-1}.

The biggest differences between both operating systems appeared in the stable balances. This could be explained with the higher animal stocking density of the organic farm (1.2 AU compared with 0.6 AU ha^{-1} agricultural land). Furthermore, loose housing stable systems with solid manure beds in cattle bearing (organic FAM farm) are more afflicted with higher NH_3 losses than stable systems with liquid manure sampling (integrated farm) (Lüttich et al., 2004). In addition, NH_3 emissions as affected by application technique during the spraying of liquid manure were taken into account in the field balance, where the NH_3 emission from the solid manure flows into the stable balance.

Assessment of the N surpluses on the two FAM farms

For the assessment of the N balances on the farm-gate or on the field level benchmarks for required, aimed and tolerable N surpluses for the agricultural or environmental sector have been published (Gutser, 2006; Düngeverordnung, 2006; Bundesregierung, 2004; BAD, 2003; Eckert et al., 1999; UBA, 1999; Frede and Dabbert, 1999; van der Ploeg et al., 1991; Gutser and Ebertseder, 2001; Gutser and Matthes, 2001; Hülsbergen and Diepenbrock, 1997; Isermann and Isermann, 1995).

The different methodological approaches and reference levels used complicate the comparison of the recommended benchmarks (i.e. farm or regional scaling, farm-gate or field basis, with or without consideration of the atmospheric deposition or NH_3 emissions).

Table 5 shows important additional information to ensure the reliable comparison of benchmarks for N balances from different farms.

The above mentioned benchmarks are frequently under criticism, since the structure of the agricultural farms is most often not or at least not adequately taken into consideration by the assessment of N balances. It is well known that the plant uptake of nitrogen from liquid or solid manure is lower than the utilisation of N in synthetic fertilisers. Necessarily, farms with animal husbandry need higher positive N balances to achieve the same yield level (and profit accordingly) than farms with only market fruit production (Gutser and Ebertseder, 2001). This becomes evident from the relationship between the animal stocking rate and the farm-gate balance shown in Figure 6. Moreover, the figure documents the fact that the establishment of both FAM farms appeared to be very practically oriented, because both farms do not contrast with the other farms of the region Pfaffenhofen.

The higher N surplus in the animal-holding farm is necessary to compensate the higher system-conditioned inevitable N losses. An optimum N utilisation is given approximately with a N excretion of 150 kg N ha^{-1} (according to an animal stocking density by approx. 1.5 AU ha^{-1}), where the N surplus, which can still be tolerated from agricultural perception, was calculated taking into account the inevitable N losses connected with this stocking density (Gutser and Ebertseder, 2001; BAD, 2003).

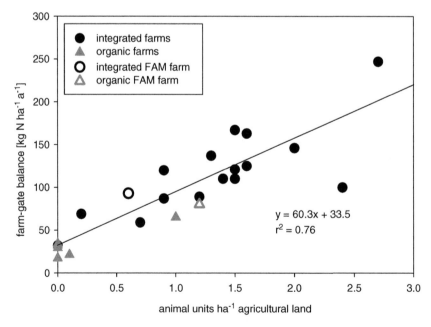

Figure 6 Relationship between the N surplus of the farm-gate balance and the animal stocking density on integrated and organic farms in the region Pfaffenhofen and on the FAM farms (Data basis REPRO: years in 1999–2001) (Gutser et al., 2003).

Table 5 Comparison of different benchmarks within the domains of agriculture and environmental protection published valuation standards for required, aimed and tolerable N surpluses (Gutser, 2006).

Source	Benchmark (kg N ha⁻¹)	Accounts for		Remarks	
		Deposition	NH₃ emission[a]		
Düngeverordnung (2006)	<60	–	–	Field balance	Independent on animal husbandry
Gutser and Ebertseder (2001) BAD (2003) 'required surpluses, unavoidable losses'	15–40	–	–	Field balance accord. to risk for N-leaching Farm-gate balance (risk for N-leaching)	Without animals up to 0.8 AU ha-1[b]
	20–45	–	–		Up to 1.4 AU ha-1
	25–55	–	–		Up to 0.8 AU ha-1
	50–95	–	++		Up to 1.4AU ha-1
	55–105	–	++		
KUL (VDLUFA, 1998) 'criteria of sustainable farms'	<30–<50	–	–	Field balance Farm-gate balance	According to risk for leaching
	<80–<100	–	++		
REPRO (Hülsbergen, 2003) 'criteria of sustainable farms'	<25	+	+	Field balance (sum reactive N)	Optimum, still tolerable range
	25–100	+	+		
Bundesregierung (2004) 'Sustainable development in BRD'	<80	+	++	Farm-gate balance Field balance	Mean values for Germany Minus 20 kg N deposition and 25 kg NH3 emission (basis 0.85AU ha⁻¹)
	<35	–	–		
UBA (1999) 'Environmental standard in German agriculture'	<50	+	–	Field balance	Mean values of agricultural sites
	<30	–	–		

[a] + = NH₃ emission from fields is included; ++ = NH₃ emissions from stable, manure storage and fields are included.
[b] AU = animal units.

The German fertilisation regulation which was amended in 2006 defines 'the good agricultural practices' of the N fertilisation in a way that the N surplus of a field balance on a farm from 2009 on may not exceed 60 kg N ha^{-1} a^{-1}. N deposition and NH_3 losses thereby remain unconsidered. On the basis of the same methodology, both FAM farms fall below this benchmark (integrated: approx. 50 kg; organic: approx. 30 kg for the whole agricultural area and 55 kg N ha^{-1} a^{-1} for the arable fields (30 kg in 1998-2000). The benchmark of this German regulation focuses on the compliance of the 'European nitrate directive'. A binding agreement on the tolerable amounts of NH_3 losses is still missing.

N surpluses are necessary for the safety of sustainable yield capacity. Gutser and Ebertseder (2001) and BAD (2003) defined benchmarks for arable soils or for the whole farm dependent on the site quality (risk of N-leaching) and on the animal stocking density. For sites with a medium N-leaching exposure and an animal density up to 0.8 AU ha^{-1} (integrated farm) the benchmark for N surpluses of the farm-gate balance is approximately 50 kg N ha^{-1} a^{-1} (including all NH_3 emissions, without N depositions), for higher animal stocking rates up to 1.4 AU ha^{-1} the corresponding value is 60 kg N ha^{-1} a^{-1}. The actual N surplus of the integrated farm is approximately 10 kg N ha^{-1} a^{-1} higher as compared to the benchmark. On the organic farm, the N surpluses met the calculated unavoidable N losses and consequently also the amount of required N surpluses for the conservation of the yield capacity of the soils.

According to the 'Commission for Sustainable Development in Germany', the aimed benchmark for N surpluses from the agricultural sector as of 2010 should be 80 kg N ha^{-1} a^{-1} (Bundesregierung, 2004). This value can be calculated with the area weighted mean farm-gate balances of all German farms. Taking the animal stocking density into account, the organic FAM farm easily meets this benchmark. The integrated farm currently has 10 kg N ha^{-1} a^{-1} of N surpluses above this claimed standard. The German Federal Environmental Agency (UBA) mentioned a mean benchmark for agricultural farms (and partly for single farms) of 30 (without deposition) or 50 kg N ha^{-1} a^{-1} (with deposition) for the N surpluses on field balance level. For integrated managed farms this N surpluses can only be achieved in the long-term at suitable locations and with animal stocking rates below 0.5 AU ha^{-1}. Among the two farms in Scheyern, the organic farm can reach this target value, however, the high NH_3 emissions (1.2 AU ha^{-1}) remain unconsidered in this calculation.

Taking into account all above mentioned restrictions and weaknesses of single assessment attempts of the N balance, the good agricultural practice of N fertilisation can be attested for both FAM farms. The N surpluses of both farms can also be tolerated in context to environmental careful land management. This is especially valid, because the results from the soil inventories showed that these N surpluses were transferred into soil by N immobilisation (see below). This immobilisation can be accepted at least in the short- and middle-term.

N losses on the two differently managed farms

As shown in Figure 5, the N surpluses from the field balance amounted to 78 kg N ha^{-1} a^{-1} for the integrated farm and 51 kg N ha^{-1} a^{-1} for the organic farm. These show a loss potential and a possible source for undesirable nutrient losses to adjacent compartments as the atmosphere and ground water. Therefore, the classification of the actual nutrient fluxes is especially important for the assessment of agricultural measures and their environmental effects.

For the N balances an atmospheric deposition of 16 kg N ha^{-1} a^{-1} was assumed. In a recent work, Gauger et al. (2002) determined a slightly higher deposition rate of 23 kg N ha^{-1} a^{-1}. In comparison with other investigation areas, these depositions are to be classified as low. Thus, an atmospheric entry of 50 kg N ha^{-1} a^{-1} was determined for the long-term investigation fields at Halle and Bad Lauchstädt in eastern Germany (Weigel et al., 2000). The values from Weigel et al. (2000) included dry, wet and gaseous N depositions. Isermann (2002) calculated a mean atmospheric N input of 30 kg N ha^{-1} a^{-1}. The high N balance of the organic FAM farm can be explained by the high N$_2$ fixation of the legume crops. Due to the substantially lower transfer of organically bound N in solid manure and compost into the subsequent arable crop, the high amounts of N fixed by legumes on the organic farm were not quantitatively utilised on a farm level. Disregarding N$_2$ fixation, the N balance of the organic farm would have been -32 kg N ha^{-1} a^{-1} (Rühling et al., 2005).

Figure 7 shows the N losses from the two FAM farms broken down into the single loss paths, where the losses were determined as follows:

- NO$_3$ leaching was determined taking the quantity of draining water and the nitrate concentrations in 1.8 m depths into account (measured in the hydrologic shafts, Matthes et al., 2001). Furthermore, the N losses via surface run-off were taken into consideration (Seiler and Hellmeier, 2001; Hellmeier, 2001). The N losses with surface run-off amounted to 1.4 and 1.7 kg N ha^{-1} a^{-1} in the organically and integrated managed soils, the losses with interflow accounted for 3.9 and 14 kg N ha^{-1} a^{-1}, and ground water formation accounted for another 10 and 11 kg N ha^{-1} a^{-1}.
- The evaluation of the NH$_3$ emission occurred by means of model calculations (Katz, 1997) supplemented with default values for NH$_3$ losses after the spreading of manure (KTBL). All the other data on NH$_3$ losses were derived from the investigations by Weber et al. (2000) and these are closely described by Ruser et al. (2007).

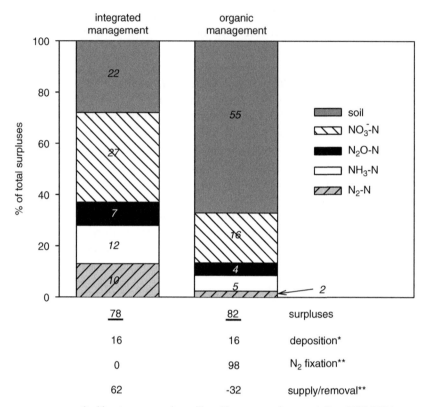

Figure 7 Fate of N surpluses on arable land of the integrated and organic
farm in Scheyern (Gutser, 2006). Numbers in italic indicate the
total amount of N loss via every single pathway in kg N ha^{-1} a^{-1}.
N surpluses of both management systems are summarised in kg
N ha^{-1} a^{-1} at the bottom.

- The N$_2$O emissions were calculated using the results of the regression
  analysis shown in Figure 3 of Chapter 2.2 (Ruser et al., 2007). The data
  entering this regression analysis were taken from Flessa et al. (2002)
  and from Sehy (2003). They varied between 1 and 10 kg N$_2$O–N ha^{-1} a^{-1}
  for fertilised soils.
- A reliable technique for the determination of the N$_2$ losses in situ was
  not available in the FAM. Therefore, the N$_2$ losses were calculated by
  subtraction of all known N losses of the N balance. It is afflicted with
  great uncertainty. Consequently, it can only roughly indicate the magnitude of N$_2$ fluxes.

From the results of the soil inventories it arose that in the integrated
managed soils 22 kg N ha^{-1} a^{-1} or 28% of the balance surplus was

immobilised in the soils, the corresponding value for the organically managed soils was $55 \, kg \, N \, ha^{-1} \, a^{-1}$ or 67%. The loss potential derived from N balances can therefore be reduced considerably by soil immobilisation. To quantify long-term N surpluses for tolerable N losses, N immobilisation may not be included in the calculation. In the organically managed soils the development of the soil N values should chiefly be pursued, because with achievement of a steady state this N accumulation potential will no more be available. Therefore, excess amounts of N would be distributed to the other loss paths as long management remains unchanged. A decrease of the N surpluses is necessary in future. This can be achieved by a decrease of the portion of legumes in the crop rotation and by the cultivation of crops with higher N efficiency (crop types or varieties richer in protein). Besides, high N surpluses in long-term will also lead to a decrease in N_2 fixation rates of the legumes, especially because the current N_2 fixation with $98 \, kg \, N \, ha^{-1} \, a^{-1}$ on arable fields is already on a very high level.

The N losses with drainage water, surface run-off, and discharge amounted to 35% of the balance surplus on the integrated and to 20% on the organic farm. Honisch et al. (2002) reported that the nitrate concentrations in the ground water were reduced by different measures from $50\text{–}60 \, mg \, NO_3 \, l^{-1}$ in 1992 on less than $30 \, mg \, NO_3 \, l^{-1}$ in 1999. This reduction is above all a result of the loss-diminishing management strategies which were successfully realised on both FAM farms. The N_2 losses of the organically managed soils seem too low ($2 \, kg \, N \, ha^{-1} \, a^{-1}$); therefore, another loss path was overrated or the input from atmospheric deposition was underestimated. N_2O production during denitrification was limited in the organically managed soils by the low supply of nitrate. Basically, the conditions for higher N_2 losses as well as for higher N_2/N_2O ratios are favourable in the organically managed soils. Relatively low nitrate concentrations and high amounts of easily available organic matter promote the reduction of N_2O to N_2, whereas the higher NO_3 concentrations in integrated managed soils favour the inhibition of N_2O-reductase and lower total N_2 losses (Blackmer and Bremner, 1978; Weier et al., 1993).

Overall the N_2O emissions (integrated farm: approx. $7 \, kg \, N_2O\text{–}N \, ha^{-1} \, a^{-1}$, organic farm: approx. $4 \, kg \, N_2O\text{–}N \, ha^{-1} \, a^{-1}$) have to be rated as very high. According to Kaiser and Ruser (2000) the location Scheyern can be considered as a high emission location. This could be possibly led back on the reduced soil tillage, because here the organic matter and the microbial activity are concentrated in the upper few centimeters of the soil. These factors enhance oxygen depletion in the soils and consequently favour denitrification. The purpose of the application of reduced soil tillage in the integrated FAM farm was the minimisation of soil erosion. This aim was unambiguously reached (Huber et al., 2005). These results clearly show, that the realisation of some of the primary purposes of the

FAM as soil conservation and reduced soil losses also means, that disadvantageous ecosystem effects are to be accepted, as for example slightly raised N balances and N surpluses, eventually increasing N_2O emissions. A significant relationship between the N balance and the N_2O emission was indicated by Kaiser and Ruser (2000) for a location in northern Germany (Figure 8). A relationship between N_2O emissions and N balance was also calculated from the measurements at the integrated plot trial after excluding the data from the potato plots (Figure 8). The slope of the regression line for Scheyern data (2.3%) is identical with the slope of the N input induced N_2O emission calculated according to the data given by Flessa et al. (2002) and by Sehy (2003). This supports the hypothesis of the high emitting location site Scheyern.

Distribution and turnover of the organic matter in soil

The results of soil inventories and C balances proved quantitative changes of the C stocks in soils. In the following the most important results of the FAM investigations on C dynamics are briefly introduced. Using very different analytical methods, these investigations all focused on the question whether and to which extent changes of the humus content

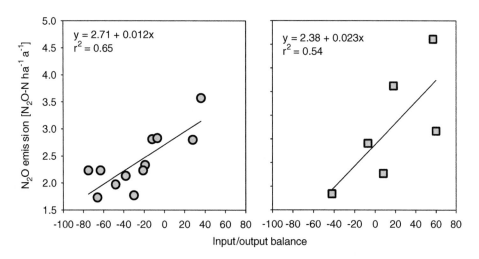

Figure 8 Mean annual N_2O emissions as affected by the input/output balance from a study site near Braunschweig (left: Kaiser and Ruser, 2000; mean of three years investigations on wheat, barley and sugar beet fields), and from experiments at the integrated crop rotation field trial of the FAM research farm (right: calculated from Ruser et al., 2001; mean of two years investigations from wheat, corn and potato fields).

can be characterised by single, easily decisive C fractions or whether they can be used as a rapid indicator for changes within the management system and C dynamics in soils.

The microbial turnover of the organic matter depends on the amount of easily available C and N compounds in soils. The management system and soil management substantially influence the amount and the quality of fresh, easily available organic matter. The availability of C compounds regulates C turnover in agricultural soils (Jenkinson, 1990; Hantschel et al., 1994). Investigations on the distribution of the organic matter in FAM soils have shown that a large part of the organic C (approximately 88%) was bound to mineral fractions (Kölbl and Kögel-Knabner, 2003). Besides, the distribution of C_{org} on different particle dimensions classes was relatively independent of respective total C_{org} contents. Approximately 55% were bound in the clay fraction, other 15% were bound in the silt fraction, where the remaining C was spread over the sand fraction or the particulate C fraction. In soils with low clay contents the essential C storage occurred in the silt fraction. According to Körschens et al. (1998) clay bound C can be considered as inert compared to other C fractions in soils.

Organic N (N_{org}) showed a similar distribution as C_{org}. On average, the clay fraction of the examined soils contained 64% of the N_{org}, only 8% were found in the sand and silt fraction (Rühling et al., 2005).

Mineral bound N_{org} and C_{org} are protected chemically and physically against decomposition. The resulting turnover rates of these fractions are very low and they can be considered as long-term storage forms in soils (Kölbl and Kögel-Knabner, 2003; Buyanovsky, et al., 1994). Consequently, management induced changes in C fractions or pools were expected chiefly in the labile soil pools. In this context, emphasis was laid on the determination of young organic matter (free particulate organic matter, free POM) and occluded C substances (occluded POM) (Kölbl and Kögel-Knabner, 2003; Kölbl et al., 2004), of microbial biomass (C_{mic}) (von Lützow et al., 2002; Wessels-Perelo and Munch, 2005) as well as of dissolved organic C (DOC) (Marx et al., 2006).

Soil texture was one of the main factors influencing the turnover of C and N. It was shown that the occluded POM fraction increases with rising clay contents in soils protecting litter or plant residues from further decomposition. This fraction accounted for up to 10% of the whole soil organic C (Kölbl and Kögel-Knabner, 2003).

Another factor that influenced the nutrient transformations was land utilisation. By means of phospholipids fatty acid patterns (PLFA) and investigations on the substrate utilisation with a Biolog© system, von Lützow and Palojärvi (1995) found a shift from microbial to fungal biomass in soils with increasing C turnover with diminished soil tillage. As compared to conventional management systems, an increase of the fungal biomass was reported from Douds et al. (1993) as well as from Pimentel et al.

(2005) in organic managed soils. This additional fungal biomass with a foremost increase of the arbuscular mykorrhiza (AM).

Comparing the share of free and occluded POM, the proportion of occluded POM increased in the order (long-term) arable field > meadow (arable until 1991) > arable (meadow until 1991) > (long-term) meadow (Figure 9). For these sensitive pools, both soils on which a rearrangement of utilisation was carried out in 1991 have not completely reached the level of the long-term utilisation until 2001. This clearly shows that the C_{org} soil contents measured in 2001 still do not represent equilibria conditions. This was also reported by Leifeld and Kögel-Knabner (2005). They valued the quantity of the organic matter (C_{org}) in ultrasonic-stable aggregates (>20 µm diameter with 22 J ml^{-1}) as an even more sensitive indicator for management changes than the shares of free and occluded POM.

Another intensively examined factor of influence was the potential yield expectation of soils. ^{15}N- and ^{13}C-enriched mustard material was mixed into the topsoil of an area with low and of an area with high yield on arable field A17 (yield mappings were carried out three preceding years

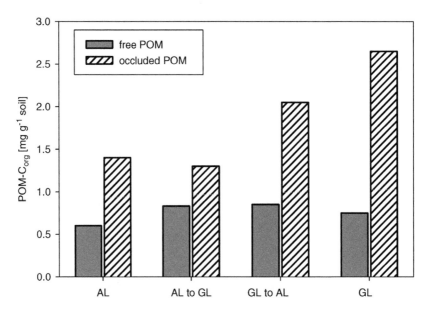

Figure 9 C_{org} concentrations of the POM fractions of a soil with different management history (long-term arable land (AL), transfer of arable (<1900–1992) to grassland (AL to GL), transfer of grassland (<1900–1992) to arable land (GL to AL) and long-term grassland (GL) according to Kölbl et al. (2004). The transfer from arable to grassland soil and vice versa was carried out in 1991; the soil sampling and analysis were conducted in 2001.

prior to mustard application). Stable isotope techniques were used to study the turnover of this material in subplots with different soil management or yield expectation. N_2O as well as the CO_2 measurements have shown that a priming effect has appeared in the soil of the high yield area. In spite of the same amounts of mustard derived C and N in the gas fluxes, the emissions from the high yield area were significantly higher as from the low yield area (Sehy, 2003; Sehy et al., 2003). In the high yield soil the organic material was quickly respired on account of a higher microbial activity (Wessels-Perelo and Munch, 2005). Besides, Wessels-Perelo and Munch (2005) could prove that the microbial biomass of both subfields differed not only concerning activity parameters (i.e. respiration or the metabolic quotient), but that they also showed distinct structural differences.

In the high yield soil higher DOC concentrations were measured and fluorescence spectrometric investigations revealed that this C was strongly converted or more humified than DOC in the low yield area soil (Steinweg, 2002; Zeller, 2006). The application of stable isotope techniques (as the green manure labelled mustard) showed that DOC, microbial C (C_{mic}) as well as microbial N (N_{mic}) represented the soil pools with the highest turnover rates. They are well suited as indicators for changes of environmental conditions or terms of utilisation, because they react within few days on changed conditions, as for example the supply of mustard (Kögel-Knabner and Munch, 2002).

The soils which differed in yield potentials also exhibited different distribution patterns of C and N in the two POM fractions. Initially, a high ^{15}N signal was measured in the free POM fraction immediately after the application of the mustard. Further on the ^{15}N enrichment in the free POM fraction decreased where the enrichment of the occluded POM fraction increased simultaneously. Degrading processes were the reason for the decrease of the ^{15}N signal of both fractions approximately 180 days after mustard application. The ^{15}N content of the high yield soil was then lower in both POM fractions. This was explained with a delay in conversion of the organic material in the low yield soil (Kögel-Knabner and Munch, 2002). It was calculated, that the turnover of fresh organic matter in the low yield soil was approximately delayed by one year. Furthermore, it was derived from these investigations that only a small part of the fresh litter reaches the mineral fraction of the soil (Kölbl and Kögel-Knabner, 2003). Consequently, these fractions are not suitable as an indicator for the detection of short-term management changes because changes of C and N pools in these fractions appear not until decades.

Basically, only modern investigation methods allowed for the differentiation between high and low yield areas of the FAM. The influence of land use (conventionally vs. precision farming) and N fertilisation which was carried out for six years could not be proven until now. As compared to conventional management an increase of the N fertilisation on the high yield fields might lead to higher soil fertility. In contrast decreased N-fertiliser amounts on the low yield fields rather lead to N depletion.

The C enrichment in the topsoils of the organic farm can be considered as positive concerning soil fertility. The better conditions for N minerali-sation in the high yield soil may lead to an easily reduced N_2 fixation on account of slightly increased N_{min} contents. The lower N_{min} contents of the soils from the low yield areas resulted in a higher N_2 fixation improving soil fertility. In the long-term qualitative differences between the sites will presumably rather decrease.

Effect of the management system on K_{CAL}- and P_{CAL}- stocks of the soils

The K^+ concentrations in the CAL extracts decreased slightly within 10 years. However, they were in the 'optimum' range in 2001 (Ebertseder et al., 2003). The stocks of K_{CAL} decreased about 140 kg K^+ ha^{-1} in the integrated and about 110 kg K^+ ha^{-1} in the organically managed soils (Table 1). Obviously, K^+ fertilisation was not necessary in both manage-ment systems due to the high clay contents of the soils, even if potato (as a fruit with a very high K^+ demand) was grown in both systems. Apart from this, a substantial part of the K^+ was returned to the fields with liq-uid manure or solid manure application. In spite of the restrictions which are well known in case of the consideration of K^+ in CAL extracts (e.g. high clay contents or longer drought of the soils reduce K^+ yield), statisti-cally significant relations between the K^+ concentration in 1991 and the change of the K^+ stock during the 10 years between the soil inventories were found on the level of grid points as well as on the field level (Figure 10).

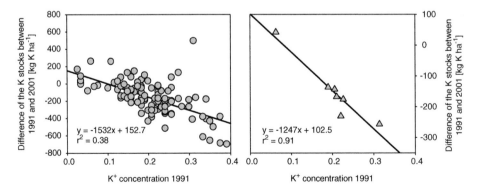

Figure 10 Relationship between the K_{CAL} concentration of the topsoils in 1991 and the change of the K_{CAL} stocks between 1991 and 2001 on the integrated farm on basis of single grid points (left) and on basis of the means of the single arable fields (right).

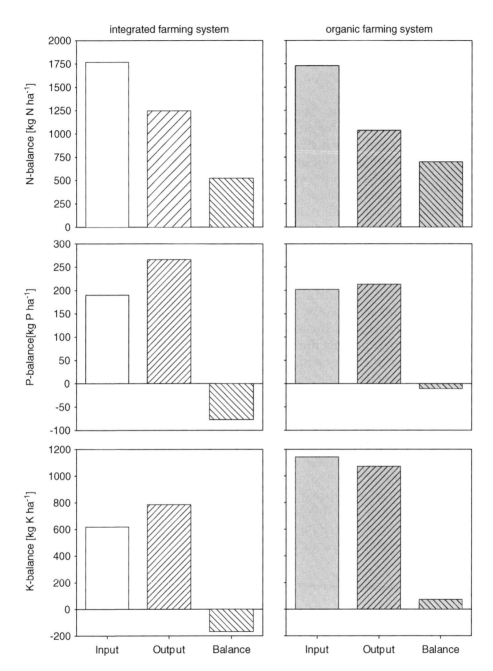

Figure 11 Input/output balances for N (arable fields), P and K (arable fields and grasslands) of the integrated farming system (left column, white bars) and of the organic farming system (right column, grey bars) during the soil inventory decade between 1991 and 2001.

The higher the K^+ supply in 1991, the stronger the decrease of the K^+ stock. A possible reason for this phenomenon might be the fact, that the K^+ concentration asymptotically approaches a limit value when K^+ fertilisation is omitted. This limit value is also reached in the subsoil (Scheffer-Schachtschabel, 1992).

The soil inventories showed that $14\,\mathrm{kg\,K^+\,ha^{-1}\,a^{-1}}$ were annually removed from the integrated managed arable soils. Using a simple input/output balance a deficit of $17\,\mathrm{kg\,K^+\,ha^{-1}\,a^{-1}}$ was calculated for the arable integrated fields (Figure 11). According to the results of the soil inventories K was depleted in the organically managed soils ($-11\,\mathrm{kg\,K^+\,ha^{-1}\,a^{-1}}$). The K^+ input/output balance proved an enrichment of $7\,\mathrm{kg\,K^+\,ha^{-1}\,a^{-1}}$ for these soils. Watson et al. (2002) as well as Vos and van der Putten (2000) reported K^+ depletions in the same scale after a five-year balance.

The P stocks decreased between 1991 and 2001 about $52\,\mathrm{kg\,P\,ha^{-1}}$ in the arable soils of the integrated and about $38\,\mathrm{kg\,P\,ha^{-1}}$ in the arable soils of the organic farm (Table 1). Despite no mineral P fertiliser was added since 1991, the P concentration in the CAL extracts can still be classified as 'optimum'. Consequently, yield depressions by the omission of P fertilisation can be excluded. Investigations on root production and distribution in a soil where P fertilisation was omitted since 14 years showed that net root formation was approximately 2-3-fold higher as compared to a P fertilised control soil (Steingrobe et al., 2001b). At the same time higher root mortality rates were determined under P-deficiency. The higher root production under P-deficiency led to a clearly better spatial development of the root system by the plant. According to Scheffer-Schachschabel (1992) the yield depression to be expected in the integrated as well as in the organically managed arable soils is negligible low with the ascertained P stocks and concentrations.

Soil inventories revealed an annual removal of $5.2\,\mathrm{kg\,P\,ha^{-1}\,a^{-1}}$ in the integrated managed arable soils, whereas the calculated P balance on the same level was $-7.7\,\mathrm{kg\,P\,ha^{-1}\,a^{-1}}$. The corresponding values for the organically managed arable soils were $3.8\,\mathrm{kg\,P\,ha^{-1}\,a^{-1}}$ and $-1.1\,\mathrm{kg\,P\,ha^{-1}\,a^{-1}}$ (cf. Table 1).

2.1.4 Conclusions

In the arable soils of the integrated farm reduced soil tillage led to the formation of a new A_p horizon with a lower thickness and higher humus content. Obviously, turnover of C compounds due to C respiration in this new A_p horizon is very intensive resulting in a relative N enrichment. The increase of the C_{org} contents in the new A_p horizon (0–10 cm) did not compensate for the C losses in 10–23.8 cm depths (old A_p thickness), so that the whole amounts of C_{org} decreased between 1991 and 2001. The stocks of N have increased in the same period.

The N stocks in the organically managed arable soils rose stronger than the C stocks, however, humus C also increased between 1991 and 2001. This disproportionate enrichment of N in the organic farming system was discussed in the context of the supply of materials with a narrow C-to-N ratio and the removal of cereal straw with a wide C-to-N ratio. It is concluded that soil fertility was enhanced.

Soil analyses and nutrient balance approaches are confirmed as suitable instruments for the optimisation of the nutrient management on farms. Soil analysis indicates the variability of the supply of nutrients. The results can be used as an indicator for the soil status, whereas nutrient balances pool the variability of the nutrient fluxes to larger units. They are to be understood as a process indicator. For the latter purpose, long-term data sets covering at least one whole crop rotation phase are required.

Modern analytical methods revealed changes in C dynamics and different C fractions. However, a reliable assessment of management-derived changes of C in soils was not possible due to the short investigation period.

Although no mineral supply of P or K has taken place since 1991 on both farms, reduced yields can be excluded as a result (i) of the high supply at the beginning of the FAM project and probably (ii) of the good buffering and soil weathering (nutrient delivery from soil) in the investigation region.

Acknowledgement

The scientific activities of the *FAM Munich Research Network on Agroecosystems* were financially supported by the German Federal Ministry of Education and Research (BMBF 0339370). Overhead costs of the Research Station Scheyern are funded by the Bavarian State Ministry for Science, Research and the Arts.

References

Al-Kaisi MM, Yin HX, Licht MA, 2005. Soil carbon and nitrogen changes as influenced by tillage and cropping systems in some Iowa soils. Agriculture Ecosystems and Environment 105, 635–647.

Angers DA, Bolinder MA, Carter MR, Gregorich EG, Drury CF, Liang BC, Voroney RP, Simard RR, Donald RG, Beyaert RP, Martel J, 1997. Impact of tillage practices on organic carbon and nitrogen storage in cool, humid soils of eastern Canada. Soil and Tillage Research 41, 191–201.

Bach M, Frede HG, 2005. Methodische Aspekte und Aussagemöglichkeiten von Stickstoff-Bilanzen. Institut für Landwirtschaft und Umwelt (Ed.), Heft 9/2005, Bonn.

BAD, 2003. Nährstoffverluste aus landwirtschaftlichen Betrieben mit einer Bewirtschaftung nach guter fachlicher Praxis. Bundesarbeitskreis Düngung (Ed.), Frankfurt/Main.

Blackmer AM, Bremner JM, 1978. Inhibitory effect of nitrate on reduction of N_2O to N_2 by soil microorganisms. Soil Biology and Biochemistry 10, 187–191.

Breitschuh G, Hochberg H, Zorn W, 2004. Nährstoffbilanzen unter Nachhaltigkeitsaspekten. In: BAD (Hrsg.) Folgen negativer Nährstoff-bilanzen in Ackerbaubetrieben, Frankfurt/Main.

Bundesregierung, 2004. Perspektiven für Deutschland – Unsere Strategie für eine nachhaltige Entwicklung. Fortschrittsbericht, Deutsche Bundesregierung, Berlin.

Buyanovsky GA, Aslam M, Wagner GH, 1994. Carbon turnover in soil physical fractions. Soil Science Society of America Journal 58, 1167–1173.

Diepenbrock W, Rost D, Hülsbergen KJ, Heine M, Deimer C, Meyer D, Heinrich J, Dubsky G, 1997–1998. Forschungsbericht zu den Projekten Informationssystem 'Agrarumwelt-indikatoren' und Betriebs-Bilanzierungsmodell REPRO. Inst. f. Acker- und Pflanzenbau, MLU Halle-Wittenberg.

Douds DD, Janke RR, Peters SE, 1993. VAM fungus spore populations and colonialization of roots of maize and soybean under conventional and low-input sustainable agriculture. Agriculture Ecosystems and Environment 43, 325–335.

Düngeverordnung, 2006. Verordnung über die Grundsätze der guten fachlichen Praxis beim Düngen vom 10.01.2006. BGBl, I, 2006.

Ebertseder T, Weinfurtner KH, Nätscher L, Kainz M, Gerl G, 2003. Ermittlung bewirtschaftungs-bedingter Veränderungen verfügbarer P- und K-Vorräte durch Bodenuntersuchungen und Bilanzierung. Vortrag VDLUFA-Kongress, Saarbrücken.

Eckert H, Breitschuh G, Sauerbeck D, 1999. Kriterien umweltverträglicher Landbewirtschaftung (KUL) – ein Verfahren zur ökologischen Bewertung von Landwirtschaftsbetrieben. Agribiological Research 52, 57–76.

Embert G, 2004. Nährstoffversorgung, Nährstoffbilanzen und Bodenfruchtbarkeit – gesetzliche Regelungen und Vorgaben. In: BAD (Ed.), Folgen negativer Nährstoffbilanzen in Ackerbaubetrieben. Bundesarbeitskreis Düngung (BAD), Frankfurt/Main, pp. 7–18.

Flessa H, Ruser R, Dörsch P, Kamp T, Jimenez MA, Munch JC, Beese F, 2002. Integrated evaluation of greenhouse gas emissions from two farming systems in southern Germany: Special consideration of soil

N_2O emissions. Agriculture Ecosystems and Environment 91, 175–189.

Frede HG, Dabbert S, 1999. Handbuch zum Gewässerschutz in der Landwirtschaft. Ecomed-Verlag.

Gauger T, Anshelm F, Schuster H, Erisman JW, Vermeulen AT, Draaijers GPJ, Bleeker A, Nagel HD, 2002. Mapping of ecosystem long-term trends in deposition loads and concentrations of air pollutants in Germany and their comparison with Critical Levels. Final Report on behalf of Federal Environmental Agency (UBA), Berlin. BMU/UBA FE-No 299 42 210. Part 1: Deposition Loads 1990–1999, Part 2: Mapping Critical Levels Exceedances.

Gutser R, 2006. Bilanzierung von Stickstoffflüssen im landwirtschaftlichen Betrieb zur Bewertung und Optimierung der Düngungsstrategien. Acta agriculturae Slovenica 87, 129–141.

Gutser R, Ebertseder T, 2001. Unvermeidbare Nährstoffverluste in der Landwirtschaft. In: Düngung; Baustein nachhaltiger Landwirtschaft. Tagung des Verbands der Landwirtschaftskammern e.V. und des Bundesarbeitskreises Düngung BAD (Eds.), Würzburg, pp. 95–114.

Gutser R, Matthes U, 2001. Gute fachliche Praxis der Düngung aus Sicht der Ökonomie und Ökologie. KTBL-Schrift 400, 91–102.

Gutser R, Reents HJ, Rühling I, Schmid H, Weinfurtner KH, 2003. Flächen- und betriebs-bezogene Indikatoren auf der Grundlage des Langzeitmonitorings. FAM-Bericht 55, 147–159.

Gutser R, Steingrobe B, Claassen N, 1998. Wurzelumsatz und – entwicklung während des Pflanzenwachstums und Kohlenstoffinput in den Boden. In: Filser J (Ed.), Schlussbericht 1993–1997, FAM-Bericht 28, pp. 15–22.

Halvorson AD, Peterson GA, Reule CA, 2002. Tillage system and crop rotation effects on dryland crop yields and soil carbon in the central Great Plains. Agronomy Journal 94, 1429–1436.

Hantschel RE, Priesack E, Hoeve R, 1994. Effects of mustard residues on the carbon and nitrogen turnover in undisturbed soil microcosms. Journal of Plant Nutrition and Soil Science 157, 319–326.

Hellmeier C, 2001. Stofftransport in der ungesättigten Zone der landwirtschaftlich genutzten Flächen in Scheyern/Oberbayern (Tertiärhügelland). PhD thesis, LMU München, ISSN 0721-1694, GSF-Bericht 07/01, p. 170.

Heuwinkel H, Locher F, Gutser R, Schmidhalter U, 2002. Variabilität der symbiontischen N_2-Fixierung: Methoden zur Abschätzung und Ursachen der Variation. FAM-Bericht 55, 69–73.

Honisch M, Hellmeier C, Weiss K, 2002. Response of surface and subsurface water quality to land use change. Geoderma 105, 277–298.

Huber B, Winterhalter M, Mallén G, Hartmann HP, Gerl G, Auerswald K, Priesack E, Seiler KP, 2005. Wasserflüsse und wassergetragene

Stoffflüsse in Agrarökosystemen. In: Osinski E, Meyer-Aurich A, Huber B, Rühling I, Gerl G, Schröder P (Eds.), Landwirtschaft und Umwelt – ein Spannungsfeld. Ergebnisse des Forschungsverbunds Agraröko-systeme München (FAM). Oekom-Verlag, München, pp. 57–98.

Hülsbergen KJ, 2003. Entwicklung und Anwendung einers Bilanzierungsmodells zur Bewertung der Nachhaltigkeit landwirts- chaftlicher Systeme. Habil.-Schrift, University Halle-Wittenberg, Shaker-Verlag, Aachen.

Hülsbergen KJ, Abraham J, Biermann S, Werner S, Hensel G, Diepenbrock W, 2000. Einsatz des Modells REPRO zur Stoff- und Energiebilanzierung im Versuchsgut Scheyern. Bericht im Rahmen des Forschungsprojekts "Langzeitmonitoring und Indikatoren" der TU München, Inst. f. Acker- und Pflanzenbau, MLU Halle-Wittenberg, pp 78.

Hülsbergen KJ, Abraham J, Küstermann B, Hensel G, 2002. Einsatz des Modells REPRO im Versuchsgut Scheyern. Forschungsbericht II, ,Langzeitmonitoring und Indikatoren' der TU München, Inst. f. Acker- und Pflanzenbau, MLU Halle-Wittenberg.

Hülsbergen KJ, Diepenbrock W, 1997. Das Modell REPRO zur Stoff- und Energiebilanzierung im Versuchsgut Scheyern. Beiträge zur 6. Wissenschaftstagung zum ökologischen Landbau, 67–74.

Isermann K, 2002. Atmosphärischen N-Einträge als unabdingbare Bestandteile der N-Bilanzen von Agrarökosystemen sowie deren tolerierbaren bzw. unvermeidbaren gasförmigen N-Emissionen darge-stellt am Beispiel Deutschlands. Workshop; N-Depositionen in Agrarökosystemen, UFZ, Halle.

Isermann K, Isermann R, 1995. Tolerierbare Nährstoffsalden der Landwirtschaft ausgerichtet an den kritischen Eintragsraten und – konzentration der naturnahen Ökosysteme. In: Umweltverträgliche Pflanzenproduktion, pp. 127–152.

Jarecki MK, Lal R, James R, 2005. Crop management effects on soil carbon sequestration on selected farmers' fields in northeastern Ohio. Soil and Tillage Research 81, 265–276.

Jenkinson DS, 1990. The turnover of organic carbon and nitrogen in soils. Philosophical Transactions of Royal Society of London, Series B, Biological Sciences 329, 361–368.

Kainz M, Weinfurtner KH, Schmid H, Danier HJ, Matthes U, Rühling I, Gerl G, Gutser R, 2003. Ermittlung bewirtschaftungsbedingter Veränderungen des C- und N-Umsatzes von Böden durch Bodenunter-suchung und Stoffbilanzierung. Vortrag VDLUFA-Kongress, Saarbrücken.

Kaiser EA, Ruser R, 2000. Nitrous oxide emissions from arable soils in Germany – An evaluation of six long-term field experiments. Journal of Plant Nutrition and Soil Science 163, 249–260.

Katz P, 1997. Cited by Menzi H, Frick R, Kaufmann R, 1997. Ammoniak Emissionen in der Schweiz. Schriftenreihe der FAL 26.

Kilian A, Gutser R, Claasen N, 1998. N_2O emissions following long-term organic fertilization at different levels. Agribiological Research 51, 27–36.

Kögel-Knabner I, Munch JC, 2002. CN-Pools und ihre Relevanz für Nährstoffnachlieferung und Strukturzustand des Bodens. FAM-Bericht 55, 35–38.

Kölbl A, Kögel-Knabner I, 2003. Isolierung und Charakterisierung der partikulären CN-Pools. FAM-Bericht 53, 67–72.

Kölbl A, Leifeld J, Kögel-Knabner I, 2004. Isolierung und Charakterisierung der C- und N-Pools. FAM-Bericht 62/II, 73–82.

Körschens M, Schulz E, 1999. Die organische Bodensubstanz – Dynamik – Reproduktion – Ökonomisch ökologisch begründete Richtwerte. UFZ-Bericht 13, pp. 46.

Körschens M, Weigel A, Schulz E, 1998. Turnover of soil organic matter (SOM) and long-term balances – Tools for evaluating sustainable productivity of soils. Journal of Plant Nutrition and Soil Science 161, 409–424.

Leifeld J, Kögel-Knabner I, 2005. Soil organic matter fractions as early indicators for carbon stock changes under different land-use? Geoderma 124, 143–155.

Leithold G, Hülsbergen KJ, Michel D, Schönmeier H, 1997. Humusbilanzierung – Methoden und Anwendung als Agrar-Umweltindikator. In: Umweltverträgliche Pflanzenproduktion – Indikatoren, Bilanzierungsansätze und ihre Einbindung in Ökobilanzen. DBU Bd. 5, 43–55.

Liang BC, McConkey BG, Campell CA, Curtin D, Lafond GP, Brandt SA, Moulin AP, 2004. Total and labile soil organic nitrogen as influenced by crop rotations and tillage in Canadian prairie soils. Biology and Fertility of Soils 39, 249–257.

Lüttich M, Dämmgen U, Eurich-Menden B, Döhler H, Osterburg B, 2004. Calculations of emissions from German agriculture – national inventory report (NIR) 2004 for 2002. Pt 2: tables, Landbauforschung Völkenrode Sonderheft 260, 33–198.

Marschner H, 1990. Mineral Nutrition of Higher Plants. Academic Press, Inc., San Diego, CA.

Marx M, Buegger F, Gattinger A, Zsolnay A, Munch JC, 2006. Determination of fate of maize and wheat exudates-C in an agricultural soil during a short-term incubation. European Journal of Soil Science *submitted*.

Matthes U, Gutser R, Gerl G, Kainz M, 2001. Stickstoffverluste durch ressourcen-schonende Bewirtschaftung – dargestellt am Beispiel des Versuchsguts Scheyern. VDLUFA-Schriftenreihe 57, 237–245.

Pimentel D, Hepperly P, Hanson J, Douds D, Seidel R, 2005. Environmental, energetic, and economic comparisons of organic and conventional farming systems. Bioscience 55, 573–582.

Rühling I, Ruser R, Kölbl A, Priesack E, Gutser R, 2005. Kohlenstoff und Stickstoff in Agrarökosystemen. In: Osinski E, Meyer-Aurich A, Huber B, Rühling I, Gerl G, Schröder P (Eds.), Landwirtschaft und Umwelt – ein Spannungsfeld. Ergebnisse des Forschungsverbunds Agrarökosysteme München (FAM). Oekom-Verlag, München, pp. 99–154.

Ruser R, Flessa H, Schilling R, Beese F, Munch JC, 2001. Effects of crop-specific field management and N fertilization on N$_2$O emissions from a fine-loamy soil. Nutrient Cycling in Agroecosystems 59, 177–191.

Ruser R, Sehy U, Weber A, Gutser R, Munch JC, 2007. Main driving variables and effect of soil management on climate or ecosystem relevant trace gas fluxes from fields of the FAM. In: Schröder P, Pfadenhauer J, Munch JC (Eds.), Perspectives for agroecosystem management – balancing environmental and socioeconomic demands, pp. 79–120.

Scheffer-Schachtschabel, 1992. Lehrbuch der Bodenkunde. 12. Auflage, Ferdinand Enke Verlag, Stuttgart.

Schmid H, Gutser R, Claasen N, 1997. Wurzelentwicklung und – umsatz von Winterweizen. Beiträge zur Wiss.-Tagg. Ökol. Landbau, Bonn, pp. 279–285.

Schmid H, Gutser R, Steingrobe B, Claassen N, 1998. Wurzelumsatz und – entwicklung während des Pflanzenwachstums und Kohlenstoffinput in den Boden. In: von Lützow M, Filser J, Kainz M, Pfadenhauer J (Eds.), FAM Jahresbericht 1997, FAM-Bericht 22, 25–30.

Sehy U, 2003. N$_2$O-Freisetzungen landwirtschaftlich genutzter Böden unter dem Einfluss von Bewirtschaftungs-, Witterungs- und Standortfaktoren. PhD thesis, TU München, Oekom Verlag, München, ISBN 3-936581-40-1, pp. 130.

Sehy U, Ruser R, Munch JC, 2003. Nitrous oxide fluxes from maize fields: Relationship to yield, site-specific fertilization and soil conditions. Agriculture Ecosystems and Environment 99, 97–111.

Seiler KP, Hellmeier C, 2001. Humic substances and agrochemicals in the discharge components of Scheyern, Freiburger Schriften zur Hydrologie 13, 367–374.

Sinowski W, 1995. Die dreidimensionale Variabilität von Bodeneigenschaften – Ausmaß, Ursachen und Interpolation. PhD thesis, TU München, Shaker Verlag, Aachen, ISBN 3-8265-0994-3, FAM-Bericht 7, pp. 158.

Steingrobe B, Schmid H, Claassen N, 2001a. Root production and root mortality of winter barley and its implication with regard to phosphate acquisition. Plant Soil 237, 239–248.

Steingrobe B, Schmid H, Claassen N, 2001b. The use of ingrowth core method for measuring root production of arable crops – influence of soil and root disturbance during installation of the bags on root ingrowth into the cores. European Journal of Agronomy 15, 143–151.

Steinweg B, 2002. Untersuchungen zur in-situ Verfügbarkeit von wasserlöslichem Humus (DOM) in Oberböden und Grundwasser-leitern. PhD thesis, TU München, Hieronymus Verlag, München, ISBN 3-89791-245-7, FAM-Bericht 52, pp. 109.

Swinnen J, van Veen JA, Merckx R, 1995. Carbon fluxes in the rhizosphere of winter wheat and spring barley with conventional vs. integrated farming. Soil Biology and Biochemistry 27, 811–820.

UBA, 1999. Entwicklung von Parametern und Kriterien als Grundlage zur Bewertung ökologischer Leistungen und Lasten der Landwirtschaft – Indikatorensysteme. UBA-Texte 42/99.

van der Ploeg R, Bach R, Efken E, 1991. Probleme einer umweltverträglichen Nährstoffver-sorgung am Beispiel der Stickstoffdüngung – Das Nitratproblem und die SCHALVO. In: Landwirtschaftliche Hochschultagung, Uni Hohenheim, 1–22.

van Noordwijk M, Brouwer G, Koning H, Meijboom FW, Grzebisz W, 1994. Production and decay of structural root material of winter wheat and sugar beet in conventional and integrated cropping systems. Agriculture Ecosystems and Environment 51, 99–113.

VDLUFA, 1998. Standpunkt 'Kriterien umweltverträglicher Landbewirtschaftung (KUL)'. VDLUFA (Ed.), VDLUFA-Infos http://www.vdlufa.de.

VDLUFA, 2004. Standpunkt 'Humusbilanzierung – Methode zur Beurteilung und Bemessung der Humusversorgung von Ackerland'. VDLUFA (Ed.), VDLUFA-Infos http://www.vdlufa.de.

VDLUFA, 2007. Standpunkt 'Nährstoffbilanzierung im Landwirts-chaftsbetrieb'. VDLUFA (Ed.), in press.

von Lützow M, Leifeld J, Kainz M, Kögel-Knabner I, Munch JC, 2002. Indications for soil organic matter quality in soil under different man-agement. Geoderma 105, 243–258.

von Lützow M, Palojärvi A, 1995. Microbial mediated C and N pools and microbial community structure. In: Kjoller A (Ed.), Final Report of the EU Environment Programme MICS. No. EV5V-CT94-0343: 20.

Vos J, van der Putten PEL, 2000. Nutrient cycling in a cropping system with potato, spring wheat, sugar beet, oats and nitrogen catch crops. I. Input and offtake of nitrogen, phosphorus and potassium. Nutrient Cycling in Agroecosystems 56, 87–97.

Watson CA, Bengtsson H, Ebbesvik M, Loes AK, Myrbeck A, Salomon E, Schroder J, Stockdale EA, 2002. A review of farm-scale nutrient budgets for organic farms as a tool for management of soil fertility. Soil Use Management 18, 264–273.

Weber A, Gutser R, Henkelmann G, Schmidhalter U, 2000. Unvermeidbare NH_3-Emissionen aus mineralischer Düngung (Harnstoff) und Pflanzenmulch unter Verwendung einer modifizierten Messtechnik. VDLUFA-Schriftenreihe 53, 175–182.

Weier KL, Doran JW, Power JF, Walters DT, 1993. Denitrification and the dinitrogen/nitrous oxide ratio as affected by water, available carbon, and nitrate. Soil Science Society of America Journal 57, 66–72.

Weigel A, Russow R, Körschens M, 2000. Quantification of airborne N-input in long-term field experiments and its validation through measurements using ^{15}N isotope dilution. Journal of Plant Nutrition and Soil Science 163, 261–265.

Weinfurtner KH, 2002. Geostatistische Auswertung der Bodeninventur 2001 des Forschungs-verbunds Agrarökosysteme München (FAM), Fraunhofer Institut für Molekularbiologie und Angewandte Ökologie, Schmallenberg.

Weinfurtner KH, 2007. Changes in nutrient status on the experimental station Klostergut Scheyern from 1991 to 2001 – Statistical and geo-statistical analysis. In: Schröder P, Pfadenhauer J, Munch JC (Eds.), Perspectives for agroecosystem management – balancing environmental and socioeconomic demands, pp. 375–406.

Wessels-Perelo L, Munch JC, 2005. Microbial immobilisation and turnover of 13C labelled substrates in two arable soils under field and laboratory conditions. Soil Biology and Biochemistry 37, 2263–2272.

Zeller K, 2006. Zeitliche Dynamik und räumliche Variabilität von wasserlöslichem Humus in Ap-Horizonten von Ackerböden mit unterschiedlichem Ertragspotenzial und unterschied-licher Bewirtschaftung. PhD thesis, TU München, http://mediatum. ub.tum.de, pp. 123.

Chapter 2.2

Main Driving Variables and Effect of Soil Management on Climate or Ecosystem-Relevant Trace Gas Fluxes from Fields of the FAM

R. Ruser, U. Sehy, A. Weber, R. Gutser and J.C. Munch

2.2.1 Production and consumption of the climate-relevant trace gases nitrous oxide and methane in soils

Nitrous oxide (N_2O) and methane (CH_4) are climate-relevant trace gases. The share of these two gases to the anthropogenic greenhouse effect was estimated to comprise 6 and 25% (IPCC, 2001). Besides the anthropogenic greenhouse effect, it was shown that N_2O is involved in stratospheric ozone destruction (Crutzen, 1981). In the stratosphere, N_2O reacts with oxygen radicals to form NO, which then catalyses the O_3 degradation.

Agricultural activities account for 75% of the anthropogenic N_2O emissions (Duxbury et al., 1993; Isermann, 1994); approximately 85% of these are derived from soil emissions (Bremner, 1997), with biological

Perspectives for Agroecosystem Management
Edited by P. Schröder, J. Pfadenhauer and J.C. Munch

nitrification and denitrification as the two most important N_2O sources in soils (Hutchinson and Mosier, 1993). Both processes are enhanced with increasing substrate supply (N concentrations). Consequently, high N_2O emissions from agricultural soils are mainly resulting from the increased N input into agricultural soils through N fertilisers, organic fertilisers and N accumulation by legumes (IPCC, 1996). To calculate N_2O emissions from agricultural soils on a national or a regional scale, the IPCC (2001) suggests an emission factor of 1.25% of the total annual N input. This linear emission factor originally was deduced by Bouwman (1996) to relate the N_2O emission to the amount of annually applied N fertiliser. On the basis of the rare data availability of field measurement series, this emission factor was developed only from field experiments on maize and grassland sites.

Duxbury et al. (1993) calculated that approximately 70% of the total anthropogenic CH_4 emission was related to agricultural activities. The most important CH_4 sources are the anaerobic production of CH_4 by methanogenic microorganisms during ruminal C fermentation, the production in flooded soils cultivated with rice, and the degassing of anaerobically produced CH_4 during manure storage in relation to the field application of the manure (IPCC, 1996).

In well-aerated soils CH_4 is oxidised to CO_2 by methanotrophic microorganisms. It was estimated that between 3 and 9% of the total anthropogenic CH_4 emission is re-oxidised annually in soils (Prather et al., 1995). N fertilisation generally reduces the potential of CH_4 oxidation in aerated soils due to the competitive effect of NH_4^+ at the active sites of the methane-monooxygenase (MMO), which is the responsible enzyme for CH_4 oxidation (Prather et al., 1995).

2.2.2 Important controlling factors for N_2O fluxes at the FAM sites

Numerous works have been published describing the effects of soil internal or external factors on the production of N_2O emissions during nitrification and denitrification (summarised by Granli and Bøckman, 1994; Bremner, 1997; Sahrawat and Keeney, 1986). In the following, we shortly describe the driving variables that were found to be the main controlling factors for N_2O emissions at the FAM study site in Scheyern.

Effect of soil temperature on the N_2O emissions from FAM sites

N_2O production rates during nitrification and denitrification rise with increasing soil temperature. Maximum emission rates have been found between 25°C and 35°C during nitrification (Haynes, 1986) and between 30°C and 50°C during denitrification (Granli and Bøckman, 1994).

Most investigations on N$_2$O emissions from arable and grassland soils at the FAM research station revealed positive relationships with the soil nitrate contents and soil moisture data of the topsoil (Flessa et al., 1995, 1996a, 2002a; Ruser et al., 1998, 2001; Sehy et al., 2003). In combination with more results from field investigations in Scheyern, these data indicate that denitrification is the most dominant process for N$_2$O production at nearly all study sites at the FAM research farm. Annually determined data of soil moisture are significantly negative correlated with soil temperature. This was supported by the investigations of Ruser (1999) on wheat, potato, corn, and fallow fields [$y = 42.8$–0.52 water-filled pore space (WFPS), $r^2 = 0.53$, $n = 598$]. High soil temperatures generally were measured when the soil moisture was far too low to induce the essential anaerobic conditions for increased N$_2$O production. We conclude that the effect of soil temperature on the N$_2$O emission rates from the FAM sites per se was negligible.

Using a semi-continuous fully automated sampling system, it was shown that daily N$_2$O fluctuations occurred less frequently at the FAM sites than was expected from literature. The only reproducible daily N$_2$O flux pattern was found after the chemical treatment of potato cabbage (Flessa et al., 2002b). The N$_2$O flux rates after this treatment were strongly correlated to the soil temperature in 2.5 cm depth with Q_{10} values >6. High Q_{10} values of the same order of magnitude for N$_2$O fluxes from soils have also been reported from Smith (1997) and from Arah and Smith (1989). The high Q_{10} values were explained with an increased denitrification activity, where the temperature effect was amplified by a low O$_2$ availability due to high soil moisture contents of the soil or due to an increased microbial O$_2$ consumption.

Effect of freezing-thawing cycles on N$_2$O fluxes from soils at the FAM sites

Trace gas flux determination at the FAM research station was carried out on an annual basis. Extremely high flux rates were found during the cold winter months. These high emission peaks, which partially exceeded the emissions maxima after N fertilisation, coincided with the thawing of frozen soil (Flessa et al., 1995). Additionally, it was shown that the N$_2$O background emission during expanded periods of severe soil frost was higher as compared to the background emission during the summer period (Ruser et al., 2001). Closed snow covers did not restrict N$_2$O fluxes (Flessa et al., 1995; Kamp et al., 1998), and they did not decrease the N$_2$O flux rates (Röver et al., 1998). These results varied distinctly from those known in regard to temperature impact on N$_2$O emission. It was generally assumed that the N$_2$O emission during the cold winter period is negligible for annual N$_2$O budgets due to the low soil temperatures. The emissions during the winter period were calculated by the interpolation of measurements in autumn and early springtime. Kaiser and Ruser (2000) summarised the results from five long-term experiments on the N_2O emissions from German agricultural study sites with annual measurement data. They calculated that approximately 50% of the total annual N_2O emission occurred beyond the growing season during the cold winter period. Currently, the following processes are discussed to contribute to the high N_2O emissions during the winter period (Flessa, 2000):

- N_2O production during denitrification and N_2O enrichment below the frozen soil layer (Burton and Beauchamp, 1994). The production of N_2O in the subsoil is favoured by the reduced O_2 diffusion into the soil due to the presence of a compact ice layer. During thawing of this ice layer, the N_2O enriched in the subsoil is liberated into the atmosphere as an emission pulse.

- Production of N_2O in the free liquid phase of the frozen soil (Röver et al., 1998). Freezing causes the concentration of salts. Consequently, a part of the soil solution remains liquid even in strongly frozen soils. Depending on the soil texture, this effect may account for between 5 and approximately 20% of the total soil water (Stähli and Stadler, 1997). Chemical or biological denitrification and linked N_2O production may be enhanced in these strongly nutrient-enriched liquid compartments.

- N_2O production during the thawing of the topsoil (Christensen and Tiedje, 1990; Christensen and Christensen, 1991; Wagner-Riddle et al., 1997). A part of the microbial biomass breaks down during the preceding frost period, and soil aggregates are disrupted resulting in an increased C and N availability during thawing. In addition to the high soil moisture, thawing water accumulates on the frozen subsoil and O_2 diffusion into the soil is further reduced and this favours denitrification. The magnitude of the N_2O emissions during soil thawing was shown to be dependent on the duration and the strength of the preceding frost period. Extremely high emissions were found after long frost periods (Flessa et al., 1995; Papen and Butterbach-Bahl, 1999). If several frost/thaw cycles occur during one winter, N_2O emission decreases from cycle to cycle due to a decreased substrate availability (Schimel and Klein, 1996; Röver et al., 1998).

All these processes can proceed simultaneously, so that their share to the in situ N_2O emission cannot be quantified exactly. Using ^{15}N-enriched nitrate and measuring the ^{15}N abundance of the N_2O emitted during soil thawing, Sehy et al. (2004) could show that denitrification was the main N_2O source in a loamy soil from a FAM site. This finding was supported by the results of a field experiment from Ruser et al. (2001), who found a strong positive correlation between the nitrate contents of the topsoil and the winter N_2O emissions from plots with different arable crops (Figure 1). The very high N_2O emission peaks during thawing at the FAM usually

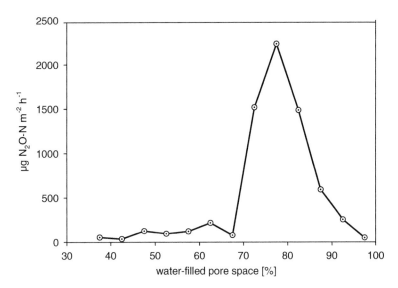

Figure 1   Effect of soil moisture on the N₂O flux rate from potato plots during the growing season. Soil moisture data were calculated in groups spanning a range of 5% WFPS each. Fluxes and moisture were measured on two differently fertilised plots in the ridge soil, in the uncompacted interrow soil and in the tractor-compacted interrow soil (Ruser et al., 1998).

occur between December and mid-March. During this time, the accumulation of N₂O in the subsoil may be less evident because the soils are not fully thawed in this period. Thawing water consistently accumulated on the remaining frost body and may have acted as an additional diffusion barrier for N₂O from the subsoil.

Effect of soil moisture and rainfall on N₂O fluxes from the soils at the FAM

Davidson (1991) modelled the N₂O emission and the (relative contribution) source strength from nitrification and denitrification (to N₂O and N₂) in soils as affected by soil moisture. Nitrification mainly occurs in well-aerated soil compartments and is consequently the main source for N₂O in dry and fresh soils. The portion of N₂O from denitrification to the total emission increases with increasing soil moisture due to the decreased diffusion of O₂ into the soil. At soil moisture contents above field capacity, N₂O is further reduced to N₂ by denitrification resulting in a decrease of the N₂O flux rates. This model was supported by field and laboratory experiments in context with the FAM. Figure 1 shows the mean N₂O emission rates from potato fields plotted against grouped soil moisture classes. Obviously, high emissions occurred at soil moisture contents above 70% WFPS. This critical value for strongly increased N_2O fluxes indicated that denitrification is the main N_2O source at high soil moisture and was also found in field investigations from Dörsch (2000). Ruser et al. (2006) also reported a threshold value of >60% WFPS and an increase of the N_2O flux rates by a factor of 40–200 above this value. These results were obtained in a laboratory experiment using undisturbed soil cores from differently compacted areas of a potato field. The magnitude of the N_2O emission was also supported by field experiments carried out by Ruser et al. (1998). In accordance with the model of Davidson (1991), Ruser et al. (2006) showed that nitrification was the main N_2O source below the mentioned threshold value of approximately 65% WFPS. Flessa et al. (1996b) measured N_2O emissions from soil cores under moisture conditions where nitrification can be considered as the sole N_2O source. Related to the amount of N nitrified, the emission accounted for approximately 0.06%.

N_2O flux rates show a very high temporal variability. Strongly increased N_2O fluxes after rainfall events have been reported from arable soils (Flessa et al., 1995; MacKenzie et al., 1997; Corre et al., 1996; Dobbie et al., 1999), from grassland soils (Clayton et al., 1994; De Klein and van Logtestjin, 1994) and from an arable soil after irrigation (Mosier and Hutchinson, 1981). Mosier et al. (1986) attributed the increased N_2O emissions after precipitation to increased soil moisture contents and an enhanced denitrification due to the decreased soil aeration. Extremely high N_2O fluxes were measured after rainfall following severe periods of drought (Firestone and Davidson, 1989; Cates and Keeney, 1987; Rudaz et al., 1991). At the FAM sites, high emission rates were induced by rewetting dry soil in summer. This effect was further amplified by low water infiltration rates in dry soil and by the high portion of silt, which promotes soil sealing. Reduced water infiltration rates apparently resulted in a surface accumulation of rain water, which directly reduced O_2 diffusion into the soil. Furthermore, it was reported in a literature review on the dynamics of dissolved organic carbon (DOC) in soils that soil drying induced an increase of microbial easily available carbon compounds in soil solution (Kalbitz et al., 2000). As indicated by the high CO_2 pulses during rewetting, these compounds are assimilated quickly after rewetting of soil. The high oxygen consumption provides anaerobic conditions which additionally stimulate N_2O production during denitrification (Flessa and Beese, 1995).

The results of all investigations on N_2O emissions at FAM sites revealed strong positive correlations between soil moisture and the N_2O flux rates. Consequently, high N_2O emission rates were found after strong rainfall events, especially enhanced by high substrate availabilities, e.g. after N fertilisation or after rewetting of dry soil in high summer (Dörsch, 2000; Kamp, 1998; Kamp et al., 2000; Ruser, 1999; Sehy, 2003). Soil moisture was identified as one of the most important driving variables for N$_2$O fluxes from arable soils. It shows a high temporal variability. By using near-continuous measurements with the field laboratory, a sampling design was developed for the reliable determination of annual N$_2$O emissions (Flessa et al., 2002b). It was shown for agroecosystems that weekly N$_2$O measurements generally are sufficient to reliably assess annual N$_2$O emissions. However, these measurements have to be completed by event-oriented measurements after strong rainfall following strong soil drying or thawing of frozen soil.

Effect of the N availability on N$_2$O emissions from FAM sites

NO$_3^-$ and NH$_4^+$ are important controlling factors for the N$_2$O production in agricultural soils. The nitrification rate is limited by the availability of NH$_4^+$, since the oxidation rate is higher as compared to the supply of NH$_4^+$. Generally, high NH$_4^+$ concentrations are present in the short term (e.g. after the application of NH$_4^+$-containing fertilisers or under very wet soil conditions). The application of NH$_4^+$ fertilisers increased N$_2$O emissions in incubation experiments; however, related to the amount of N added, the cumulative emission accounted for only between 0.1 and 0.45% (Bremner and Blackmer, 1981; Goodroad and Keeney, 1984). A similar low NH$_4^+$–N-inducing N$_2$O emission of 0.06% has also been reported from a microcosm experiment with a loamy soil from the FAM research station (Flessa et al., 1996b). For low and medium soil moisture, Ruser (1999) calculated emission factors for the nitrification-derived N$_2$O between 0.1 and 0.2% for a soil from the crop rotation plot trial of the integrated FAM farm. As the NO$_3^-$ concentrations in unfertilised soils at the FAM were very low, it can be assumed that nitrification was the main source for the annual background emissions in Scheyern.

Positive correlations between the soil nitrate contents and the N$_2$O emissions have numerously been reported from study sites where denitrification was shown to be the main N$_2$O source (Ambus and Christensen, 1995; Thornton and Valente, 1996; Smith et al., 1998). Besides its direct concentration effect during denitrification, the nitrate concentration also influences the N$_2$O/N$_2$ ratio. High NO$_3^-$ concentrations were shown to inhibit the N$_2$O reductase activity (Blackmer and Bremner, 1978; Weier et al., 1993), thus increasing the net N$_2$O emission. As compared to the N$_2$O fluxes, Ruser et al. (2006) found that N$_2$ was negligible following the addition of nitrate in a usual dose for potato production.

Normally, the temporal variability of the N$_2$O emission can only be explained poorly by using the soil nitrate contents. This is due to the fact that the limiting factors for N$_2$O production and emission in soils vary in the short term and that they might amplify or attenuate each other (Robertson, 1994). This conclusion is also valid for the trace gas investigations at FAM sites. The soil nitrate contents often correlated with the N$_2$O flux rates where the coefficient of determination in the majority of cases was below 50%. Definite better results were obtained when the single data were aggregated on an annual basis. Ruser et al. (2001) could explain 81% of the variability of the annual N_2O emissions from differently managed fields with the mean annual soil nitrate contents (Figure 2(C)).

These data were measured on wheat, maize, potato, and fallow fields, where two N fertiliser treatments were included in the measurement design. Each point represents the mean of two sampling years with similar mean air temperatures and amounts of total rainfall. Such high coefficients of determination also resulted from other temporal aggregations of the N_2O emission (vegetation period, $r^2 = 0.74$, Figure 2(A); winter period, $r^2 = 0.93$, Figure 2(B)). As compared to calculations with single data sets, Sehy (2003) also reported better correlations, when the soil nitrate contents and the flux data were aggregated to annual data sets (data not shown). The outcomes of Sehy (2003) also indicated the strong effect of soil moisture on N_2O emission, and they reflected the paramount importance of denitrification for N_2O production in the investigated soils. The strong relationships between soil nitrate contents and N_2O emissions (Figure 2) clearly demonstrate that the soil nitrate is the most important factor to reduce N_2O emissions from arable soils. The nitrate concentrations can at least be partly controlled by a farmer. As demonstrated by the split data sets in Figures 2(A) and 2(B), the above-stated effect is valid for the vegetation period as well as for the winter period. Consequently, the temporal and the spatial synchronisation of plant demand and N supply is the most effective measure for a significant reduction of N_2O emissions from arable soils. This was also confirmed by the work of Gutser et al. (2000). They fertilised each of two summer wheat sites with 150 kg N ha^{-1} in a threefold split dose. In comparison with the site which received a relatively high amount of N during the beginning phase of the vegetation period (70 kg N ha^{-1}), the site with only 30 kg N ha^{-1} at this stage emitted approximately 700 g ha^{-1} less N_2O–N during the complete vegetation period. The higher N_2O emission in the treatment with high N amounts at the beginning of the vegetation period can be explained by a lower N uptake of the wheat plants and by the higher soil moisture.

Gutser et al. (2000) showed that probably due to the lower nitrate concentrations, the addition of a nitrification inhibitor reduced the N_2O emissions from different crops as compared to treatments fertilised with ammonium nitrate by approximately 16%. Simultaneously, it was possible to reduce the frequency of N applications from three to two doses per year by the addition of nitrification inhibitors. The plant yield was not negatively influenced by this reduction, which also minimises the amount of diesel and consequently the CO_2 production from fossil energy sources.

One of the main aims of precision farming is the optimisation of the spatial N fertiliser distribution. Sehy et al. (2003) reported a N_2O reduction

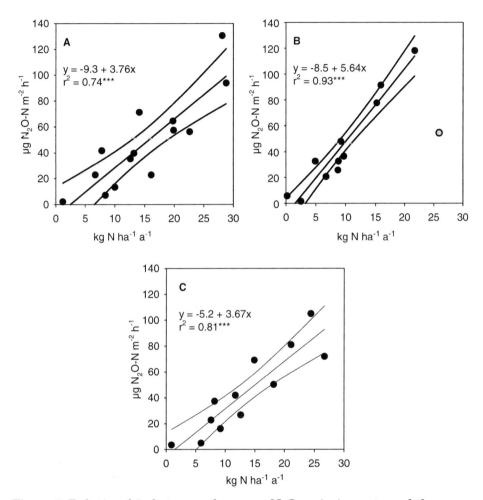

Figure 2 Relationship between the mean N₂O emission rate and the mean soil nitrate contents of the topsoil (0–30 cm depth) from different crops during the cropping season (A), during the winter period (B), and for annual data (C) (Ruser et al., 2001). During the winter season, one outlier (grey circle) was excluded from statistical calculations.

potential by the use of precision farming of approximately 10% as compared to a conventionally managed system. This was achieved by increased N use efficiency through the reduction of the amounts of N fertiliser given to areas with long-term low plant yield. The results of Sehy et al. (2003) also showed that besides short-term effects, the N₂O emission also reflects the N mineralisation potential of a soil. The emission from the high-yield area was 7.8 kg N₂O–N ha⁻¹ a⁻¹, whereas the emission from the conventionally managed low-yield area was only $3.7 \, kg \, N_2O\text{–}N \, ha^{-1} \, a^{-1}$. Further investigations in these two areas generally showed distinct differences in C and N mineralisation potentials, and this was explained by the higher soil moisture in the high-yield soil (Sehy, 2003) and by differences in the microbial community structure of soil in the two yield areas (Wessels-Perelo, 2003).

Several trace gas studies have shown that the cultivation of some crop types is afflicted with very high N_2O emissions. This crop type effect was demonstrated for potato production at FAM sites (Ruser et al., 2001) and for sugar beets, potatoes, rapeseed, and broccoli in external investigations (Kaiser et al., 1998; Smith et al., 1998; Dobbie et al., 1999; Kaiser and Ruser, 2000). All these investigations included N_2O flux measurements from wheat sites under good agricultural practices. As compared to the N_2O emissions from the wheat sites, the higher emissions from the above-mentioned crop types were explained by a higher N_2O production during denitrification due to the higher nitrate contents of the soil during the vegetation period and after harvest. The high nitrate contents after harvest and during the winter period might be a result of a decreased N uptake prior to harvest and of the low C-to-N ratio of these crop types.

Effect of N fertilisation on N_2O emissions from FAM sites

The availability of mineral N is one of the most important controlling factors for N_2O production during nitrification and denitrification in soils. Several investigations on the effect of N fertilisation on N_2O fluxes from arable soils have shown that N_2O emission increases with increasing fertiliser amounts (summarised by Eichner, 1990; Granli and Bøckman, 1994; Bouwman, 1996). Eichner (1990) tried to correlate the N_2O emissions with the type of fertiliser, which was problematic due to the lack of annual data sets. Therefore, this approach primarily images short-term effects. Using annual data sets, Bouwman (1996) calculated a N fertiliser–related annual emission factor of 1.25% and an additional background emission of $1.0 \, kg \, N_2O\text{–}N \, ha^{-1} \, a^{-1}$ from unfertilised soils. This calculation was based on results from measurements on grassland and cornfields. Bouwman (1996) explicitly mentioned that annual data are essential for a reliable estimation of annual N_2O emissions from arable soils.

Kaiser and Ruser (2000) and Ruser et al. (2001) found for five long-term investigation sites in Germany including results of the FAM studies that the assessment of the N_2O emission with the amount of N fertiliser is inadequate. Much more reliable results were obtained by the guidelines of the IPCC (2001), considering all N inputs into the soil. In independent investigations, Flessa et al. (2002a) and Sehy (2003) explained between 61 and 67% of the variability of the annual N_2O emission with the total annual N input. Figure 3 summarises the results of these two investigations. When the N_2O emission was plotted against the amount

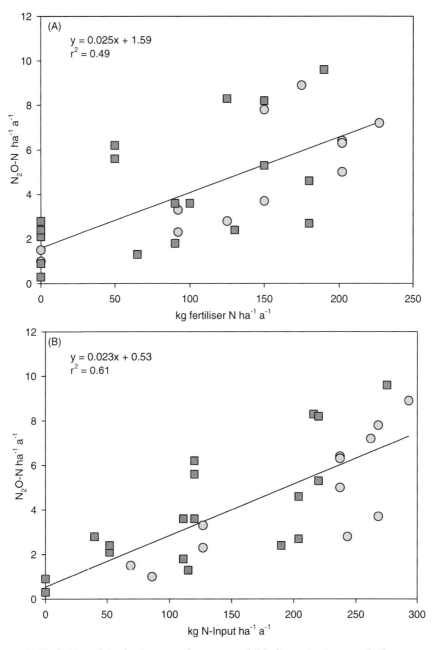

Figure 3 Relationship between the annual N$_2$O emission and the annual amount of N fertiliser (A) and between the annual N$_2$O emission and the amount of total N input (B). Data from Flessa et al. (2002a) (dark grey squares) and from Sehy (2003) (bright grey circles) were combined.

of N fertiliser, the natural background emission was strongly overestimated with $1.59 \, \text{kg} \, N_2O{-}N \, ha^{-1} \, a^{-1}$. Reasons for this overestimation were the non-consideration of the high N_2 fixation rates by legumes (clover-lucerne-mixture and lupine data), of the N mineralisation after mulching of fallow fields and of the mineralisation of crop residues with a narrow C-to-N ratio during autumn and winter (potato crop).

If the total N input was taken into consideration instead of the N fertiliser amount only, the resulting natural background emission received a much more realistic dimension ($0.53 \, \text{kg} \, N_2O{-}N \, ha^{-1} \, a^{-1}$) and the variability of the annual N_2O emission could be explained with 61% to a higher extent (Figure 3). Related to the annual N input, 2.3% are emitted as N_2O within 1 year. This emission factor is approximately twice of that suggested by the IPCC (2001). On a national and international basis, it was shown that the N_2O emission at the study site in Scheyern is on a very high level. Consequently, the FAM study site must be considered as a high emitting location (Kaiser and Ruser 2000, Figure 4). The reason for this phenomenon is still unclear. Kaiser and Ruser (2000) concluded from N_2O

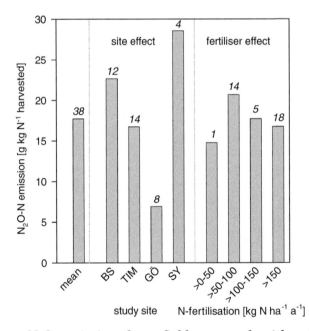

Figure 4 Mean N_2O emission from fields cropped with winter cereals related to one unit of N grain yield, calculated for all available data (mean), for different study sites (BS = Braunschweig, TIM = Timmerlah, GÖ = Göttingen, SY = Scheyern), and for ranges of N fertiliser applied. The bar labels represent the number of data sets available for each bar (Kaiser and Ruser, 2000).

measurements at four German cereal sites that the higher N_2O emissions from the study site in Scheyern were a result of the distinct and more frequently occurring freezing/thawing and drying/rewetting events. Another reason may be the generally high textural portion of silt in the FAM soils, and this results in a decreased O_2 diffusion due to soil sealing after rainfall events, thus favouring the creation of anoxic microsites and denitrification. The concentration of microbial easily available carbon in the topsoil as a result of reduced soil management and the subsequent O_2 consumption was also stated as one reason for the higher N_2O emissions from the FAM sites.

So far, only few researchers have related the N_2O emission to plant yield or specific yield parameters. Kaiser and Ruser (2000) related the annual N_2O emissions from cereal fields to the N uptake of the grains. The reduction of N fertiliser did not have a statistically significant effect on the yield-related N_2O emission (Figure 4). Despite the enormous potential of measures to reduce N_2O emissions from arable fields (e.g. synchronisation of plant demand with spatial and temporal N supply), these results show that the production of foods and animal feed will always be connected with a certain yield-specific N_2O emission. The production in organic farming systems cannot prevent such N_2O emissions. Sehy (2003) calculated an emission factor of 0.82 kg $N_2O–N\,t^{-1}$ for wheat grain at an organically managed field and factors between 0.51 and 0.62 kg $N_2O–N\,t^{-1}$ from conventionally managed fields.

The establishment of fallow sites significantly reduced the N_2O emission from the FAM fields. In 1991, 9.0 ha of steep fields or fields with low yield expectations were transformed from arable use to fallow. The ammonium and nitrate contents of unfertilised fallow soils are extremely low. It can be concluded that the substrate availability for the main processes of N_2O production in soils is limited in such soils. As demonstrated in Figure 2, the N_2O emission was strongly correlated to the nitrate contents of the topsoil (Ruser et al., 2001; Sehy, 2003). Consequently, the N_2O emission from unfertilised fallow and wheat sites was lower as compared to the fertilised arable sites. It varied between 0.3 and 0.9 kg $N_2O–N\,ha^{-1}\,a^{-1}$ (Kamp, 1998; Stolz, 1997; Ruser et al., 2001). In regression analysis of the N induced N_2O emission at the FAM sites, an intercept of 0.53 kg $N_2O–N\,ha^{-1}\,a^{-1}$ for unfertilised soils was computed (Figure 3). As long as the fertilisation intensity at the remaining arable and grassland sites is not intensified, the establishment of unfertilised fallow sites at locations with long-term low yield expectations efficiently reduces the N_2O emission on a farm level. The mean N input at arable sites of the integrated FAM farm was 256 kg $N\,ha^{-1}\,a^{-1}$ (140 kg N synthetic fertiliser, 39 kg N organic fertiliser, 45 kg N from crop residues from cash crops and 32 kg N from intercrops). In accordance to the regression analysis shown in Figure 3, the input of 256 kg $N\,ha^{-1}\,a^{-1}$ produces a N_2O emission of 6.41 kg $N_2O–N\,ha^{-1}\,a^{-1}$. In contrast, the establishment of fallow sites with one mulching measure per year resulted in an emission of only 1.45 kg N_2O–N ha^{-1} a^{-1}. This measure reduced N_2O emission on a farm level by 7% as compared to soil management in 1990 (prior to the establishment of fallows).

Effect of soil compaction and loosening on the N_2O emissions from FAM sites

Figure 5 shows the macro-pore volume of soil under different crop types at the integrated crop rotation plot trial plotted against the bulk density in the topsoil (5–10 cm depths).

The portion of fine and smaller medium pores (<0.2–167 µm) was more or less solely dependent on texture characteristics, whereas the effect of soil compaction was negligible (data not shown). In contrast, the portion of macro-pores (>1200 µm) was found to be strongly correlated to the bulk density of a soil. The portion of freely draining macro-pores is crucial for the aeration of a soil (Stepniewski et al., 1994; Xu et al., 1994). Horton et al. (1994) reported a sharp decrease of water infiltration with increasing bulk density, resulting in water logging after heavy rainfall events. All these factors increase N_2O emissions derived from denitrification with increasing soil compaction (Yamulki and Jarvis, 2002;

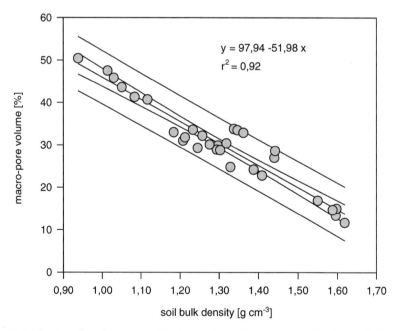

Figure 5 Relationship between the bulk density of an arable topsoil (5–10 cm depth) and the macro-pore volume of soil cores collected from wheat, corn, potato, and fallow fields of the conventional FAM farm.

Table 1 Cumulative N_2O emissions from potato fields as affected by soil compaction, macro-pore volume, and N fertilisation during the growing season (with permission from Ruser, 1999).

Bulk density (g cm^{-3})	Macro-pore volume (Vol. %)	N_2O emission (kg N ha^{-1})	
		50	150
1.05[a]	43.6[a]	1.88[a]	3.59[a]
1.26[b]	32.2[b]	2.87[a]	5.07[b]
1.56[c]	13.9[c]	18.25[b]	50.35[c]

Different letters (a–c) indicate statistically significant differences between the treatments within one column (Student–Newman–Keuls Test, $\alpha < 0.05$).

Hansen et al., 1993; Bakken et al., 1987). Trace gas investigations on potato plots at the FAM research station revealed high gaseous losses of N fertiliser from tractor-compacted areas. Related to the amount of N fertiliser about one third was emitted during the vegetation period from the strongly compacted tractor tram lines (Table 1, Ruser et al., 1998).

Soil sampling during the vegetation period showed that potato roots did not penetrate into these tractor-compacted areas. An appreciable uptake of nitrate and soil water by the potato plants from the compacted areas can therefore be excluded. Thus, optimum conditions for an increased N_2O production during denitrification were given (high nitrate and moisture contents, low nitrate and moisture removal). In 1998 and 1999, the chair of plant production and breeding of the TU Munich conducted an experiment on the fertilisation of potatoes on the integrated FAM farm. The main results were that the application of N into the potato ridges and the temporal splitting of the N fertilisation did not negatively influence the most relevant yield parameters (Maidl et al., 2002). Both measures contributed very efficiently to the minimisation of the nitrate concentrations in the tractor-compacted areas and the interrow areas. Both fertilisation measures are likely to reduce the N_2O emissions from potato-cropped soils. However, the final confirmation of these findings has to be verified by field examinations.

As shown in Table 1 and as reported from another potato-cultivated soil (Smith et al., 1998), soil loosening decreased N_2O emissions. This can be explained with an increase in the volume of quickly draining macropores. The bulk density in the interrow soil was 1.26 g cm^{-3}, this value was representative for the bulk density of topsoils under wheat and corn crops at the same site (Ruser, 1999). Due to the increased pore volume in the loose ridge soil, the critical threshold value of >60% WFPS is rarely exceeded after rainfall events as compared to the interrow soil. Generally, it can be concluded that a loose soil structure effectively prevents from N_2O emissions from arable soils. Therefore, the use of broad tyres as practised on both FAM farms helps to minimise N_2O emissions as compared to conventional agriculture.

2.2.3 Important controlling factors for CH_4 fluxes at the FAM sites

Effect of soil moisture on the CH_4 fluxes from FAM soils
    Soil moisture was shown to be the most relevant controlling factor for the CH_4 uptake from soils at the FAM research station. All investigations showed strong negative correlations between soil moisture contents and the CH_4 uptake. This relation is illustrated by Figure 6. Increasing soil

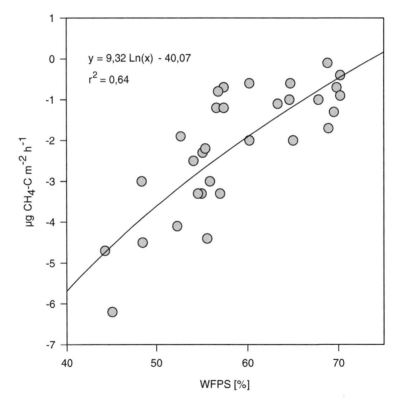

Figure 6 Relationship between the soil moisture (0–30 cm depth) and the mean CH_4 flux rates from plots with different crops (wheat, corn, potato, and fallow) and N fertilisation. Data from Ruser (1999) were calculated as means from 6 months periods each. One period represents the growing season, the second one represents the winter season.

moisture reduces the diffusion of CH$_4$ into the soil, and thus the substrate availability decreases (Crill, 1991; Castro et al., 1994; Smith et al., 2000). It was reported from laboratory experiments that CH$_4$ oxidation decreases at soil moisture below 20% WFPS (Del Grosso et al., 2000; Adamsen and King, 1993), whereas no CH$_4$ oxidation could be observed in air-dried soil (Nesbit and Breitenbeck, 1992). Highest CH$_4$ uptake rates at the FAM sites were measured during summer (Kamp, 1998; Ruser, 1999; Dörsch, 2000) because such low levels of soil moisture (<20% WFPS) were not reached. Similar temporal patterns of the CH$_4$ fluxes with maximum uptake rates in summer have also been reported from other arable study sites (Ambus and Christensen, 1995; Dobbie and Smith, 1996).

CH$_4$ uptake quickly decreases with increasing soil moisture contents. Figure 7 shows the CH$_4$ flux rates measured on potato fields during the cropping season plotted against the WFPS. The mean CH$_4$ uptake between 20 and 40% WFPS was 4.1 µg CH$_4$–C m^{-2} h^{-1}. At soil moisture contents between 40 and 60% WFPS, the mean CH$_4$ uptake rate was reduced by approximately 60% and soil moisture above 80% WFPS. The latest was only measured in the tractor-compacted interrow soil, inducing a net CH$_4$ release. In contrast, the high moisture contents in winter did not result in a net CH$_4$ release, as CH$_4$ production was strongly correlated to the soil temperature (Ruser et al., 1998). Prior to the investigation on the potato plots, measurements were carried out in wheat crop on the same plots for approximately 1 year. During this year, no CH$_4$ emission was observed (Ruser, 1999). Although CH$_4$ production in soils is a strongly anaerobic process, these results showed that methanogenic microorganisms are able to survive in well-aerated soils over a long time and that they can be activated by short-term induced anaerobiosis. This was supported by the determination of archaea in the potato-cropped soil (Gattinger et al., 2002).

Effect of soil compaction on the CH$_4$ fluxes at the FAM sites
    The cumulative CH$_4$ fluxes of differently fertilised and differently compacted areas of potato fields are shown in Table 2. Soil compaction reduces the portion of freely draining macro-pores which are mainly responsible for the aeration of a soil (Figure 5) and consequently for the CH$_4$ supply in the soil. Therefore, diffusion of atmospheric CH$_4$ into the soil decreases with increasing bulk density. Hansen et al. (1993) reported that the CH$_4$ uptake of a tractor-compacted soil with a bulk density of 1.30 g cm^{-3} was 50% lower as compared to an uncompacted soil with a bulk density of 1.21 g cm^{-3}. Stepniewski et al. (1994) determined gas diffusion as affected by the bulk density and by the air-filled porosity. With increasing bulk density, the redox potential decreased simultaneously with decreasing O$_2$ diffusion rates and increasing soil moisture contents. In 15 cm soil depth, the redox

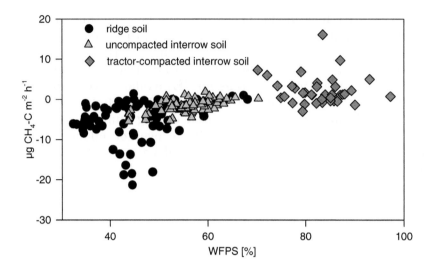

Figure 7 Relationship between the water-filled pore space and the CH_4
fluxes from differently compacted areas of potato fields during
the growing season (Ruser et al., 1998).

Table 2 Cumulative CH_4–C fluxes from different areas of potato fields as
affected by bulk density and N fertilisation during the growing
season (with permission from Ruser, 1999).

Area	Bulk density ($g\ cm^{-3}$)	N fertilisation ($50\ kg\ N\ ha^{-1}$)	N fertilisation ($150\ kg\ N\ ha^{-1}$)
		$g\ CH_4$–$C\ ha^{-1}$	
Ridge	1.05	−143.1[a]	−98.1[a]
Uncompacted interrow	1.26	−25.4[b]	−29.9[b]
Tractor-compacted interrow	1.56	+71.8[c]	+56.0[c]

Statistically significant differences between the compacted areas are designated with
different letters (a–c) ($p < 0.05$, Student–Newman–Keuls Test).

potential was $<0\ mV$ in a compacted treatment and $+200\ mV$ in the
uncompacted treatment. Besides the reduction of CH_4 uptake rate, it was
shown that soil compaction in conjunction with high amounts of rainfall at
a FAM potato site annihilated the sink function of the soil and turned it
into a source (Table 2). This means that CH_4 production in the strongly
compacted soil exceeded CH_4 oxidation thus resulting in net emission
during the growing season (Ruser et al., 1998). The CH_4 release indicated

the poor O$_2$ availability in the wet compacted soil. The anaerobic decomposition processes obviously not only reached the redox level for nitrate reduction, as indicated by the high N$_2$O emission (Table 1), but transiently proceeded even further to the stage of CH$_4$ formation.

Effect of N fertilisation on the CH$_4$ fluxes at the FAM sites

N fertilisation normally reduces the CH$_4$ uptake in soils. This was reported for forest soils (Schnell and King, 1994), grassland soils (Mosier et al., 1991), and arable soils (Steudler et al., 1989; Hütsch et al., 1993; Mosier et al., 1996). This reduction was explained by the fact that the CH$_4$-oxidising enzyme MMO of methanotrophic bacteria can also oxidise other substrates. The competitive inhibition of NH$_4^+$ or of ammonia (NH$_3$) at the active enzyme sites was considered to be responsible for this effect (Knowles, 1993; Dunfield and Knowles, 1995). Another explanation reported by Schnell and King (1994) is the toxic effect of nitrite enriched during ammonium oxidation. Flessa et al. (1996b) demonstrated for a Scheyern soil that the inhibition of the CH$_4$ uptake was higher following NH$_4^+$ fertiliser application as compared to the addition of NO$_3^-$. However, the CH$_4$ uptake in the nitrate treatment was also affected and 21% lower as in the unfertilised control treatment. The authors expected that the reduction in the nitrate treatment was probably due to the re-mineralisation of microbial transformed nitrate to ammonium.

The annual CH$_4$ uptake at the FAM sites was very low. The rates varied between 80 and 567 g CH$_4$–C ha^{-1} a^{-1}. The CH$_4$ oxidation capacity seems to be strongly constricted due to the long-term intense agricultural use at the FAM sites with high fertiliser amounts. This long-term inhibition was reported by Bronson and Mosier (1993) and Hütsch (1996). Comparing CH$_4$ uptake from different ecosystems, arable soils showed the lowest CH$_4$ uptake (Ambus and Christensen, 1995). Ojima et al. (1993) reported a reduction of the CH$_4$ uptake between 30 and 75% after N fertilisation of arable soils and after the conversion of grassland to arable soil. Long-term N fertilisation can reduce CH$_4$ oxidation in a way that the inhibitory effect cannot be observed after single N fertilisation measures.

Effect of farming system on the atmospheric pollution with climate-relevant trace gases

We calculated the atmospheric burden of the two farming systems of the FAM research station (the integrated farm and the organic farm). All general conditions and applied emission factors which have not been determined by our working group (N$_2$O, CH$_4$ and CO$_2$ emissions during the production of synthetic N fertiliser, CO$_2$ emission during the exploration and the consumption of diesel fuel, N$_2$O and CH$_4$ release from animal husbandry and from the storage of animal excrements) were explained in detail by Flessa et al. (2002a). In contrast to Flessa et al.

(2002a) we used the emission factor shown in Figure 3 for the N_2O emissions from arable, grassland, and fallow soils because it additionally included the emission data from Sehy (2003). Furthermore, some input data were slightly modified, and the data for the dynamics of the carbon contents in the soils were partially also taken into consideration.

Figure 8 shows the portion of the different emission sources to the total atmospheric pollution as affected by the two management systems realised at the FAM research station. The portion of the N_2O emission from the arable sites to the total atmospheric burden was 53% for the integrated farm and 38% for the organic managed farm. This distinct difference can be explained by the higher N fertilisation level on the fields of the integrated farm. Emission from the production of synthetic fertilisers did not account for in the organic farming system due to the abandonment of synthetic resources. The organic farming system was afflicted with higher amounts of diesel-derived CO_2 equivalents.The reasons were the generally deeper soil management and the higher frequency of mechanical weed control measures. This was also reflected in the emission data shown in Figure 8, where the portion of diesel (exploration and consumption) was 12% for the integrated farm and 27% for the organic farm.

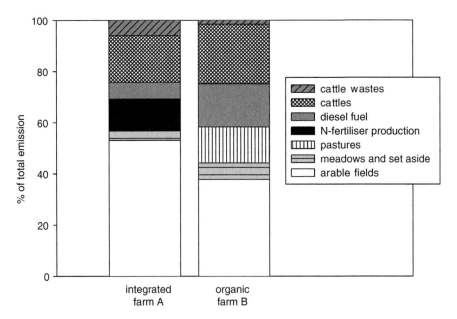

Figure 8 Portion of the different greenhouse-relevant trace gas sources to the total emission (base: CO_2 equivalents) of the integrated and of the organic farm system at the FAM research station (Flessa et al., 2002a, modified). Changes in humus C are not considered.

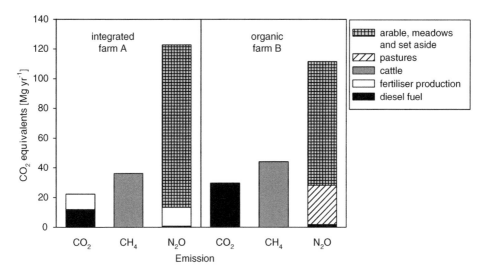

Figure 9 Total CO_2, CH_4 and N_2O emissions from the agricultural production of the two FAM farming systems and the contribution of different sources to these emissions. CH_4 uptake by soils and C release/consumption by humus are not considered.

Figure 9 shows the portion of the different sources to the total emission of each single trace gas. With 192 t CO_2 equivalents a^{-1} for the integrated farm and 187 t a^{-1} for the organic farm, the total atmospheric burden was very similar for both farming systems. Assuming that the change of the humus contents proceeded linearly between 1991 and 2001, additional 13 t a^{-1} has to be taken into account for the integrated farm, whereas 21 t a^{-1} has to be subtracted from the value of the organic farm. For both farming systems, N_2O accounted for more than 60% of the total emission. Animal husbandry was 0.9 AU ha^{-1} on the integrated farm and 0.6 AU ha^{-1} on the organic farm. In the surrounding of the FAM research station, many specialised farms are located with animal husbandry between 2 and 3 AU ha^{-1} (Ruser et al., 2007). For these farms, it can be supposed that the share of the single trace gases to the total emission shifts towards a higher portion of ruminal CH_4.

On the two FAM farms, different crop rotations were established and thus different management measures were realised that do not allow for the direct comparison of the two farming systems. The integrated fields were installed on more fertile soils due to the thicker loess layers as compared to the organic fields. Therefore, the two farming systems should be compared on the basis of two general yield-related emission factors.

– The total emission related to plant production (animal husbandry not included) was 4284 kg CO_2 equivalents $ha^{-1} a^{-1}$ from the integrated farm. This value was 25% higher as compared to the emissions from the organic farm (3211 kg ha^{-1} a^{-1}). In contrast the mean grain yield on all wheat fields was 55.9 and 29.6 dt ha^{-1} on the integrated and on the organic farm, respectively. The yield-related emission was 770 g kg^{-1} on the integrated farm and 1080 g kg^{-1} on the organic farm.

– The total emissions per cattle unit from enteric fermentation and waste storage were about 25% higher for the integrated farm (1.6 t AU^{-1}) than from the organic farm (1.3 t AU^{-1}). This difference was attributed to higher emissions from waste management on the integrated farm (liquid manure). However, emissions were nearly identical when related to beef production, as animal productivity (mean daily increase in weight) was about 20% higher for the integrated farm ($\approx$1200 g d^{-1}) than for the organic farm ($\approx$950 g d^{-1}). If greenhouse gas emission from feed production is included, total emissions per unit beef are probably higher for the more extensive cattle management than for the intensive bull fattening. Similar conclusions were drawn by Subak (1997), who compared greenhouse gas emissions from extensive and intensive beef production systems in Europe.

Both calculations clearly show that organic farming does not result in a reduced emission of trace gases when related to yield parameters.

2.2.4 Ammonia emissions from agricultural land use

NH$_3$ belongs to the group of trace gases that indirectly contribute to the greenhouse effect. Its natural atmospheric concentration is <100 ppb. It is highly reactive. NH$_3$ pollution was shown to be the main reason for eutrophication and acidification of natural ecosystems (Bergkvist and Folkeson, 1992; Jongebreur and Voorburg, 1992). NH$_3$ is also a precursor for secondary aerosols, a part of the NH$_3$ produced in stables is already solid bound in the stable and distributed over large distances as fine particulate matter (Thorne, 2002). According to Ahlgrimm (1995), approximately 700 000 t NH$_3$ was annually emitted in Germany. In Europe about 95% of the NH$_3$ emission originated from agricultural activities with a mean emission density of approximately 25 kg N ha^{-1} a^{-1} (ECETOC, 1994; Klaassen, 1994). Ross et al. (2002) modelled the NH$_3$ emission from a dairy farm with liquid manure storage. The calculated NH$_3$ emission ranged between 27 (scenario with preventing measures) and 107 kg N ha^{-1} a^{-1} ('worst case' scenario without any preventing measure). These calculations clearly reflect the high emission potential and the preventing capabilities in context to NH$_3$.

Seventy-six per cent of the total NH_3 emitted in Germany is derived from animal husbandry with $245\,kt\,a^{-1}$ from stables and excrement storage, $46\,kt\,a^{-1}$ from animal grazing on pastures, and $212\,kt\,a^{-1}$ from the application of organic fertilisers as liquid or solid manure (ECETOC, 1994). The NH_3 emission from the agricultural sector, other than emission from animal husbandry, was almost completely associated to the application of synthetic fertilisers.

Numerous investigations have focused on NH_3 emissions from animal husbandry during the storage and the application of organic fertilisers (Groot Koerkamp et al., 1998; Bouwman et al., 1997; Sommer et al., 1993; Döhler, 1991; Amberger, 1990). Collateral investigations at the chair of plant nutrition (TU Weihenstephan) have mainly focused on unavoidable NH_3 losses in context to agricultural land use which occurred after the application of NH_4^+-containing fertilisers or after the mulching of plant materials. The investigations have been carried out at the research station of the TU-Munich Weihenstephan in Dürnast. Dürnast is located about 25 km southeast from the FAM research station. The total annual rainfall ($\approx 800\,mm\,a^{-1}$), the mean annual air temperature ($7.4°C$) and the annual rainfall distribution (not shown) were almost identical for both study sites. The experimental soils were also in good agreement with the soils in Scheyern. All investigations were carried out on carbonate-free Dystric Eutrochrepts derived from loess with a pH value <6.5. The various similarities ensure the direct comparability of the two study sites Scheyern and Dürnast, allowing the transfer of the Dürnast findings to the FAM sites.

NH_3 emissions after the application of synthetic N fertilisers

NH_3 losses after the application of synthetic N fertilisers are affected by many factors such as the form and the amount of the N applied, soil pH, land use system, application technique, rainfall, and temperature (Misselbrook et al., 2004). Among the controlling factors, soil pH is often reported to be one of the main influencing variables (He et al., 1999; Sommer and Ersboll, 1996). NH_3 volatilisation increases with increasing soil pH (pK_A value of $NH_4^+/NH_3 = 9.3$). The soils of the two research stations are predominantly decalcified. Soil pH values vary slightly in the weakly acidic range. To stabilise this pH value, lime is applicated regularly. Despite the weakly acidic conditions, high NH_3 emissions can occur due to the enrichment of NH_4^+ and the rise of pH during urea hydrolysis in wet 'microsites' (Fenn and Richards, 1986). Under extreme conditions such as during water logging in paddy soils, the NH_3 loss after the application of urea may account for up to 50% (De Datta, 1995). In principle, high NH_3 can be expected with increasing amounts of N fertiliser and after the surface application without any further measure to incorporate the fertiliser into the soil (Fox and Piekielek, 1987). In a laboratory experiment, Vermoesen et al. (1996) determined NH_3 losses of up to 30% of the urea N applied. These investigations are of special relevance for the FAM because ammonium nitrate urea solution (AUN) was the most frequently used fertiliser type on the integrated FAM farm. Table 3 shows the NH_3 emissions after the application of different NH_4^+ containing synthetic fertilisers (Weber et al., 2000). As explained in detail by Weber et al. (2000), the experimental design also took into account the diurnal variations of the fluxes due to temperature fluctuations as reported by Chinkin et al. (2003).

Temporally, the NH_3 volatilisation from all investigated treatments varied strongly. The maximum loss accounted for 5.5% of the fertiliser applied in the second experimental year (Table 3). For the fertilisers used in the FAM [AUN and calcareous ammonium nitrate (CAN)], the highest NH_3 emission was measured with 1.2 kg N ha^{-1} after the addition of CAN at the end of April 2000. The highest emission measured after the application of AUN was 0.72 kg N ha^{-1} or 0.9% of the fertiliser applied at the end of May 1999 (Figure 10). The NH_3 emissions from the sites fertilised with CAN show a distinct inter-annual variation. This was especially obvious when comparing the fluxes after fertilisation at the end of April.

Highest NH_3 volatilisation rates were determined after the application of granulated urea. This is in accordance to investigations on the effect of N fertiliser amount and type on the NH_3 emissions from arable

Table 3 NH_3–N losses from wheat fields and environmental conditions after the broadcast surface application of different synthetic N fertilisers.

Year	Application date	Fertiliser type (% of total N)				Mean air temperature (°C)	Sum of rainfall (mm)
		U	UAS	AUN	CAN		
1999	End of March	0.6	–	0.2	<0.1	7.7	11.1
	Begin of April	−0.1	–	−0.1	<0.1	7.7	26.1
	End of April	0.5	–	0.4	0.1	13.3	0.5
	End of May	2.6	–	0.9	0.1	19.1	49.5
	Mid June	−0.1	–	−0.1	−0.1	12.4	22.1
2000	End of March	0.6	0.3	–	<0.1	5.7	41.7
	Begin of April	0.2	0.2	–	0.2	8.0	32.3
	End of April	5.5	4.6	–	1.5	15.4	4.8
	Begin of June	1.4	1.3	–	0.7	18.5	9.5

N application was carried out homogeneously at different application dates with 80 kg N ha^{-1}; the flux measurement period covered 10 days (from Weber et al., 2000). U = granulated urea, UAS = urea ammonium sulphate, AUN = ammonium nitrate urea solution, CAN = calcareous ammonium nitrate, – = not determined.

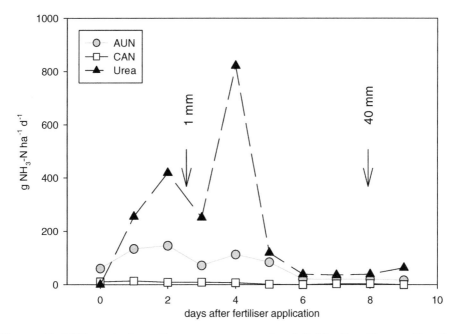

Figure 10 NH_3 emissions from a summer wheat field after the application
of different N fertilisers. Rainfall occurrences are indicated by
an arrow and the amount.

and grassland sites (Bouwman et al., 2002; Herrmann et al., 2001; van der
Weerden and Jarvis, 1997; Sommer et al., 2004; He et al., 1999). However,
the magnitude of the NH_3 volatilisation differed strongly among these
investigations. During the investigations of Weber et al. (2000), the highest NH_3 emission accounted for 5.5% of the urea N applied, whereas other
researchers reported values of up to 48% on maize fields and of 20% on
wheat sites (Cai et al., 2002). A range between 12 and 46% after urea addition was also reported by van der Weerden and Jarvis (1997), where the
application of NH_4NO_3 or CAN resulted in NH_3 emissions of <1%. The
direct comparison of the single data sets is restricted due to the fact that
the measurement periods were not identically and strongly dependent on
the timing of N fertilisation as shown in Table 3. The experiments of
Weber et al. (2000) showed that the investigation period should cover at
least 10 days after fertilisation, whereas in some cases more than 12 days
were necessary to reliably determine the NH_3 fluxes. Only few authors
tried to quantify or at least to estimate annual NH_3 emissions from soils.
Eckard et al. (2003) reported a NH_3 emission of 17, 32, and 57 kg NH_3–N
ha^{-1} a^{-1} from unfertilised, NH_4NO_3-fertilised, and urea-fertilised pastures,
respectively. Bouwman et al. (2002) calculated the global fertiliser-induced
NH_3 release from soils, and this accounted for 18% in developing countries

and 7% in developed countries. The higher emissions in the developing countries were explained by the lower costs for urea fertilisers resulting in significantly higher portion of these fertilisers in the total fertiliser use and by the higher temperatures after fertiliser application.

The main observation from the work of Weber et al. (2000) was that temperature and rainfall were two of the most important influencing variables for the NH_3 losses after synthetic fertiliser application in the region of the FAM station. The following reasons may explain this phenomenon:

– After urea-containing fertilisers, the NH_4^+ concentration as reactant for NH_3 production is dependent on the original NH_4^+ concentration and on the additional supply of NH_4^+. Thus NH_3 volatilisation is controlled by activity of urease, which shows the typical enzyme temperature dependence. In case of high temperatures, the urea is quantitatively converted to NH_4^+ within a few days (Beline et al., 1998; Muck and Richards, 1980). The increasing effect of enhanced temperatures on the urea hydrolysis in soils was demonstrated by Häni et al. (1986). During the hydrolysis urea is converted to NH_4^+ and HCO_3^-. The production of HCO_3^- results in an additional alkalinisation of the environment, and this enhances NH_3 volatilisation in soil compartments with a low pH buffer capacity.

– The NH_3 partial pressure exponentially increases with (soil) temperature (Farquhar et al., 1980), whereas NH_3 solubility in water decreases disproportionately (Sommer et al., 1991). Both effects increase NH_3 release from soils with increasing temperature.

Furthermore, it has to be mentioned in conjunction with the soil temperature that the NH_3 losses in early spring were negligible as long as the soil temperatures were below 10°C and as long as the amount of rainfall did not atypically increase. Increased NH_3 volatilisation rates with rising soil and air temperatures have been reported by many researchers (He et al., 1999; Misselbrook et al., 2004; Cai et al., 2002; Bouwman et al., 2002; Roelle and Aneja, 2002).

A rainfall event of approximately 1 mm in 2 days after application of urea reduced NH_3 emission by nearly 50% (Figure 10). This clearly demonstrates that soil moisture and rainfall also are important factors for the NH_3 release after fertilisation. Strongly increased NH_3 fluxes after N fertilisation in conjunction with rainfall events have often been reported (Al-Kanani et al., 1991; Burch and Fox, 1989; Denmead et al., 1987), where extremely enhanced fluxes occurred after rewetting of dry soil in midsummer (Roelle and Aneja, 2002). Varying the soil moisture, Sigunga et al. (2002) observed highest NH_3 volatilisation after N fertilisation from vertisols at 80% of the water-holding capacity. They concluded that this was the optimum soil moisture for urea hydrolysis. In contrast, Priebe and Blackmer (1989) demonstrated the dislocation of urea fertiliser from the soil surface into the soil during rainfall. This transport into the soil reduced the surface contact of the fertiliser with the atmosphere thus reducing NH_3 volatilisation (Singuna et al., 2002). The low amount of only 1 mm rainfall 2 days after fertilisation may only have marginally contributed to the translocation of urea or NH_4^+ (Figure 10). However, obviously the amount was adequate to reduce the NH_3 flux rate from 400 to 200 g N $ha^{-1} d^{-1}$. The following factors may also have contributed to this 50% reduction:

  - the low amount of rainfall may have resulted in dissolving of fertiliser aggregates and thus in a more homogenous spatial distribution of urea / NH_4^+ within the pH buffering soil system or in a less alkalisation due to a more effective buffering.
  - the reduction of the pH due to acid precipitation in compartments with low buffering capacity at the interlayer soil /atmosphere may also have contributed to this decrease. The mean pH value of rainfall or wet precipitation in Germany varied between 4.35 and 4.90 for the observation period 1982–1998 (UBA, 1999). Under the conditions of low soil buffering, already low amounts of acid rain may significantly decrease soil pH at the interlayer, since pH is presented as logarithm.

NH$_3$ emissions after mulching of plant materials

Figure 11 shows the NH_3 fluxes after mulching of legumes plants. The flux rate increased approximately 5 days after mulching of the pea-vetches mixture. Rana and Mastrorilli (1998) also reported a lag phase for the increase of NH_3 emissions after the mulching of plants. Mulching mechanically breaks up the plant material so that cell and plant saps are released. The main share of the water in plants is located in the cytoplasm with a pH between 7.0 and 7.5 and in the phloem with pH values between 7.8 and 8.0 (Marschner, 1990). Thus, it can be expected that the liquid phase of the mulch immediately after mulching has a pH value >7.0 as long as the main part of the plant material has no soil contact. If the moisture content of the mulch material is sufficient, the microbial decomposition releases NH_4^+ resulting in increased NH_3 emissions. As shown in Figure 11, rainfall events were decisive for the NH_3 flux rates following mulching. During the first period of the measurement period, rainfall significantly increased NH_3 emissions from the legumes site (pea-vetches mixture) and from the oil radish sites. We suppose that the mulch material loosely lying on the soil surface dries very fast, thus disabling a further microbial decomposition until the next rainfall event, which then provides sufficient moisture for microbial activity. During 2 weeks of investigation, 3.4 kg NH_3–N ha^{-1} was emitted from a clover-grass mixture mulch. This accounted for 6% of the total N applied with the mulch (Table 4, Weber et al., 2000). The corresponding emission from a clover-grass mixture with lower total N content and a higher C-to-N ratio was 1.7 kg N ha^{-1}.

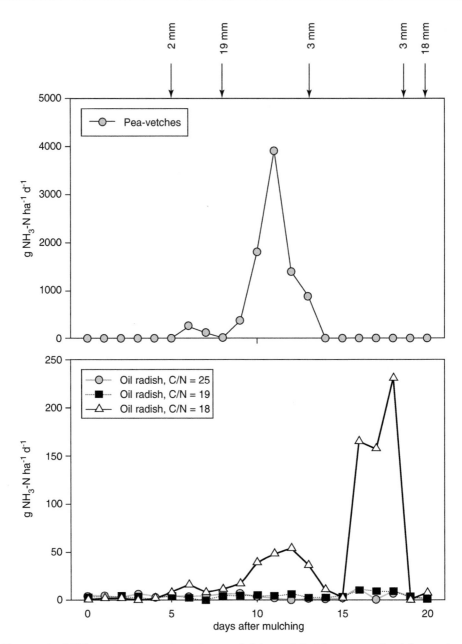

Figure 11 NH$_3$ emissions after the mulching of fields cropped with a mixture of pea and vetches (top) and NH$_3$ emission of three plots
cropped with differently fertilised oil radish, the latter resulting
in different C-to-N ratios of the plants (bottom). Rainfall events
are indicated by the daily amounts and arrows.

Table 4 N fertilisation, N content and C-to-N ratios of the plant biomass and NH_3–N losses after mulching of clover/grass mixtures, of a pea-vetches mixture and of oil radish sites (Weber et al., 2000).

Crop type		N fertilisation (kg N ha^{-1})	Plant N content (% dry mass)	C/N ratio	NH_3 emissions (kg N ha^{-1})	NH_3–N as % of plant N (%)
Main crop	Clover/ grass mixture A	0	4.7	8	3.4[a]	6
Main crop	Clover/ grass mixture B	0	4.2	10	1.7[a]	3
Intercrop	Peas-vetches	0	3.6	11	9.5[b]	4
Intercrop	Oil radish	40	1.6	25	0.1[b]	<1
Intercrop	Oil radish	80	2.1	19	0.2[b]	<1
Intercrop	Oil radish	120	2.2	18	1.3[b]	1.2

[a] Measurement period = 2 weeks.
[b] Measurement period = 3 weeks.

For the investigations with the intercrops shown in Table 4, a significant negative correlation between the C-to-N ratios of the mulch material and the NH_3 emission was found as well as a positive relationship between the NH_3 emissions and the N contents of the mulch (Weber et al., 2000).

The results from Weber et al. (2000) are in good accordance to investigations from Larsson et al. (1998), who determined the NH_3 emissions from sites with grass mulch (1.15 and 2.12% N content) and from an lucerne site (4.33% N content) over a period of 3 months. Irrespective of the higher total NH_3 emissions between 17 and 39% of the mulch N, NH_3 emission significantly increased with total N content of the mulch. Glasener and Palm (1995) also reported a negative correlation between the NH_3 emission and the total N content. Moreover, they found a negative effect of the lignin content of the mulch on the NH_3 emission.

The high NH_3 volatilisation after mulching of plants (primarily with high N contents or low C-to-N ratios) was one of the important results from the investigations on unavoidable NH_3 fluxes. The emissions accounted for 3–6% of the N bound to the mulch materials. Total emissions reached a maximum level of 9.5 kg N ha^{-1} during a 3-week investigation. For the calculation of N balances on field and on farm scale, these high N losses as well as the emission from animal husbandry and from the storage and application of excrements must be taken into account.

Currently, it is assumed that these high NH_3 losses are limited to mulched fields, where the mulch remains on the surface (Glasener and Palm, 1995). Indeed, Rana and Mastrorilli (1998) found strongly increased NH_3 emissions after the ploughing of beans. Consequently, future investigations on unavoidable NH_3 losses should focus more closely on the handling of intercrops and crop residues.

Environmentally sound management practices and related adverse effects – Reduced soil management to minimise soil erosion might increase N_2O emissions

To minimise soil erosion in this hilly landscape, the practice of reduced soil management was established already at the beginning of the FAM research period. The necessity of call for action was clearly demonstrated by the erosion events in the autumn of 1990 (>9 t ha^{-1} a^{-1}). The implementation of reduced soil management practices successfully reduced soil erosion to a rate below 200 kg ha^{-1} a^{-1} (Auerswald et al., 2003). Otherwise, the reduced soil management results in the concentration of easily available carbon in the upper centimetres of the topsoil. The microbial consumption of this carbon induces an increased O_2 depletion in soil. Thus, N_2O production during denitrification is enhanced, also assisting higher N_2O emission rates. Numerous investigations have shown that reduced soil management significantly increased N_2O emissions as compared to ploughed soils (Burford et al., 1981; Linn and Doran, 1984; Aulakh et al., 1984). The N_2O emissions at the FAM study sites, being higher than the emissions from other national and international long-term N_2O study sites (2.3% related to the N applied), may be a result of the reduced soil management practices. Further investigations on the plot trial of the integrated FAM farm may add helpful information on the effect of different soil management strategies (three different strategies).

Injection of liquid manure to reduce NH_3 emissions

The injection or at least soil management after liquid manure applications in arable soils is a good agricultural practice because it efficiently reduces NH_3 losses as compared to the broadcast application (Horlacher and Marschner, 1990). The application of liquid manure also increases the N_2O emissions due to the high amounts of easily available N (Thompson et al., 1987; Dosch and Gutser, 1996). Flessa and Beese (2000) investigated the effect of the liquid manure application technique (control versus broadcast application versus injection) on the N_2O emissions in a microcosm study. During the observation period of 9 weeks, N_2O emissions from the surface application treatment accounted for 0.2% of the manure N. The corresponding value for the injection treatment was 3.3%. Additionally, the source function of the soil for atmospheric CH_4 (-12 g ha^{-1} CH_4–C in the control treatment) was turned into a net source where 39 g CH_4–C g ha^{-1}

was emitted after the injection treatment. The CH_4 and the high N_2O flux rates temporally coincided with strongly reduced redox potentials (approximately -200 mV) in the injection slots. These results prove the outstanding importance of denitrification to the total N_2O emissions from the soils at the FAM research farms. Obviously, the injection induced strongly increasing anaerobic conditions favourable not only for denitrification but also for the net production of CH_4 in the injection treatment. Similar effects from experiments on N_2O emissions from arable fields following liquid manure were reported by Dosch and Gutser (1996) and by Velthof et al. (1997), where the increased N_2O emission after incorporation of liquid manure was also explained by the increased N_2O production during denitrification. To reduce N_2O losses after the incorporation of liquid manure, soil moisture during application should not exceed 60% WFPS, since this value was found to be the critical value for increased N_2O emissions from denitrification (Flessa and Beese, 2000). As a further effective measure to reduce N_2O emissions, a nitrification inhibitor should be added to the liquid manure (Dendooven et al., 1988). Fermentation or separation of liquid manure decreases the dry matter content of the remaining liquid, and thus it also reduces the amount of microbial easily available carbon. These treatments minimise O_2 consumption in soils after the application of the liquid. They were shown to effectively reduce N_2O fluxes as compared to fields treated with untreated liquid manure (Dosch and Gutser, 1996).

Acknowledgements

The scientific activities of the FAM Munich Research Network on Agroecosystems were financially supported by the German Federal Ministry of Education and Research (BMBF 0339370). Overhead costs of Research Station Scheyern were funded by the Bavarian State Ministry for Science, Research and the Arts.

References

Adamsen AP, King GM, 1993. Methane consumption in temperate and subarctic forest soils: rates, vertical zonation, and responses to water and nitrogen. Applied and Environmental Microbiology 59, 485–490.

Ahlgrimm HJ, 1995. Beitrag der Landwirtschaft zur Emission klimarelevanter Spurengase – Möglichkeiten zur Reduktion? Landbaufor- schung Völkenrode, 45/4, 191–204.

Al-Kanani T, MacKenzie AF, Barthakur NN, 1991. Soil water and ammonia volatilization relationships with surface-applied nitrogen fertilizer solutions. Soil Science Society of America Journal 55, 1761–1766.

Amberger A, 1990. Ammonia emissions during and after land spreading of slurry. In: Nielsen VC, Voorburg JH, L'Hermite P (Eds.), Odour and Ammonia Emissions from Livestock Farming. Elsevier Applied Science, London, pp. 126–131.

Ambus P, Christensen S, 1995. Spatial and seasonal nitrous oxide and methane fluxes in Danish forest-, grassland-, and agroecosystems. Journal of Environmental Quality 24, 993–1001.

Arah JRM, Smith KA, 1989. Steady-state denitrification in aggregated soils: A mathematical model. Journal of Soil Science 40, 139–149.

Auerswald K, Kainz M, Fiener P, 2003. Erosion potential of organic versus conventional farming evaluated by USLE modelling of cropping statistics for agricultural districts in Bavaria. Soil Use and Management 19, 305–311.

Aulakh MS, Rennie DA, Paul EA, 1984. Gaseous nitrogen losses from soils under zero-till as compared with conventional-till management systems. Journal of Environmental Quality 13, 130–136.

Bakken LR, Børresen T, Njøs A, 1987. Effect of soil compaction by tractor traffic on soil structure, denitrification, and yield of wheat (Triticum aestivum L.). Journal of Soil Science 38, 541–552.

Beline F, Martinez J, Marol C, Guiraud G, 1998. Nitrogen transformations during anaerobically stored ^{15}N-labelled pig slurry. Bioresource technology 64, 83–88.

Bergkvist B, Folkeson L, 1992. Soil acidification and element fluxes of a Fagus sylvatica forest as influenced by simulated nitrogen deposition. Water, Air, & Soil Pollution 65, 111–113.

Blackmer AM, Bremner JM, 1978. Inhibitory effect of nitrate on reduction of N_2O to N_2 by soil microorganisms. Soil Biology & Biochemistry 10, 187–191.

Bouwman AF, 1996. Direct emissions of nitrous oxide from agricultural soils. Nutrient Cycling in Agroecosystems 46, 53–70.

Bouwman AF, Boumans LJM, Batjes NH, 2002. Estimation of global NH_3 volatilization loss from synthetic fertilizers and animal manure applied to arable lands and grasslands. Global Biogeochemical Cycles 16, doi: 10.1029/2000GB001389.

Bouwman AF, Lee DS, Asman WAH, Dentener FJ, van der Hoek KW, Olivier GJ, 1997. A global high-resolution emissions inventory for ammonia. Global Biogeochemical Cycles 11, 561–587.

Bremner JM, 1997. Sources of nitrous oxide in soils. Nutrient Cycling in Agroecosystems 49, 7–16.

Bremner JM, Blackmer AM, 1981. Terrestrial nitrification as a source of atmospheric nitrous oxide. In: Delwiche CC (Ed.), Denitrification, Nitrification, and Atmospheric Nitrous Oxide. John Wiley and Sons, New York, pp. 151–170.

Bronson KF, Mosier AR, 1993. Nitrous oxide emissions and methane consumption in wheat and corn-cropped systems in northeastern Colorado. In: Harper LA, Moiser AR, Duxbury JM, Rolston DE (Eds.), Agricultural Ecosystems Effects on Trace Gases and Global Climate Change. ASA Special Publication No 55, American Society of Agronomy, Madison, WI, pp. 133–144.

Burch JA, Fox RH, 1989. The effect of temperature and initial soil moisture on the volatilization of ammonia from surface-applied urea. Soil Science 147, 311–318.

Burford JR, Dowdell RJ, Crees R, 1981. Emissions of nitrous oxide to the atmosphere from direct-drilled and ploughed clay soils. Journal of the Science of Food and Agriculture 32, 219–223.

Burton DL, Beauchamp EG, 1994. Profile nitrous oxide and carbon dioxide concentrations in a soil subjected to freezing. Soil Science Society of America Journal 58, 115–122.

Cai GX, Chen DL, Ding H, Pachoslki A, Fan XH, Zhu ZL, 2002. Nitrogen losses from fertilizers applied to maize, wheat and rice in the North China Plain. Nutrient Cycling in Agroecosystems 63, 187–195.

Castro MS, Melillo JM, Steudler PA, Chapman JW, 1994. Soil moisture as a predictor of methane uptake by temperate forest soils. Canadian Journal of Forest Research 24, 1805–1810.

Cates RL, Keeney DR, 1987. Nitrous oxide production throughout the year from fertilized and manured fields. Journal of Environmental Quality 4, 443–447.

Chinkin LR, Ryan PA, Coe DL. 2003. Recommended improvements to the CMU ammonia emission inventory model for use by LADCO. Prepared Lake Michigan Air Directors Consortium.

Christensen S, Christensen BT, 1991. Organic matter available for denitrification in different soil fractions: effect of freeze/thaw cycles and straw disposal. Journal of Soil Science 42, 637–647.

Christensen S, Tiedje JM, 1990. Brief and vigorous N_2O production by soil at spring thaw. Journal of Soil Science 41, 1–4.

Clayton H, Arah JRM, Smith KA, 1994. Measurements of nitrous oxide emissions from fertilized grassland using closed chambers. Journal of Geophysical Research 99, 16.599–16.607.

Corre MD, van Kessel C, Pennock DJ, 1996. Landscape and seasonal patterns of nitrous oxide emissions in a semiarid region. Soil Science Society of America Journal 60, 1806–1815.

Crill PM, 1991. Seasonal patterns of methane uptake and carbon dioxide release by a temperate woodland soil. Global Biogeochemical Cycles 5, 319–334.

Crutzen PJ, 1981. Atmospheric chemical processes of the oxides of nitrogen including nitrous oxide. In: Delwiche CC (Ed.), Denitrification,

Nitrification and Atmospheric N_2O. John Wiley & Sons Ltd, Chichester, 17–44.

Davidson EA, 1991. Fluxes of nitrous oxide and nitric oxide from terrestrial ecosystems. In: Rogers JE, Whitman WB (Eds.), Microbial Production and Consumption of Greenhouse Gases: Methane, Nitrogen Oxides and Halomethanes. American Society for Microbiology, Washington DC, pp. 219–235.

De Datta SK, 1995. Nitrogen transformation in wetland rice ecosystems. Fertilizer Research 42, 193–203

De Klein CA, van Logtestjin RS, 1994. Denitrification and N_2O emission from urine-affected grassland soil. Plant and Soil 163, 235–242.

Del Grosso SJ, Parton WJ, Mosier AR, Ojima DS, Potter CS, Borken W, Brumme R, Butterbach-Bahl K, Crill PM, Dobbie K, Smith KA, 2000. General CH_4 oxidation model and comparison of CH_4 oxidation in natural and managed systems. Global Biogeochemical Cycles 14, 999–1020.

Dendooven L, Bonhomme E, Merckx R, Vlassak K, 1998. N dynamics and sources of N_2O production following pig slurry application to a loamy soil. Biology and Fertility of Soils 26, 224–228.

Denmead OT, Nulsen R, Thurtell GW, 1987. Ammonia exchange over a corn crop. Soil Science Society of America Journal 42, 840–842.

Dobbie KE, McTaggart IP, Smith KA, 1999. Nitrous oxide emissions from intensive agricultural systems: variations between crops and seasons, key variables, and mean emission factors. Journal of Geophysical Research 104: 26891–26899.

Dobbie KE, Smith KA, 1996. Comparison of CH_4 oxidation rates in woodland, arable and set aside soils. Soil Biology & Biochemistry 28, 1357–1365.

Döhler H, 1991. Laboratory and field experiments for estimating ammonia losses from pig and cattle slurry following application. In: Nielsen VC, Voorburg JH, L'Hermite P (Eds.), Odour and Ammonia Emissions from Livestock Farming. Elsevier Applied Science, London, England, pp. 132–140.

Dörsch P, 2000. Nitrous oxide and methane fluxes in differentially managed agricultural soils of a hilly landscape in Southern Germany. Fakultät für Landwirtschaft und Gartenbau, PhD thesis, TU München, Hieronymus-Verlag, München, ISBN 3-89791-126-4, FAM-Bericht 44, 226 pp.

Dosch P, Gutser R, 1996. Reducing N losses (NH_3, N_2O, N_2) and immobilization from slurry through optimized application techniques. Fertilizer Research 43, 165–171.

Dunfield PF, Knowles R, 1995. Kinetics of inhibition of methane oxidation by nitrate, nitrite and ammonium in a humisol. Applied and Environmental Microbiology 61, 3129–3135.

Duxbury JM, Harper LA, Mosier AR, 1993. Contribution of agroecosystems to global climate change. In: Harper LA, Mosier AR, Duxbury JM, Rolston JM (Eds.), Agricultural Ecosystems Effects on Trace Gases and Global Climate Change. ASA Special Publication 55. ASA, CSSA, and SSSA, Madison, WI, pp. 1–18.

ECETOC, 1994. Ammonia emissions to air in western Europe. Technical report Nr. 62, European Centre for ecotoxicology, Brussel, Belgium.

Eckard RJ, Chen D, White RE, Chapman DF, 2003. Gaseous nitrogen losses from temperate perennial grass and clover dairy pastures in south-eastern Australia. Australian Journal of Agricultural Research 54, 561–570.

Eichner MJ, 1990. Nitrous oxide emissions from fertilized soils: Summary of available data. Journal of Environmental Quality 19, 272–280.

Farquhar GD, Firth PM, Wetselaar R, Weir B, 1980. On the gaseous exchange of ammonia between leaves and the environment: determination of the ammonia compensation point. Plant Physiology 66, 710–714.

Fenn LB, Richards J, 1986. Ammonia loss from surface applied urea-acid products. Fertilizer Research 9, 265–275.

Firestone MK, Davidson EA, 1989. Microbial basis of NO and N_2O production and consumption in soil. In: Andreae MO, Schimel DS (Eds.), Exchange of Trace Gases Between Terrestrial Ecosystems and the Atmosphere. John Wiley and Sons Ltd., Chichester, pp. 7–21.

Flessa H, 2000. Steuerung der N_2O- und CH_4-Spurengasflüsse aus oder in Böden durch Standort- und Nutzungsfaktoren. Habilitationsschrift an der Fakultät für Forstwissenschaft und Waldökologie, Georg-August-Universität Göttingen, 254 pp.

Flessa H, Beese F, 1995. Effects of sugarbeet residues on soil redox potential and nitrous oxide emission. Soil Science Society of America Journal 59, 1044–1051.

Flessa H, Beese F, 2000. Laboratory estimates of trace gas emissions following surface application and injection of cattle slurry. Journal of Environmental Quality 29, 262–268.

Flessa H, Dörsch P, Beese F, 1995. Seasonal variation of N_2O and CH_4 fluxes in differently managed arable soils in Southern Germany. Journal of Geophysical Research 100, 23115–23124.

Flessa H, Dörsch P, Beese F, König H, Bouwman AF, 1996a. Influence of cattle wastes on nitrous oxide and methane fluxes in pasture land. Journal of Environmental Quality 25, 1366–1370.

Flessa H, Pfau W, Dörsch P, Beese F, 1996b. The influence of nitrate and ammonium fertilization on N_2O release and CH_4 uptake of a well-drained topsoil demonstrated by a soil microcosm experiment. Journal of Plant Nutrition and Soil Science 159, 499–503.

Flessa H, Ruser R, Dörsch P, Kamp T, Jimenez MA, Munch JC, Beese F, 2002a. Integrated evaluation of greenhouse gas emissions from two farming systems in southern Germany: Special consideration of soil N_2O emissions. Agriculture, Ecosystems & Environment 91, 175–189.

Flessa H, Ruser R, Schilling R, Loftfield N, Munch JC, Kaiser EA, Beese F, 2002b. N_2O and CH_4-fluxes in potato fields: automated management, management effects and temporal variation. Geoderma 105, 307–325.

Fox RH, Piekielek WP, 1987. Comparison of surface application methods of nitrogen solution to no-till. Journal of Fertilizer Issues 4, 7–12.

Gattinger A, Ruser R, Schloter M, Munch JC, 2002. Microbial community structure varies in different soil zones of a potato field. Journal of Plant Nutrition and Soil Science 165, 421–428.

Glasener KM, Palm CA, 1995. Ammonia volatilization from tropical legume mulches and green manures on unlimed and limed soils. Plant and Soil 177, 33–41.

Goodroad LL, Keeney DR, 1984. Nitrous oxide emissions from soils during thawing. Can. Journal of Soil Science 64, 187–194.

Granli T, Bøckman OC, 1994. Nitrous oxide from agriculture. Norwegian Journal of Agricultural Sciences, suppl. 12.

Groot Koerkamp PWG, Metz JHM, Uenk GH, Phillips VR, Holden MR, Sneath RW, Short JL, White RP, Hartung J, Seedorf J, Schroeder M, Linkert KH, Pedersen S, Takai H, Johnsen JO, Wathes CM, 1998. Concentrations and emissions of ammonia in livestock buildings in Northern Europe. Journal of Agricultural Engineering Research 70, 79–95.

Gutser R, Linzmeier W, Kilian A, 2000. N_2O-Emissionen aus landwirtschaftlich genutzten Flächen in Abhängigkeit der N-Düngung und des N-Potentials der Böden. VDLUFA-Schriftenreihe 55, 190199.

Häni H, Calame F, Neyroud J, 1986. Einfluss von Boden, Pflanzendecke und Witterung auf die Ammoniakverflüchtigung. FAC-Tagung 1988: Stickstoff in Landwirtschaft, Luft und Umwelt. Schriftenreihe der FAC Liebefeld 7, 77–89.

Hansen S, Mæhlum JE, Bakken LR, 1993. N_2O and CH_4 fluxes in soil influenced by fertilization and tractor traffic. Soil Biology & Biochemistry 25, 621–623.

Haynes RJ, 1986. Nitrification. In: Haynes RJ, Orlando FL (Eds.), Mineral Nitrogen in the Plant-Soil System. Academic Press, New York, pp. 127–165.

He ZL, Alva AK, Calvert DV, Banks DJ, 1999. Ammonia volatilization from different fertilizer sources and effects of temperature and soil pH. Soil Science 164, 750–758.

Herrmann B, Jones SK, Fuhrer J, Feller U, Neftel A, 2001. N budget and NH_3 exchange of a grass/clover crop at two levels of N application. Plant and Soil 235, 243–252.

Horlacher D, Marschner H, 1990. Schätzrahmen zur Beurteilung von Ammoniakverlusten nach Ausbringung von Rinderflüssigmist. Z. Pflanzenern. Bodenk. 153, 107–115.

Horton R, Ankeny MD, Allmaras RR, 1994. Effects of compaction on soil hydraulic properties. In: Soane BD, can Ouwerkerk C (Eds.), Soil Compaction in Crop Production. Elsevier Science Publishers, Amsterdam, pp. 141–165.

Hutchinson GL, Mosier AR, 1993. Processes for production and consumption of gaseous nitrogen oxides in soil. In: Delwiche CC (Ed.) Denitrification, nitrification and atmospheric N_2O. John Wiley & Sons Ltd, Chichester, 79–94.

Hütsch BW, 1996. Methane oxidation in soils of two long-term fertilization experiments in Germany. Soil Biology & Biochemistry 28, 773–782.

Hütsch BW, Webster CP, Powlson DS, 1993. Long-term effects of nitrogen fertilization on methane oxidation in soil of the Broadbalk wheat experiment. Soil Biology & Biochemistry 25, 1307–1315.

IPCC, 1996. Climate Change 1995, the science of climate change. Contribution of working group I to the second assessment report of the Intergovernmental Panel on Climate Change. Cambridge University Press, UK.

IPCC, 2001. Climate Change 2001: The Scientific Basis. Cambridge University Press, UK.

Isermann K, 1994. Agriculture's share in the emission of trace gases affecting the climate and some case-oriented proposals for sufficiently reducing this share. Environmental Pollution 83, 95–111.

Jongebreur AA, Voorburg JH, 1992. The role of ammonia in acidification. Perspectives for the prevention and reduction of emissions from livestock operations. Studies in Environmental Science 50, 55–64.

Kaiser EA, Kohrs K, Kücke M, Schnug E, Heinemeyer O, Munch JC, 1998. Nitrous oxide release from arable soil: importance of N-fertilization, crops and temporal variation. Soil Biology & Biochemistry 30, 1553–1563.

Kaiser EA, Ruser R, 2000. Nitrous oxide emissions from arable soils in Germany – An evaluation of six long-term field experiments. Journal of Plant Nutrition and Soil Science 163, 249–260.

Kalbitz K, Solinger S, Park JH, Michalzik B, Matzner E, 2000. Controls on the dynamics of dissolved organic matter in soils: a review. Soil Science 165, 277–304.

Kamp T, 1998. Freiland- und Laboruntersuchungen zur N_2O-Freisetzung eines landwirtschaftlich genutzten Bodens unter definierten Temperaturbedingungen. PhD thesis, Westfälische Wilhelms-Universität Münster.

Kamp T, Steindl H, Hantschel RE, Beese F, Munch JC, 1998. Nitrous oxide emissions from a fallow and wheat field as affected by increased soil temperatures. Biology and Fertility of Soils 27, 307–314.

Kamp T, Steindl H, Munch JC, 2000. Monitoring trace gas fluxes (N_2O, CH_4) from different soils under the same climatic conditions and the same agricultural management. Phyton 41, 119–130.

Klaassen G, 1994. Options and costs of controlling ammonia emissions in Europe. European Review of Agricultural Economics 21, 219–240.

Knowles R, 1993. Methane; Processes of production and consumption. In: Harper LA, Moiser AR, Duxbury JM, Rolston DE (Eds.), Agricultural Ecosystems Effects on Trace Gases and Global Climate Change. ASA Special Publication No 55, American Society of Agronomy, Madison, WI, pp. 145–156.

Larsson L, Ferm M, Kasimir-Klemedtsson A, Klemedtsson L, 1998. Ammonia and nitrous oxide emissions from grass and alfalfa mulches. Nutrient Cycling in Agroecosystems 51, 41–46.

Linn DM, Doran JW, 1984. Effect of water-filled pore space on carbon dioxide and nitrous oxide production in tilled and nontilled soils. Soil Science Society of America Journal 48, 1267–1272.

MacKenzie AF, Fan MX, Cadrin F, 1997. Nitrous oxide emission as affected by tillage, corn-soybean-alfalfa rotations and nitrogen fertilization. Can. Journal of Soil Science 77, 145–152.

Maidl FX, Brunner H, Sticksel E, 2002. Potato uptake and recovery of nitrogen ^{15}N-enriched ammonium nitrate. Geoderma 105, 167–177.

Marschner H, 1990. Mineral Nutrition of Higher Plants. Fourth Printing, Academic Press Inc, San Diego, CA.

Misselbrook TH, Sutton MA, Scholefield D, 2004. A simple process-based model for estimating ammonia emissions from agricultural land after fertilizer applications. Soil Use and Management 20, 365–372.

Mosier AR, Guenzi WD, Schweizer EE, 1986. Soil losses of dinitrogen and nitrous oxide from irrigated crops in northeastern Colorado. Soil Science Society of America Journal 50, 344–348.

Mosier AR, Hutchinson GL, 1981. Nitrous oxide emissions from cropped fields. Journal of Environmental Quality 10, 169–173.

Mosier AR, Parton WJ, Valentine DW, Ojima DS, Schimel DS, Delgado JA, 1996. CH_4 and N_2O fluxes in the Colorado shortgrass steppe: 2. impact of landscape and nitrogen addition. Global Biogeochemical Cycles 10, 387–399.

Mosier AR, Schimel DS, Valentine DW, Bronson K, Parton W, 1991. Methane and nitrous oxide fluxes in native, fertilized and cultivated grassland. Nature 350, 330–332.

Muck RE, Richards BK, 1980. Losses of manurial N in free-stall barns. Agricultural Manure 7, 65–93.

Nesbit SP, Breitenbeck GA, 1992. A laboratory study of factors influencing methane uptake by soils. Agriculture, Ecosystem & Environment 41, 39–54.

Ojima, DS, Valentine DW, Mosier AR, Parton WJ, and DS Schimel, 1993. Effect of land use change on methane oxidation in temperate forest and grassland soils. Chemosphere 26, 675–685.

Papen H, Butterbach-Bahl K, 1999. A 3-year continuous record of nitrogen trace gas fluxes from untreated and limed soil of a N-saturated spruce and beech forest ecosystem in Germany; 1. N_2O emissions. Journal of Geophysical Research 104, 18487–18503.

Prather M, Derwent R, Ehhalt D, Fraser P, Sanhueza E, Zhou X, 1995. Other trace gases and atmospheric chemistry. In: Houghton JT, Meira Filho LG, Bruce J, Lee H, Callander BA, Haites E, Harris N, Maskell K (Eds.), Climate Change 1994: Radiative Forcing of Climate Change and an Evaluation of the IPCC IS92 Emission Scenarios. Cambridge University Press, Cambridge, United Kingdom and New York, NY, USA, pp. 73–126.

Priebe DL, Blackmer AM, 1989. Soil moisture content at the time of application as a factor affecting losses of nitrogen from surface-applied urea. Journal of Fertilizer Issues 6, 62–67.

Rana G, Mastrorilli M, 1998. Ammonia emissions from fields treated with green manure in a Mediterranean climate. Agricultural and Forest Meteorology 90, 265–274.

Robertson K, 1994. Nitrous oxide emission in relation to soil factors at low to intermediate moisture levels. Journal of Environmental Quality 23, 805–809.

Roelle PA, Aneja VP, 2002. Characterization of ammonia emissions from soils in the upper coastal plain, North Carolina. Atmospheric Environment 36, 1087–1097.

Ross CA, Scholefield D, Jarvis SC, 2002. A model of ammonia volatilisation from a dairy farm: an examination of abatement strategies. Nutrient Cycling in Agroecosystems 64, 273–281.

Röver M, Heinemeyer O, Kaiser EA, 1998. Microbial induced nitrous oxide emissions from an arable soil during winter. Soil Biology & Biochemistry 30, 1859–1865.

Rudaz AO, Davidson EA, Firestone MK, 1991. Sources of nitrous oxide production following wetting of dry soil. FEMS Microbiology Ecology 85, 117–124.

Ruser R, 1999. Freisetzung und Verbrauch der klimarelevanten Spurengase N_2O und CH_4 eines landwirtschaftlich genutzten Bodens in Abhängigkeit von Kultur und N-Düngung, unter besonderer Berücksichtigung des Kartoffelanbaus. PhD thesis, TU München, Hieronimus-Verlag, München, ISBN 3-89791-034-9, FAM-Bericht 36, 124 pp.

Ruser R, Flessa H, Russow R, Schmidt G, Buegger F, Munch JC, 2006. Emissions of N_2O, N_2 and CO_2 from soil fertilized with nitrate: effect of compaction, soil moisture and rewetting. Soil Biology & Biochemistry 38, 263–274.

Ruser R, Flessa H, Schilling R, Beese F, Munch JC. 2001. Effects of crop-specific field management and N fertilization on N_2O emissions from a fine-loamy soil. Nutrient Cycling in Agroecosystems 59, 177–191.

Ruser R, Flessa H, Schilling R, Steindl H, Beese F, 1998. Effects of soil compaction and fertilization on N_2O and CH_4 fluxes in potato fields. Soil Science Society of America Journal 62, 1587–1595.

Ruser R, Gerl G, Kainz M, Ebertseder T, Reents HJ, Schmid H, Munch JC, Gutser R, 2007. Effect of management system on the N-, C-, P- and K-fluxes from FAM soils. In: Schröder P, Pfadenhauer J, Munch JC (Eds.), Perspectives for agroecosystem management – balancing environmental and socioeconomic demands, pp. 41–78.

Sahrawat KL, Keeney DR, 1986. Nitrous oxide emission from soils. Advances in Soil Science 4, 103–148.

Schimel JP, Klein JS, 1996. Microbial response to freeze-thaw cycles in tundra and taiga soils. Soil Biology & Biochemistry 28, 1061–1066.

Schnell S, King GM, 1994. Mechanistic analysis of ammonium inhibition of atmospheric methane consumption in forest soils. Applied Environmental Microbiology 60, 3514–3521.

Sehy U, 2003. N_2O-Freisetzungen landwirtschaftlich genutzter Böden unter dem Einfluss von Bewirtschaftungs-, Witterungs- und Standortfaktoren. PhD thesis, TU München, Oekom Verlag München, ISBN 3-936581-40-1, 130 pp.

Sehy U, Dyckmans J, Ruser R, Munch JC, 2004. Adding dissolved organic carbon to simulate freeze-thaw related N_2O emissions from soil. Journal of Plant Nutrition and Soil Science 167, 471–478.

Sehy U, Ruser R, Munch JC, 2003. Nitrous oxide fluxes from maize fields: relationship to yield, site-specific fertilization, and soil conditions. Agriculture, Ecosystems & Environment 99, 97–111.

Sigunga DO, Janssen BH, Oenema O, 2002. Ammonia volatilization from vertisols. European Journal of Soil Science 53, 195–202.

Smith KA, 1997. Soils and the greenhouse effect. Soil Use and Management 13, 229.

Smith KA, Dobbie KE, Ball BC, Bakken LR, Sitaula BK, Hansen S, Brumme R, Borken W, Christensen S, Priemé A, Fowler D, MacDonald JA, Skiba U, Klemedtsson L, Kasimir-Klemedtsson A, Degorska A, Orlanski P, 2000. Oxidation of atmospheric methane in N. European soils, comparison with other ecosystems, and uncertainties in the global terrestrial sink. Global Change Biology 6,791–803.

Smith KA, McTaggart IP, Dobbie KE, Conen F, 1998. Emissions of N_2O from Scottish agricultural soils, as a function of fertilizer N. Nutrient Cycling in Agroecosystems 52, 123–130.

Sommer SG, Christensen BT, Nielsen NE, Schjørring JK, 1993. Ammonia volatilization during storage of cattle and pig slurry: Effect of surface cover. Journal of Agricultural Science 121, 63–71.

Sommer SG, Ersboll AK, 1996. Effect of air flow rate, lime amendments, and chemical soil properties on the volatilization of ammonia from fertilizers applied to sandy soils. Biology and Fertility of Soils 21, 53–60.

Sommer SG, Olesen JE, Christensen BT, 1991. Effects of temperature, wind speed and air humidity on ammonia volatilization from surface applied cattle slurry. Journal of Agricultural Science 117, 91–100.

Sommer SG, Schjørring JK, Denmead OT, 2004. Ammonia emission from mineral fertilizers and fertilized crops. Advances in Agronomy 82, 557–662.

Stähli M, Stadtler D, 1997. Measurement of water and solute dynamics in freezing soil columns with time domain reflectometry. Journal of Hydrology 195, 352–369.

Stepniewski W, Glinski J, Ball BC, 1994. Effects of compaction on soil aeration. In: Soane BD, van Ouwerkerk C. (Eds.), Soil Compaction in Crop Production. Elsevier Science Publishers, Amsterdam, pp. 167–189.

Steudler PA, Bowden RD, Mellilo JM, Aber JD, 1989. Influence of nitrogen fertilization on methane uptake in temperate forest soils. Nature 341, 314–316.

Stolz B, 1997. Kohlendioxid-, Lachgas- und Methanemissionen nach dem Umbruch einer Kleerotationsbrache sowie der Zusammenhang von Nitratreduktaseaktivität und Lachgasemissionen in Böden. Diploma work Ludwig-Maximilians-Universität München.

Subak S, 1997. Full cycle emissions from extensive and intensive beef production in Europe. In: Adger WN, Pettnella D, Whitby M (Eds.), Climate Change Mitigation and European Land-use Policies. CAB International, Oxon, UK, pp. 145–157.

Thompson RB, Ryden JC, Lockyer DR, 1987. Fate of nitrogen in cattle slurry following application or injection to grassland. Journal of Soil Science 38, 689–700.

Thorne PS, 2002. Air quality issues. In: Merchant JA, Ross RF (Eds.), Iowa Concentrated Animal Feeding Operations Air Quality Study, Final report, Iowa State University and the University of Iowa Study Group, pp. 35–44.

Thornton FC, Valente RJ, 1996. Soil emissions of nitric oxide and nitrous oxide from no-till corn. Soil Science Society of America Journal 60, 1127–1133.

UBA, 1999. Jahresbericht 1998 aus dem Messnetz des Umweltbundesamts. UBA-Texte 66, ISSN 0722-186X.

van der Weerden TJ, Jarvis SC, 1997. Ammonia emission factors for N fertilizers applied to two contrasting grassland soils. Environmental Pollution 95, 205–211.

Velthof GL, Oenema O, Postma R, van Beusichem ML, 1997. Effects of type and amount of applied nitrogen fertilizer on nitrous oxide fluxes from intensively managed grassland. Nutrient Cycling in Agroecosystems 46, 257–267.

Vermoesen A, Demeyer P, Hofman G, van Cleemput O, 1996. Ammonia volatilization from mineral N-fertilizers, influenced by pH, temperature and soil moisture content. In: Diekkrüger B, Heinemeyer O, Nieder R (Eds.), Transactions of the 9th Nitrogen Workshop, TU-Braunschweig, pp. 573–576.

Wagner-Riddle C, Thurtell GW, Kidd GK, Beauchamp EG, Sweetman R, 1997. Estimates of nitrous oxide emissions from agricultural fields over 28 months. Canadian Journal of Soil Science 77, 135–144.

Weber A, Gutser R, Henkelmann G, Schmidhalter U, 2000. Unvermeidbare NH_3-Emissionen aus mineralischer Düngung (Harnstoff) und Pflanzenmulch unter Verwendung einer modifizierten Messtechnik. VDLUFA-Schriftenreihe 53, 175–182.

Weier KL, Doran JW, Power JF, Walters DT, 1993. Denitrification and the dinitrogen/nitrous oxide ratio as affected by water, available carbon, and nitrate. Soil Science Society of America Journal 57, 66–72.

Wessels-Perelo L, 2003. Microbial immobilisation and turnover of ^{13}C and ^{15}N labelled substrates and microbial diversity in two arable soils under field and laboratory conditions. Phd thesis at the TU-University of Munich, 93 pp.

Xu X, Nieber L, Gupta SC, 1994. Compaction effect on the gas diffusion coefficient in soils. Soil Science Society of America Journal 56, 1743–1750.

Yamulki S, Jarvis SC, 2002. Short-term effect of tillage and compaction on nitrous oxide, nitric oxide, nitrogen dioxide, methane and carbon dioxide fluxes from grassland. Biology and Fertility of Soils 36, 224–231.

Chapter 2.3

Precision Farming – Adaptation of Land Use Management to Small Scale Heterogeneity

U. Schmidhalter, F.-X. Maidl, H. Heuwinkel, M. Demmel,
H. Auernhammer, P.O. Noack and M. Rothmund

2.3.1 Basics and objective

Site-specific farming can contribute in many ways to long-term sustainability of agriculture production, confirming the intuitive idea that precision agriculture should reduce environmental loadings by applying inputs such as fertilisers only where they are needed, when they are needed (Bongiovanni and Lowenberg-Deboer, 2004), and in site-specific amounts. Site-specific crop management aims at optimising agriculture production by managing both the crop and the soil with an eye towards the different conditions found in each field.

Site-specific management requires detailed information about the heterogeneity of fields to adapt soil cultivation, seeding, fertilising, and fungicide and herbicide application to the locally varying conditions. Previously existing soil and plant information seldom matches the requirements either with respect to the intensity of the required information or with respect to the quality of the derived maps to delineate management units. Conventional methods are too costly and time-consuming. Preferably fast, non-contacting and non-destructive methods should be available to obtain the required information. Management recommendations corresponding to within-field site-specific characteristics are rarely available (Robert, 2001). Implementing the knowledge gained in sound management practices is clearly lagging. In this respect some differences between the two land-use systems studied by the FAM project (integrated farming and organic farming) have to be emphasised: Organic farming generally lacks the potential for a short-term on-the-go reaction especially with respect to fertiliser application. Further on, in organic farming, nitrogen, the key nutrient for agricultural production, will mainly enter crop rotation via the symbiotic legume-rhizobium N_2 fixation, i.e. a biological process that already strongly interacts with site-specific conditions as well as organic manure, that delivers plant-available N also strongly in accordance with site-specific conditions.

The spatial and temporal variability of soil water and nitrogen supply capabilities, as well as the spatial and temporal changes in plant nitrogen uptake on the field and farm levels, require different (fertiliser) management strategies to obtain economically and ecologically reasonable yields. This report focuses on recent developments to characterise the spatial and temporal variability of soil water, soil nitrogen, plant nitrogen uptake, biomass development, and yield more efficiently, with the aim to optimise inputs relative to the site-specific yield potential. In the integrated farm, site-specific crop management approaches have been designed and tested to optimise agricultural production, while for the organic farm appropriate management options are still investigated.

2.3.2 Development of methods to characterise spatial variability of soils, crops, and yield

Soils and crops are not uniform but vary according to the spatial location. To get information on their spatial variability and their local distribution, soil, crop, and yield parameters have to be measured on as many locations in the field as possible. These locations have to be defined by position information.

To get such geo-referenced information all over the field, traditional sampling strategies would require a large amount of time and labour in the field as well as in the lab. Therefore, the development and availability of continuously and on-the-go measurement systems and accurate and reliable positioning systems as well as electronic communication systems are prerequisites to detect small-scale heterogeneity.

Technical development – Prerequisites for precision farming

Developments of electronic devices and information technique in all industrial and private sectors made these technologies also available for agriculture. Some applications could be integrated in the FAM project and other related projects of the Technical University of Munich with little modifications, e.g. receiver for the global satellite navigation systems. Others needed and some still need a lot of effort for development and evaluation such as sensor systems for soil and crop parameters.

Geo-referencing and communication

Although radio navigation systems are available since World War II, reliable, affordable, and easy-to-operate position detection and navigation render possible not until global satellite navigation systems were operable.

With the increasing number of sensor and actuator systems, the electronic communication on agricultural machines became very important (Auernhammer, 1997). Although starting with point-to-point connections, the need for bus system-based communication system increased with the complexity of the systems. Therefore, development and standardisation of an agricultural bus system started very early and reached the level of International Standardisation Organisation ISO (Ehrl, 2005).

Positioning systems for agricultural use. At the end of the 80th different positioning systems have been evaluated for the use in agriculture, especially with yield sensor systems and for tractor guidance. Most of them were based on the radio navigation principle with either active or passive beacons (Searcy et al., 1989). The accuracy was sufficient for the investigated applications, but the necessary ground-based infrastructure made a widespread use in agriculture impossible.

With the development and operability of the Global Positions System – NAVigation System by Time and Range (GPS-NAVSTAR) of the USA and the GLObal NAvigation Satellite System (GLONASS) of Russia – satellite positioning and navigation systems covering the whole world without any individual user-owned infrastructure became available not only for military use but also for civil use.

Both systems work on the same principles (Figure 1).

The system is based on three segments. The space segment consists of more than 24 satellites on six orbital planes at an altitude of 20 183 km. These satellites are continuously sending messages that identify the satellite vehicle (sv). It provides the positioning, timing and ranging data, satellite status, and orbit parameters to the user. The master control station, monitor, and uploading stations (control segment) control the space segment. The user segment consists of all equipment that receives and tracks the satellite signals. Time synchronisation with the satellite clocks enables the GPS receivers to calculate the distances between satellite and

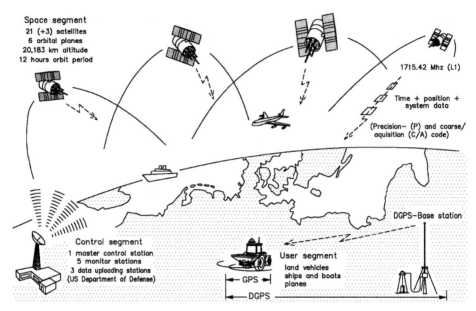

Figure 1 Principle of the global satellite navigation systems GPS NAVSTAR.

receiver by run-time measurement. Based on these 'pseudo ranges' to more than three (or more than two for two-dimensional) satellites and the information on the positions of the satellites, the receiver can calculate the position of the receiver antenna. The GPS NAVSTAR provides accuracies of absolute position detection within ±10–15 m.

For higher accuracy a differential global positioning system is needed. DGPS additionally uses a base station (or a network of base stations) that calculates the errors of the run-time and pseudo-range measurement by comparing the calculated position with the real position of the antenna. With this correction information, normally transferred via radio link, the positioning accuracy of the mobile receivers can be reduced to ±2–5 cm depending on the used DGPS principle (see the section 'Accuracy of GPS and DGPS').

Accuracy of GPS and DGPS

Beginning with the first application of GPS in agriculture in 1990 (Auernhammer et al., 1991), the first integration in local yield detection systems on combine harvesters in 1991 (Auernhammer et al., 1993), the availability of pseudo range correction DGPS in 1992 (Auernhammer et al., 1994), the declaration of full operational capability (FOC) on 17th June 1995 and the switching off of selective availability (SA) for civil users on 2nd May 2000, a lot of investigations have been made to define the accuracy of GPS and DGPS for different applications and configurations.

On one hand, technical developments and system developments increased the accuracy of GPS and DGPS equipment continuously. On the other hand, investigations of the positioning accuracy became more and more difficult with increasing system accuracies, especially in dynamic applications like on agricultural vehicles and machines (Steinmayr et al., 2000; Stempfhuber, 2001; Ehrl et al., 2003).

Summarising the published investigations of different GPS and DGPS configurations following positioning accuracy can be determined (Table 1).

From Table 1 it can be concluded that at present GPS and DGPS technology offers all ranges of accuracies to fulfil the requirement of most agricultural applications. The investment for the systems increases with increasing accuracy.

Communication systems for agricultural equipment

Sophisticated farming, especially precision farming, integrates a variety of computerised equipment and tools. Until now most of these mechatronic systems are based on specific controllers with integrated man–machine interfaces and point-to-point connections to the actuators on the machines. These circumstances often result in tractor cabins filled with a high number of controllers for different machines, which lead to a confusion of the tractor driver, increase the operating errors, and decrease the acceptance of electronic control systems (Ehrl et al., 2003).

To overcome these problems, compatible electronic communication systems are needed like in other fields of application (e.g. industry automation). For that purpose BUS Systems for mobile applications have been developed and first introduced to automotive applications. But also for agricultural electronics two communication standards have been developed, both using controller area network (CAN) (Auernhammer, 2002).

– German 'Landwirtschaftliches BUS-System' (LBS), codified as DIN 9684/2–5, is based on the 11-bit identifier of CAN V2.0A. It connects a maximum of 16 controllers, including the terminal; data transfer

Table 1 General positioning accuracy of different GPS and DGPS configurations.

GPS or DGPS configuration	Absolute positioning error range (m)	Pass-to-pass error range[a] (m)
Standard GPS, single frequency	$< \pm 10$	$< \pm 5$
DGPS, pseudo range correction	$< \pm 3$	$< \pm 1$
DGPS, 2-frequency, code smoothing	$< \pm 0.5$	$< \pm 0.1$
RTK DGPS, real-time kinematik	$< \pm 0.05$	$< \pm 0.02$

[a] Pass-to-pass error = error between two points within a short time (<15 min).

speed is 125 kB s^{-1}. It was only used by a small number of German agricultural equipment manufacturers and is not further supported.
– ISO 11783 standard works with the extended identifier of CAN V2.0B, 256 kb s^{-1} data transfer speed, and is able to connect a maximum of 32 controllers. Its detailed structure using the ISO/OSI layer model and 13 parts of special definitions tries to cover all requests of agricultural tractor–implement combinations. Most agricultural equipment manufacturers have declared to support the standard and integrate it in their electronic development (Figure 2).

Machine guidance

A fast-spreading application of precision farming technologies to optimise the production process are navigation systems based on the satellite navigation system GPS NAVSTAR. A wide range of configurations from guidance up to automatic steering are available. The basic principles of the different systems are identical. With the satellite positioning system, the positions of a first trace in the field are stored and parallel tracks in a distance defined by the machine user are calculated.

Using guidance systems a display shows the driver the new/next reference track or the deviation from the new/next reference track and the driver tries to follow with his vehicle (tractor, combine, sprayer) as exactly as possible along this virtual line.

Automatic steering systems on agricultural machines use an additional solenoid steering valve or a small electric servo drive at the steering wheel to navigate the vehicle along the reference line.

The accuracy of the satellite-based guidance and steering systems depends on the used GPS systems and DGPS correction principles and services. The accuracy of guidance systems is also influenced by the skill of the driver to steer the vehicle along the displayed track.

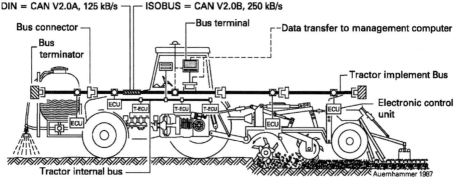

T-ECU Tractor internal Electronic Control Unit

Figure 2 Communication between tractor and implement using ISO 11783 (Auernhammer, 1997).

With the correction services exempt of charges like BEACON or WAAS/EGNOS, normally a pass-to-pass accuracy (within a time period of 15 min) between 10 and 50 cm can be reached. An accuracy between 5 and 10 cm can be reached using chargeable satellite-based correction services. To reach this accuracy, additional sensors to compensate rolling and tilting and their integration in the navigation controller are necessary.

The highest accuracy with errors between 1 and 5 cm can only be reached using real-time kinematik differential DGPS (RTK DGPS) with separate base stations within a distance <25 km and the integration of rolling, tilting, and jawing sensors. This level of accuracy is only used with automatic steering systems. The prices of the systems increase with their accuracy and with the level of automation (Demmel, 2002).

Manual operated guidance systems can be favourably used for the application of fertilisers and herbicides, especially if there are no tramlines available. Gaps and overlapping can be prevented. Especially in corn these systems avoid the counting of rows during turning at the headland. List prices (in 2006) vary between 2000 € and 6000 €.

Depending on the chosen (bought) accuracy, automatic steering systems are used for tillage, seedbed preparation, seeding, planting, and cultivation. Harvesting machines such as combine harvesters or self-propelled mowers can also be controlled. Overlapping, which occurs especially with large working width, can be minimised. Time for turning can be reduced driving every second pass first and filling the gaps later. Workload relieving and increase in performance of drivers especially at long-lasting workdays and under bad conditions (dust, fog, darkness) are enormous. Automatic steering systems cost between 10 000 € and 45 000 € and can create correction service fees up to 2000 € per year.

Soil Sensors

Site-specific management requires detailed information about the locally varying soil conditions. The lack of high spatial resolution topsoil data is a serious limitation to the establishment of site-specific soil and crop management. Simple methods to detect important soil properties would facilitate the development of optimised management strategies. In some countries, previously existing soil information from conventional soil coring may be available. However, such data seldom match the requirements either with respect to the intensity of the soil sampling or with respect to the quality of the derived maps to delineate management units. Additionally, conventional methods are too costly and time-consuming. Preferably fast, non-destructive and non-contacting methods should be available to obtain the required information.

Soil properties such as clay, organic matter, or plant-available water capacity are important factors of soil fertility. Classical methods to determine such parameters are lengthy, space consuming, and laborious. More rapid and inexpensive methods would be valuable in obtaining such information. With the recent advancement in non-destructive proximal or remote-sensing techniques, this goal of characterising field-site characteristics using high spatial resolution soil data seems to be within reach.

Non-destructive principles to sense soil properties – Mapping soils by apparent electrical conductivity measurements

Electromagnetic induction represents a fast non-contacting method to get information about field heterogeneity of soil texture and soil water content. Measurements of the apparent electrical conductivity represent the influence of several factors, including soil texture and organic matter content, soil salinity, soil water content, and soil bulk density. Whereas the influence of salinity plays normally a minor role under temperate conditions, information about clay content (de Jong et al., 1979) and water content (Kachanoski et al., 1988) can be derived. See also Sommer et al. (2007), this volume.

Determination of soil texture and soil carbon content by NIRS

Owing to the rapid progress in data processing during the last decade, the use of near-infrared reflectance spectroscopy in chemical, biological, and agricultural sciences has been enhanced. Near-infrared reflectance spectroscopy has already been demonstrated as an accurate method to obtain valuable information on soil texture and organic matter (Stenberg et al., 1995; Ben-Dor and Banin, 1995). Studies of soils encompassing very different origins and composition are rare and are addressed in this work.

Mapping soil surface properties by aerial reflectance measurements

The spatial variability of topsoil texture and organic matter across fields can be studied using field spectroscopy and airborne hyper spectral imagery with the aim of improving soil-mapping procedures. Organic matter and clay content are correlated with spectral properties. Topsoil reflectance (330–2500 nm) can be measured in the field using airborne sensors such as the HyMap sensor for recording hyper spectral images (420–2480 nm, 128 channels) of bare soil fields. Using partial least square regression (PLSR) allows developing and calibrating models that establish a quantitative relationship between the spectra and soil parameters.

Mapping available soil water by aerial thermography or proximal reflectance measurements

Many applications in fields such as hydrology, meteorology, and agriculture require mapping of soil moisture, since the amount and status of water in soils impact crop growth. This requires reliable techniques to perform accurate soil water content measurements with minimal soil disturbance.

Crop growth depends on soil attributes. It should be feasible to use the crop stand condition as a bio-indicator of soil productivity. Biomass is one of the important parameters to differentiate crop stand conditions. For regions with negative water balance during the growing season, the site-specific availability of soil water is the main limiting soil resource. Biomass production and transpiration/evapotranspiration are related to each other linearly under water-deficit conditions (Schmidhalter and Oertli, 1991; Funk and Maidl, 1997). From this general relationship, it can be concluded that biomass production is related to the available soil water, particularly under water-deficit conditions (Selige and Schmidhalter, 2001; Brunner, 1998). Furthermore, because differences in evapotranspiration can be reflected by different canopy temperatures of crop stands, a feed-forward soil–crop response mechanism would allow the plant-available water capacity to be inferred from the surface temperature of sensitive crops such as winter wheat during specific, so-called 'bio-indicative', crop development stages. Thus, surface temperatures recorded by remotely sensed thermography could allow the pattern of plant-available water in fields to be detected (Selige and Schmidhalter, 2001).

Biomass sensors

Remote sensing has a great potential for characterising the effect of stresses on plants. Understanding the factors that influence the reflectance signal will greatly enhance the quality of the data and the potential for detecting stresses (Major et al., 2003).

Previous research has shown that spectral measurements can indirectly describe biomass, nitrogen concentration, and nitrogen uptake. In the past mostly hand-held spectrometers were used for this purpose. Reflectance measurements have been widely used in order to estimate the N status of plants. Leaf reflectance in the visible region is driven primarily by chlorophyll absorption, in the near-infrared region by leaf structure and in the short-wave infrared by water absorption. Primarily, leaf reflectance and absorption of light, the amount of leaves and the reflectance of the soil surface, or other background determine canopy reflectance.

Usually, biomass yield and nutrient status of canopies are assessed manually by cutting biomass, followed by laboratory analysis. Nevertheless, these methods are time-consuming, labour-intensive, costly, and of limited value for numerous examinations required within heterogeneous fields. Thus, different tools and sensors have been developed to replace manual biomass determination. These devices are used in direct contact with leaves or in the close-up range of plants and are applicable on agricultural fields.

The terms biomass yield and nutrient status of canopies are charac-terised by different parameters and need definition. A crop is composed of above-ground (leaves, blades, stems, ears, flowers, etc.) and sub-surface (roots, sprouts, etc.) material. The general aim of biomass sensors is the detection of above-ground biomass yield. In this context, fresh matter yield is defined as biomass per unit ground area, and is composed of dry matter yield and water content. Analogous, dry matter yield is defined as dry matter per unit ground area. Further parameters of a canopy with agronomic relevance are the concentration and content of photosynthetic active pigments (e.g. chlorophyll) or nutrients such as nitrogen. A concen-tration indicates the constituent per unit dry matter. In contrary, a content describes the constituent on an areal resolution (per unit soil or leaf area). When multiplying N concentration and dry matter yield, N uptake or N con-tent is achieved, an areal parameter. Nevertheless, a comparable N uptake is achieved with different structured canopies: Either a high dry matter yield is in association with a low N concentration or a low dry matter yield is asso-ciated with a high N concentration. A further parameter of plant canopies is the leaf area index that indicates the leaf area per unit ground area.

Some sensors are only valuable in detecting specific parameters such as nutrient concentration (nitrate test; SPAD-Meter, Hydro-N-Tester), bio-mass yield (Pendulum-Meter), or leaf area index (LAI-2000 plant canopy analyzer). Other systems (reflectance sensors (YARA N-Sensor), laser sen-sors (MiniVeg N; Planto sensor)) generate measurement values that cor-relate to different parameters of plant growth.

Tools used in the FAM research project and other important tech-niques for detecting canopies are shown in this chapter. The devices can be divided into three groups: chemical, mechanical, and optical devices. A chemical test is the nitrate-N test. The so-called pendulum-meter is a mechanical device. Owing to their measurement principle, the optical sen-sors can be further divided into passive (N-Sensor) and active sensors (SPAD-Meter, N-Tester, MiniVeg N, Planto N-Sensor, YARA N-Sensor *ALS*). Passive sensors rely on a minimal level and quality of irradiance, whereas active sensors use their own light source.

Nitrate test

The nitrate test determines the nitrate-N concentration in the sap of growing plants. This parameter is an indicator of N supply from soil. A specific gripper (Figure 3) is used for pressing out the sap at the stem basis of plants. Generally, 30 stems are needed to derive an adequate amount of sap (Baumgärtel, 2001). After contacting the sap with specific test strips, a colour reaction appears, which is dependent on the value of nitrate-N concentration. The comparison with a colour chart leads to the def-inition of N fertiliser demand. If the actual values of nitrate-N concentration in the plant sap are lower than those necessary to achieve maximal yield,

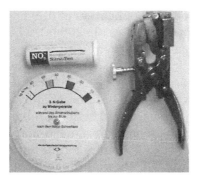

Figure 3 Nitrate-N-test: gripper and test strip (by courtesy of Sächsische Landesanstalt für Landwirtschaft, Leipzig, Germany). (For colour version of this figure, please see page 423 in colour plate section.)

a nitrogen topdressing is necessary. The method only indicates the actual value of nitrate-N concentration, thus the measurements have to be repeated within 10–14 days to define the starting point of N deficiency, i.e. the timing of N topdressing. Nevertheless, the method is characterised by some disadvantages. The nitrate-N concentration in the sap of plants is strongly dependent on the water supply. So the derived values of nitrate-N test show a great variation, even in the course of a day (MacKerron et al., 1995). A further limitation is that the measurements are not applicable for site-specific N fertilisation within heterogeneous fields.

Pendulum-Meter (Crop-Meter)

The pendular sensor (Pendulum-Meter, Crop-Meter) is a physical pendular, attached in the front of a tractor (Figure 6). The mechanical device is a passive method for deriving information of the plant biomass status. With forward motion of the tractor, the pendular is guided through the canopy in a defined height and moves in accordance with the resistance of the plants. Nevertheless, a correction procedure is necessary for varying depth of lanes as well as for unforeseen tractor movements. Then, the degree of deflection of the pendular depends mainly on the mass of the plants. In this context, the angle of deflection proved to be highly correlated to the biomass yield of the canopy (Ehlert et al., 2004b).

Some limitations of the pendular sensor are evident. The application of the device is impractical for small plants such as during the period of tillering of cereals. Thus, Ehlert et al. (2004b) used the Crop-Meter starting from the third nitrogen dressing in winter wheat, at EC 37. As only plant mass leads to a deflection of the pendular, no information about the nutrient status of the canopy is available.

Nevertheless, the readings can be derived independent of external weather conditions. The detection of biomass yield enables the use of the

Crop-Meter for a site-specific application of fungicides and growth regula-
tors (Dammer et al., 2001; Ehlert et al., 2004a), where reduced amounts in
areas with low plant mass seem promising.

SPAD-Meter (or N-Tester)

The hand-held devices SPAD-Meter (Figure 4) and N-Tester (Co.
YARA) are identical in their construction and measurement principle.
Both optical tools are active sensors for detecting chlorophyll concentration
of leaves. An internal radiation source emits light and the transmission
through a leaf is measured in the red (650 nm) as well as near-infrared
(920 nm) spectral regions. For the measurements, a single leaf is fixed in
a defined position between two arms and the light source is activated with
manual pressing. With the SPAD-Meter, every single measurement value
is seen. The N-Tester depends on measurements of at least 30 leaves for
deriving results.

As chlorophyll concentration of leaves varies within the horizontal
profile of a canopy as well as within the horizontal and vertical areas of a
leaf, the youngest and fully developed leaves have to be measured and the
detection must be performed in the middle of the leaf. The derived measurement values display a relative information of the chlorophyll concentration of leaves. This parameter is closely correlated to the N concentration
(Ercoli et al., 1993; Bredemeier and Schmidhalter, 2001; Schächtl, 2004).
Thus, the N concentration of leaves is gained in an indirect way. As an
enhanced N supply leads to an increasing N concentration, the values of
N-Tester measurements are in accordance with mineral N dressings
(Table 2). So N-Tester measurements enable the definition of N demand of
plants.

Figure 4 SPAD-Meter (by courtesy of YARA GmbH & Co. KG, Dülmen,
        Germany). (For colour version of this figure, please see page 424
        in colour plate section.)

Table 2 N-Tester values dependent on the nitrogen supply of two winter wheat cultivars (EC 37, Schächtl, 2004).

	N amount (kg N ha^{-1})				
Cultivar	0	60	120	180	Ø
Flair	509	584	623	658	594
Orestis	427	502	525	604	514

The results are independent of external conditions. So time of day as well as the detection of wet leaves display no problem. But an important effect is due to different cultivars. Varieties often differ in chlorophyll concentration and greenness, even though their values of N concentration are comparable (Maidl et al., 2001). Table 2 gives an example of a cultivar with dark-green coloured leaves (Flair, high chlorophyll concentration) and with light-green coloured leaves (Orestis, low chlorophyll concentration). The N concentration of both cultivars was comparable and increased with the amount of mineral N fertiliser. Without considering the cultivar effects, measurements with the N-Tester would recommend a comparable N application for un-fertilised cv. Flair and cv. Orestis that already received 120 kg N ha^{-1}.

A problem of SPAD-Meter and N-Tester measurements is the impossible differentiation between the lack of sulphur or nitrogen, as a shortage of both nutrients leads to a reduction in chlorophyll concentration. Also, water deficiency as well as drought stress impairs the derived measurement values (Martinez and Guiamet, 2004). However, nutrient deficiencies other than nitrogen and water stress influence similarly all non-destructive passive and active nutrient sensors. Both hand-held devices are only suited for point measurements on single leaves. Thus, a high areal solution within a heterogeneous field is very time-consuming.

LAI-2000 plant canopy analyzer

The LAI-2000 plant canopy analyzer (Figure 5) enables a detection of leaf area index in the field without destroying plants. For the passive method, a minimal intensity of irradiance is essential. Light intensity is measured at the top as well as the bottom of the canopy. The reduction of radiation on the way through the canopy is proportional to the number of leaf layers and correlated to the leaf area index. Two different measurement techniques are available. When using one sensor, the measurements have to be performed in a defined sequential course at the top and the bottom of the canopy. With two sensors, a simultaneous detection of radiation at the top and the bottom of the canopy is enabled. Approximately eight measurements are needed for robust values. Especially for plants with a

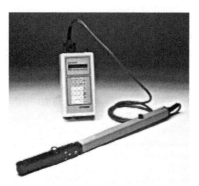

Figure 5 LAI-2000 plant canopy analyzer (by courtesy of LI-COR Biosciences GmbH, Bad Homburg, Germany). (For colour version of this
figure, please see page 424 in colour plate section.)

Figure 6 Crop-Meter (Ehlert, 2004; by courtesy of agrocom. GmbH & Co.
Agrar-system KG, Bielefeld, Germany. (For colour version of this
figure, please see page 424 in colour plate section.)

large row width such as maize, potatoes, and sugar beet, a systematic
placement of the sensor is essential, between as well as within the rows
(Figure 5).

The knowledge of leaf area index enables a derivation of biomass yield
(Schächtl and Maidl, 2002; Figure 12). Nevertheless, the correlation
between leaf area index and dry matter yield depends on the effect of
canopy architecture with different cultivars.

The application of the LAI-2000 plant canopy analyzer is limited for
defined external weather conditions. The readings are only valuable for
situation with diffuse irradiance and no direct sunlight like in the early
morning or late afternoon period or for cloudy conditions. Wet canopies

should be avoided as raindrops on the lens lead to a refraction of light. The sensor is only suited to gain point information, thus the application is limited within heterogeneous fields.

Reflectance measurements

Reflectance spectra of canopies are detected with a spectrometer (Figure 7, hand-held device). The measurement method is an optical passive tool as direct sunlight or a minimal intensity of diffuse irradiance is required for deriving a reflection signature. In order to obtain the degree of reflection, the reflectance of the crops as well as the irradiance is gathered. Typical reflectance spectra of winter wheat canopies with different nitrogen supply are given in Figure 13. A characteristic feature for green plants is the trend in the visible (400–700 nm) and near-infrared (>700 nm) regions. Blue (400–500 nm) and red (600–700 nm) wavelengths of the incoming light are highly absorbed by green plants, thus the reflection in these wavelengths is low. In these regions, plant pigments (carotinoids, chlorophylls) exhibit absorption maxima. The low absorption and therefore the higher degree of reflection at green wavelengths (500–600 nm)

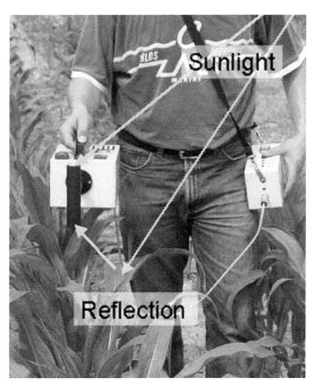

Figure 7 Hand-held spectrometer (by courtesy of tec5 AG, Oberursel, Germany).

lead to the typical green colour of the plants. In the near-infrared region (>700 nm), a steep rise in the reflectance spectra appears in the crossover between red and near-infrared wavelengths, leading to the so-called red-edge of reflectance spectra.

A varying N supply is associated with changes in reflectance signatures of plants (Figure 13), due to altered biomass yield as well as values of concentration of absorbing constituents. With increasing chlorophyll concentration, the intensity of absorption is reinforced in blue and red wavelengths, whereas the reflection is diminished. Furthermore, an enhanced amount of biomass leads to a higher intensity of reflection in the near-infrared spectral region. Accordingly, the local minimum in reflectance at about 670 nm is enlarged with increasing N supply, and the red-infrared edge of spectra shifts towards longer wavelengths (Liebler et al., 2001).

But numerous studies revealed the influence of diurnal variation of external effect such as weather conditions, cloudiness, and solar angle. As a result, the performance of a vegetation index is subjected to diurnal variation. A vegetation index, suitable for an application in plant production, should be sensitive to crop parameter and insensitive to environmental parameters. Sticksel et al. (2004) found that all vegetation indices varied significantly between morning, noon, and afternoon measurements (Table 3). RVI, IRG, and GR were more subjected to a diurnal effect than other indices. By far, REIP was least affected by a diurnal factor. N fertiliser treatment strongly affected all vegetation indices (Table 4).

In general, increasing amounts of applied N resulted in increasing index values. It is important to notice that REIP, IRG, and IRI differed significantly for all tested N amounts, while for NDVI, RVI, and GR, a saturation effect was observed at the highest level of biomass formation (Table 4).

Reflectance measurements are influenced by the external conditions. In contrast to handheld spectrometers, oblique oligo-view sensors such as

Table 3 Absolute and relative values (in brackets) of different vegetation indices as affected by time of day (Sticksel et al., 2001).

Time of day	Vegetation index					
	REIP	NDVI	RVI	IRG	IRI	GR
Morning	725.5	0.914	28.9	10.45	1.45	2.68
	(99.9)	(98.4)	(92.6)	(96.3)	(100)	(97.1)
Noon	725.2	0.858	20.7	8.96	1.41	2.13
	(99.9)	(92.4)	(66.3)	(83.3)	(97.2)	(77.2)
Afternoon	725.9	0.920	31.2	10.84	1.45	2.76
	(100)	(100)	(100)	(100)	(100)	(100)

Table 4 Vegetation indices as affected by N fertiliser treatment (Sticksel et al., 2004).

N (kg ha^{-1})	Vegetation index					
	REIP	NDVI	RVI	IRG	IRI	GR
0	721.2 a	0.804 a	13.2 a	6.08 a	1.27 a	2.00 a
100	725.1 b	0.912 b	27.0 b	9.88 b	1.44 b	2.64 b
160	727.6 c	0.940 c	34.3 c	12.16 c	1.52 c	2.72 c
220	728.3 d	0.940 c	34.5 c	12.49 c	1.53 c	2.79 c

Values followed by different letters in a column are significantly different at $p < 0.05\%$.

the YARA N-Sensor allow for measurements being highly independent of daytime, azimuth angle, and cloudiness (Mistele et al., 2004).

Measurements, however, cannot be performed at very low zenith angles. Such limitations do not exist for the newly developed active principle implemented in the YARA N-Sensor *ALS*. This sensor has its own radiation source and measurements can also be conducted reliably at night.

Reflectance measurements are influenced by soil reflectance. Thus, at very early growth stages with a reduced ground cover measurements reflect a mixed signal between soil and plant. In contrast to reflectance measurements in the Nadir position, sensors with oblique oligo-view optics allow earlier measurements. Thus, reliable assessment of biomass and nitrogen content can be obtained as early as EC 28 in wheat and at the growth stage EC 14 in maize (Liebler, 2003; Schächtl, 2004; Mistele et al., 2004; Mistele, 2006). Structure and composition (leaves, stems, flowers, ears, awns, etc.) of a canopy may further influence reflectance measurements.

Calculation of vegetation indices allows reducing effects of influencing factors. These parameters are mathematical combinations of various wavelengths in different regions of the reflectance spectra. The vegetation indices show strong correlations to different parameters of biomass growth and nutrient status. An example is the vegetation index REIP ('red edge inflection point') that indicates the shifting point at the red-infrared of reflectance spectra. An increasing nitrogen supply of the canopy leads to a shift in REIP values towards longer wavelengths (Liebler, 2003; Figure 14). Nevertheless, sensor measurements have to be corrected for cultivar effects, mainly due to different canopy architecture. For cultivars with a planophile growth habit (e.g. cv. Pegassos), REIP values reach a saturation level at a lower biomass yield than when measuring varieties with a more erectophile growth habit (e.g. cv. Xanthos).

The principles of passive and active reflectance measurements are already available in a commercially available product, the 'N-Sensor' of the company YARA (Figure 8). The device is mounted on the roof of a tractor.

Figure 8 N-Sensor on tractor roof (by courtesy of YARA GmbH & Co. KG, Dülmen, Germany). (For colour version of this figure, please see page 425 in colour plate section.)

Simultaneously, the canopy is scanned in four vertical directions at an oblique view from the top of the tractor and an additional sensor detects the incoming irradiance in vertical direction. The N-Sensor is applicable for on-the-go measurements of nutrient status within heterogeneous fields. The plant information can be used to guide amount and distribution of nitrogen fertiliser dressings in an online-mode (Reusch, 1997; Link et al., 2002).

Fluorescence measurements

Another technique to monitor the nutritional status of plants by means of non-destructive and remote measurements is based on the fluorescence of plant pigments such as chlorophyll. The use of chlorophyll fluorescence in plant physiology studies in not new, since this method has been used for many years as a tool for photosynthesis research and for stress detection in plants. Laser-induced chlorophyll fluorescence is the optical emission from chlorophyll molecules that have excited to a higher energy level by the absorption of electromagnetic radiation. Changes in the chlorophyll concentration can be detected on the basis of changes in the plant's fluorescence spectra. FAM tested newly developed sensors to describe the nitrogen content and biomass of crop stands under field conditions.

The fluorescence sensor MiniVeg N (Figure 9, hand-held device; Figure 10, tractor-mounted sensor) is an active optical sensor using the measuring principle of laser-induced chlorophyll fluorescence. Core of the device is an internal laser diode (red light laser), inducing the chlorophyll molecules in plant cells to emit fluorescence light. The intensity of fluorescence light is detected with highly sensitive optical components at

Figure 9 Hand-held fluorescence sensor (by courtesy of Fritzmeier Umwelttechnik GmbH & Co., Germany).

Figure 10 MiniVeg N (by courtesy of Fritzmeier Umwelttechnik GmbH & Co., Germany).

the wavelengths of 690 nm (red; F690) and 730 nm (near-infrared; F730) and the vegetation index ratio is calculated (F690/F730).

The emission of fluorescence light is due to the induction of chlorophyll molecules by laser light leading to an excessive level of energy. Nevertheless, other chlorophyll molecules may selectively re-absorb the emitted fluorescence light during the path through a leaf (Agati et al., 1993). Owing to the partial overlapping of absorption spectra and fluorescence spectra of chlorophyll at about 670 nm, the degree of re-absorption is maximised in the red region and of minor importance in near-infrared wavelengths. Thus, red fluorescence light is highly re-absorbed, whereas the radiation in the near-infrared region passes the leaves nearly uninfluenced. The degree of re-absorption depends on the chlorophyll concentration. At higher values of chlorophyll concentration, the re-absorption at 690 nm increases disproportionally. Thus, the intensity of detected chlorophyll fluorescence at 690 nm (F690) decreases with augmenting chlorophyll concentration, whereas in the range of 730 nm (F730) only small changes are detected. Therefore, the vegetation index ratio is negatively related to the chlorophyll concentration of plants (Sticksel et al., 2001; Figure 15). Owing to the strong correlation between chlorophyll concentration and N concentration (Ercoli et al., 1993), the ratio describes the N supply of plants.

Very recently, a fluorescence sensor is available as tractor-mounted device in the front of a tractor (Figure 10). The laser-induced chlorophyll fluorescence readings of canopies can be performed independent of external conditions, even during the night (Schmidhalter et al., 2004; Schächtl et al., 2005). As only chlorophyll molecules are induced to emit fluorescence light, no effects of soil reflection have to be considered. Thus, the application of the sensor is already practicable at the tillering of cereals, under conditions with low LAI ground cover of plants. Furthermore, a detection in row cultivars such as maize, potatoes, and sugar beet seems promising.

An alternative laser-induced chlorophyll fluorescence sensor is the tractor-based Planto N-Sensor (Figure 11). The principle is similar to the MiniVeg N-sensor. However, whereas the MiniVeg N-sensor measures in close contact with the plant canopy, the Planto N-Sensor is mounted on the roof of the tractor and measures at about 3 m distance from the plant canopy (Schmidhalter et al., 2004). The Planto N-Sensor has a scanning function that allows to scan the biomass independent of the detection of the chlorophyll content and augments the detected area. The unique system allows to determine independently biomass and chlorophyll density (Bredemeier and Schmidhalter, 2003, 2005). The scanned area, however, is considerably smaller for laser sensors as compared to that of reflectance-based sensors. Fluorescence measurements are influenced by temperature. An integrated temperature sensor, however, allows considering such an influence (Bredemeier and Schmidhalter, 2003; Schächtl et al., 2005). Although light intensity seems to have little influence on laser-induced chlorophyll measurements particularly under controlled conditions (Bredemeier and Schmidhalter, 2003), such an influence has to be considered, particularly at clear sunny days (Blesse and Schmidhalter, unpublished; Maidl and Limbrunner, unpublished).

Figure 11 Laser-induced chlorophyll fluorescence sensor Planto N-Sensor (Planto GmbH, Leipzig, Germany, by courtesy of Technical University of Munich). (For colour version of this figure, please see page 425 in colour plate section.)

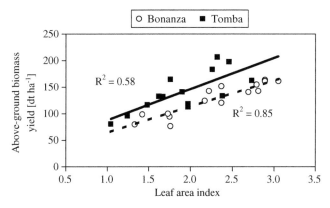

Figure 12 Correlation between leaf area index and above-ground dry matter yield for two potato cultivars (Schächtl and Maidl, 2002).

Yield measurement systems

To detect the yield of agricultural crops, measurement systems working directly on the harvesting equipment have been developed and evaluated. To realise local yield detection, continuously working measurement

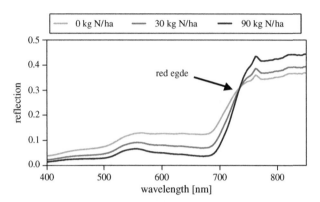

Figure 13 Reflectance spectra of wheat canopies (EC 32) with different N supply (Liebler et al., 2001).

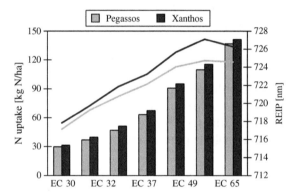

Figure 14 REIP (lines) und N uptake (bars) for two winter wheat cultivars during vegetation period (Liebler, 2003).

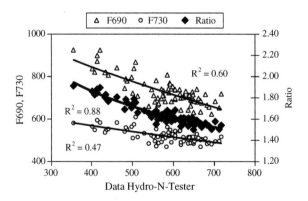

Figure 15 Correlation of F690 (triangles), F730 (circles), and ratio (squares) on chlorophyll concentration (Sticksel et al., 2001).

systems are necessary. Counter weighing of the total mass of crops harvested on a field can be used to calibrate and control the systems.

Components for most local yield measurement systems include (Figure 16):
- Mass flow sensor
- Sensor for working capacity (speed and working width)
- Position detection system (usually DGPS)
- Processing, monitoring, and data storing unit
- Data transfer to office computer

The mapping process of geo-referenced yield data is normally realised on the office computer with analysis and mapping software.

Yield detection for combinable crops
For yield measurement within the combine harvester, several meters, also referred to as yield sensors, have been developed and introduced into practical use. They work on the continuous flow principle and are installed in the upper part or on the head of the clean grain elevator. The available measuring systems are based upon two different measuring principles (Figure 17).

With the *volume measurement principle*, the corn flow is registered according to its volume via the specific weight (hl-weight or density) into mass flow. The volume is registered by determining the corn volumes on the elevator paddles.

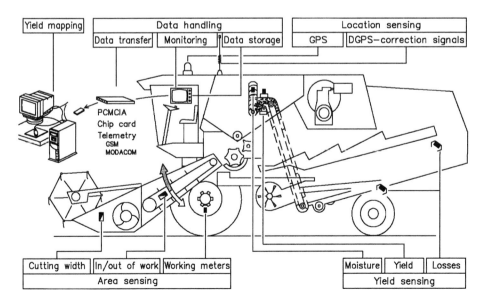

Figure 16 Components for local yield detection in a combine harvester.

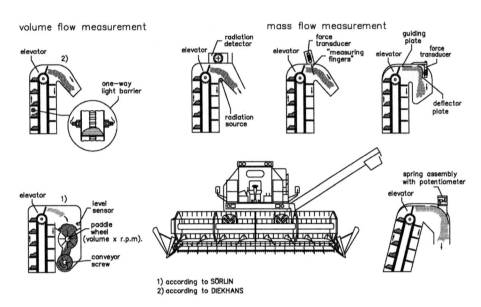

Figure 17 Yield measurement systems for combine harvesters.

Two systems on the market (Quantimeter of Claas and Ceres 2 of RDS) operate with a light barrier in the upper part of the feed-flow side of the grain elevator (left side of Figure 17). The corn conveyed by the elevator paddles interrupts the light beam. From the length of the dark phase and from calibration functions, the height and hence the volume of the corn charge on the paddles are calculated. The zero tare value is served by the darkening rate when the elevator is running empty. A tilt sensor is designed to compensate the influence of a non-uniform loading of the elevator paddles on a side slope. The hl-weight, which has to be determined with a beam balance, is used by the evaluation electronics to deduce the mass flow (t h^{-1}). As in all other measuring systems, this is converted into the aereal yield (t ha^{-1}) by being offset against harvested area produced from entered cutting width and measured threshing distance (wheel sensor). In addition, the harvested area is used to determine the area output (ha h^{-1}). Like with all other systems, the mass flow and the yield can be adjusted to standard moisture through the use of a continuously working moisture sensor.

In *determining the mass* of the corn flow, one relies on either the force/impetus measurement principle or the absorption of gamma rays by mass in a radiometric measuring system.

One measurement system on the market (Massey Ferguson Flowcontrol) operates according to the radiometric principle. The corn discharged from the elevator paddles passes through the region between a weakly radioactive source (Americium 241, activity 35 MBq) and a radiation sensor. As it does so, radiation is absorbed. The degree of absorption corresponds to the areal weight of the corn in the region of the measuring window. The material velocity, which is deduced from the elevator speed, is used to calculate mass flow. Today similar systems are also used in food processing.

A number of yield measurement systems developed in the USA use the force/impetus measurement and are likewise fitted on the elevator head in the discharge path of the corn. The sensor consists either of a baffle plate, which is fitted to a force-measuring cell (AgLeader, Case, Deutz-Fahr, picture top-right), or of a curved plate fitted to a spring element measuring the displacement way (John Deere), or of a curved plate mounted in patented geometry to a force measurement cell to compensate varying friction force (New Holland). Corn hitting the baffle plate or the curved plate causes a force effect to the bending bar, the spring element or the load cell, which is electrically sensed with strain gauges or the displacement sensor. Since this impetus is the product of mass and velocity, it is possible to calculate the mass flow. The material velocity is deduced, in turn, from the elevator velocity.

Besides the sensor element, all systems consist of processing, monitoring, and data storage units in the cab and have the possibility to integrate a moisture sensor. Most of the systems are factory-installed accessories; some products can be retrofitted to a selection of combine types (Demmel, 2001).

Yield detection for forage crops

Typical crop rotations in many regions of the world, especially in Western Europe, do not only consist of combinable crops. To get information on the variability of yields of different kinds of crops, yield measurement systems for other harvesting equipment than combine harvesters are necessary.

Since 1990, research on mass flow sensors for forage choppers have been reported. First results were published by Vansichen and De Baerdemaeker (1993). In 1995, Auernhammer et al. presented the results of a comparison of mass flow measurement systems based on the clearance between upper and lower feed rolls (volumetric), the power consumption of the cutter drum, the power consumption of the blower, and a radiometric measurement system (mass flow) in the spout (Figure 18).

As in yield measuring systems for combines, the mass flow is converted into the area yield (t ha^{-1}) by being offset against harvested area produced from entered cutting width and measured harvesting distance (wheel sensor). In addition, the harvested area is used to determine the area output (ha h^{-1}). Continuous yield sensor readings (often 1 Hz) are combined with information from the satellite positioning system for geo-referencing.

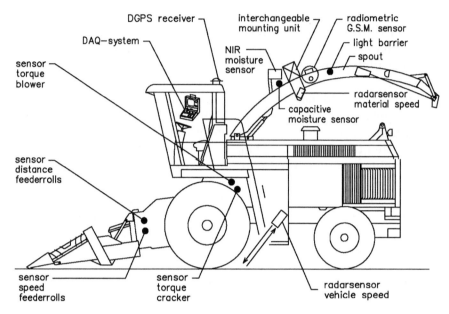

Figure 18 Possible sensor configuration for mass flow detection in a forage
          chopper.

As an alternative for mass flow detection in the forage chopper,
Missotten et al. (1997) developed and evaluated the friction-compensated
curved plate in the spout.

Since 2005, Claas and John Deere have been selling mass flow and
yield detection systems based on the measurement of the displacement of
the feeder rolls (volume flow system) as options for their self-propelled forage choppers.

Research and development on systems for local yield detection in
round balers, square balers, and self-loading trailers did not result in
products on the agricultural equipment market (Auernhammer and
Rottmeier, 1990; Behme et al., 1997).

Since 2000, three research groups have published their work on the
development of mass flow measurement technology for tractor-mounted
grass mowers. The systems were based on belt weighing technique, force,
and torque measurement. Until now, none of the developments is available
on the market (Kumhala et al., 2001; Demmel et al., 2002; Wild et al., 2004).

Yield detection for root crops

To determine the geo-referenced yield of root crops such as potatoes
and sugar beet, different measurement principles have been integrated
into harvesting equipment, tested, and evaluated (Figure 19).

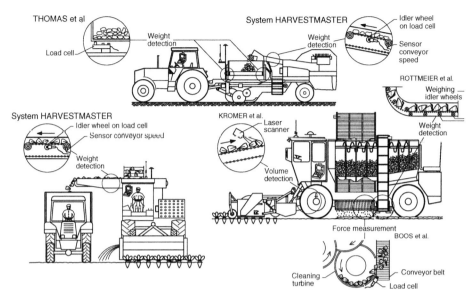

Figure 19 Sensor applications to determine mass flow and yield in root crop harvesting equipment.

A number of authors report that they have successfully developed and evaluated conveyor belt weighing systems combined with GPS positioning systems and data processing units (Campbell et al., 1994; Walter et al., 1996; Demmel and Auernhammer, 1999).

Godwin and Wheller (1997) used a trailer equipped with load cells to obtain yield data based on the mass accumulation rate.

Kromer and Degen (1998) tried to deduce sugar beet yield by estimating the volume of the beet using a laser scanner.

Hennes et al. (2002) adapted the friction-compensated curved plate principle to the conditions at a rotating cleaning turbine of a sugar beet harvester.

In the USA a number of potato custom harvesters are using the commercially available Harvestmaster HM-500 yield measurement system (conveyor weighing principle) to control the load of the transport trucks to avoid fines for overload.

Yield detection for other crops (not common to Western Europe)

Also for crops not common to Western Europe, continuously working mass flow and yield measurement systems able to deliver data for geo-referenced yield detection and yield mapping have been developed.

In the USA and Australia, yield measurement systems for self-propelled cotton pickers and cotton strippers have been intensively developed and evaluated. The cotton flow is determined with optical or microwave sensors and the sensor readings are converted into mass flow by calibration algorithms (comparable to optical mass flow measurement systems in combines; Durrance et al., 1998; Searcy et al., 1989). First systems are available commercially (John Deere, AgLeader, Microtrac, Farmscan, AGRIplan).

Further sensor research and developments tried to make systems available to continuously detect mass flow and yield of sugar cane (Cox et al., 1998; Benjamin et al., 2001), peanuts (Perry et al., 1998), grape (Tisseyre et al., 2001), pea (Glancey et al., 1997), and tomato (Pelletier and Upadhyaya, 1998).

Accuracy of yield detection in harvesters

Combinable crops: Extensive studies on the measuring accuracy of the individual measuring systems were carried out in the years 1991–1995 (Demmel, 2001). They were supplemented by joint test bench trials of all four systems in 2000 and 2001 (Kormann et al., 1998; Demmel, 2001).

The level of accuracy in practical use was determined by counter weighing the grain tank loads on calibrated platform scales. The measuring systems were examined, in part, on different combine harvester models with different grain types in lightly to medium cropped land (Table 5).

The mean relative error represents the measure of the calibration quality. It should ideally measure zero, or at least close to zero. This requirement was successfully achieved by all meters. The standard deviation (s) is the measure of the measuring accuracy. It indicates the range of error within around two thirds of all measurements. Despite the different measuring

Table 5 Errors of yield measuring systems for combine harvesters in practical use (Demmel, 2001).

Meter manufacturer	Period of study, total area, number of grain tank loads	Combine harvester models, grain types	Relative calibration errors (%)	Std. dev. of the relative error (%)
CERES 2 RDS	3 years, 140 ha, 179 tank loads	3 combine models, 4 grain types	−0.14	±3.43
FLOW CONTROL MASSEY FERGUSON	2 years, 140 ha, 132 tank loads	2 combine models, 2 grain types	−1.01	±4.07
YM 2000 AGLEADER LH565 LH AGRO	3 years, 130 ha, 182 tank loads	3 combine models, 4 grain types	−1.83	±4.06

Table 6 Errors of yield measuring systems for combine harvesters at different throughputs – 2000/2001 test bench studies, flat-standing position, 10, 15, 20, 25, and 30 t h^{-1} throughput, 5 repetitions/treatment, n = 25/m, reference mass/treatment 1 t, winter wheat (Demmel, 2001).

Meter manufacturer	Relative calibration error (%)	Std. dev. of the relative error (%)
CERES 2 RDS	−0.57	±5.50
FLOWCONTROL MASSEY FERGUSON	−1.64	±3.02
YM 2000 AGLEADER LH 565 LH AGRO	−1.71	±3.65
QUANTIMETER CLAAS	−2.71	±1.72
PRO SERIES 2000 RDS	−3.89	±5.54
GREENSTAR JOHN DEERE	−2.89	±2.81
FIELDSTAR (Force) DRONNINGBORG/AGCO	−0.22	±1.52

principles, all measuring systems are characterised by approximately equal ranges of error between ±3.5 and ±4%.

In the test bench studies, the accuracy of the measuring systems was intended to be determined under identical, clearly defined conditions. Particular considerations were given to the effect of different throughput levels and of transverse and longitudinal tilts (Table 6).

When the measuring accuracy of the various yield measuring systems is checked in the test bench under flat conditions at different throughputs, mean calibration errors <3% are obtained. Only at lower throughputs (10 t h^{-1}) do occur larger deviations (3–10%). This indicates that the calibration curves plotted in the instruments are not yet optimally matched to low throughputs.

The standard deviations vary at the individual throughput levels between 0.5 and 3%, across all throughputs they varied between 2 and 6%.

Lateral and longitudinal tilts of the combine harvesters at constant throughputs (20 t h^{-1}) exert a very much greater influence upon the accuracy of the meters (Table 7).

The least reaction to tilt influences is exhibited by the radiometric measuring system. The two volumetric measuring systems are equipped, for compensation of this influence, with one or two axle tilt sensors. Nevertheless, the errors caused by lateral and longitudinal tilts cannot successfully be compensated under all conditions. In this regard, the force measuring systems occupy a middle position between radiometric and volumetric meters.

Table 7 Errors of yield measuring systems for combine harvesters at different tilts – 2000/2001 test bench studies, 20 throughput, 5, 10, and 13° of lateral tilt to the left and to the right and longitudinal tilt forward and back, as well as combinations thereof, 5 repetitions/treatment, $n = 60/m$, reference mass/treatment 1 t, winter wheat (Demmel, 2001).

Meter manufacturer	Relative calibration error (%)	Std. dev. of the relative error (%)
CERES 2 RDS	−3.38	±8.07
FLOWCONTROL MASSEY FERGUSON	−1.11	±2.17
YM 2000 AGLEADER LH 565 LH AGRO	−0.24	±4.31
QUANTIMETER CLAAS	−0.91	±3.74
PRO SERIES 2000 RDS	−0.90	±11.73
GREENSTAR JOHN DEERE	−1.36	±3.37
FIELDSTAR (Force) DRONNINGBORG/AGCO	−0.02	±2.38

Table 8 Accuracy of mass flow and yield measurement systems for forage choppers.

Measurement principle	System placement	Author	Evaluation extent	Measured accuracy
Gamma ray absorption	Spout	Auernhammer et al. (1997)	24 field, 416 loads	Avg = −0.5%, SD = 3.3%
Friction compensated curved plate	Spout	Missotten et al. (1997)	1 field, 9 loads	Avg = +0.0%, SD = 2.7%
Force measurement	Blower wall	Barnett and Shinners (1998)	n.c.	Avg = n.c., SD = <12%
Ultrasonic absorption	Spout	Barnett and Shinners (1998)	n.c.	Avg = n.c., SD = n.c.
Feeder roll displacement	Feeding rolls	Ehlert (1999)	1 field, 41 loads	Avg = −1.0%, SD = 8.2%
Laser surface scanning	Spout	Schmittmann et al. (2000)	n.c.	n.c.

Forage

Based on a very high number of control weights of trailer loads during the evaluation of the gamma ray absorption-based measurement system for a self-propelled forage chopper, Auernhammer et al. (1997) reported on an accuracy (standard deviation of relative error) of $\pm 3.3\%$ after optimisation of the meter in 1993 and 1994. Table 8 contrasts the results of the evaluation of different measurement principles to detect mass flow and yield in forage choppers published by different authors.

Root crops

The compilation of publications about studies on the accuracy of measurement systems for root crops show that the majority of the authors have used the conveyor weighing system HM 500 from Harvestmaster (Table 9).

Although integrated in very different harvesting equipment, the error level seems to be similar. Demmel and Auernhammer (1999) tested this particular meter in a trailed one-row bunker hopper potato harvester and in a self-propelled six-row side loading sugar beet harvester. They reported on an accuracy (standard deviation of relative error) of 4.9 and 2.2%, respectively.

Research and development needs

Continuously working mass flow and yield measurement systems able to deliver data for yield mapping are available or in development for most harvesting technologies and machines. Nevertheless, this first generation of sensors and meters are in most cases retrofit solutions for existing machine systems. Therefore, a number of compromises have been made to integrate them. In many cases these compromises negatively influence the operability and in some cases they also reduce the measurement accuracy.

One aim for the future must be to optimise the application of mass flow and yield sensors by integrating them in new machine designs. Second, the accuracy has to be increased and stabilised, especially under worse and changing operating conditions. At least the operability must be facilitated, especially calibration effort and complexity.

Detection/determination of site-specific soil variability

Soil variability can be determined by soil sampling and analysis, which cannot be the chosen method for site-specific farming (see above). Here, non-destructive principles to sense soil properties are introduced. They can be separated into proximal and remote sensing, for both of them two methods are outlined.

Electromagnetic induction measurements to survey the spatial variability of soils

Measurements of the electromagnetic induction by EM38 were calibrated and validated on different levels, on the field level and on the farm level (Schmidhalter et al., 2001b; Schmidhalter, 2001; Heil and Schmidhalter, 2003; Sommer et al., 2007, this volume), and a survey was conducted within geographic regions of various origins.

Table 9 Accuracy of mass flow and yield measurement systems for potato and sugar beet harvesters.

Measurement principle	Harvester type, Crop	Author	Evaluation extent	Measured accuracy
Mass accumulation system 'Silsoe'	Trailer sugar beet, potatoes	Godwin and Wheeler (1997)	1 field, 15 loads	Avg = −1.1%, SD = 4.0%
Basket weighing system 'Tifton'	Trailed two-row basket combine, peanuts	Durrance et al. (1998)	2 fields, 40 loads	Avg = +0.2%, SD = 3.1%
Conveyor weighing 'Harvestmaster'	Trailed two-row side loading, potatoes	Rawlins et al. (1995)	1 field, 48 loads	Avg = n.c., SD = 4.9%
Conveyor weighing 'Harvestmaster'	Trailed six-row side loading, sugar beet	Hall et al. (1997)	1 field, 99 loads	Avg = −1.0%, SD = 2.2%
Conveyor weighing 'Harvestmaster'	Trailed one-row bunker hopper, potatoes	Demmel et al. (1998)	2 fields, 77 loads	Avg = −1.3%, SD = 4.1%
Conveyor weighing 'Harvestmaster'	Self-propelled six-row side loading, sugar beet	Demmel et al. (1998)	2 fields, 39 loads	Avg = +1.0%, SD = 3.7%
Conveyor weighing 'Rottmeier'	Self-propelled six-row tanker, sugar beet	Demmel et al. (1998)	5 fields, 23 loads	Avg = 2.1%, SD = 5.6%
Force curved plate system 'Leuven'	Self-propelled tanker loader, sugar beet	Broos et al. (1998)	1 field, 19 loads	Avg = 0.4%, SD = 1.6%
Laser optical volume system 'Bonn'	Self-propelled cleaner loader, sugar beet	Kromer and Degen (1998)	2 fields, 15 loads	Avg = n.c., SD = 4.0%

Calibration was performed for individual soil horizons with detailed investigations of soil texture, soil water content, and the electrical conductivity of the soil solution. Clay content and water content in 0–90 cm soil depth were the parameters most closely related to the apparent electrical conductivity with R^2 values between 0.31 and 0.67 for clay and 0.31 and 0.64 for water content. Other soil parameters such as silt and sand content or the electrical conductivity of the soil solution were in general not related to the apparent electrical conductivity. Values of the electrical conductivity in the horizontal and vertical modes correlated with each other ($R^2 = 0.93$). The results point out that relevant information for site-specific management can be obtained by this non-contacting method.

A further segmentation of the data in different soil groups improved the relationships significantly to R^2 values higher than 0.67 for clay, silt, and sand (Heil and Schmidhalter, 2003). By this way, soil water content at field capacity could be determined with adjusted R^2 values higher than 0.89.

Principles developed within the FAM project were further introduced and adopted to the German-wide precision farming project Preagro where 2800 ha of arable land in largely different geographic and climatic zones were mapped by electromagnetic induction. Data of the apparent electrical conductivity (ECa) were compared to various other information sources (national soil inventory, yield maps, and spectral information from airborne remote sensing) (Neudecker et al., 2001). Multi-temporal measurements showed comparable patterns in ECa over time. Zones of different soil substrates could be better delineated by electromagnetic induction than by the previously existing information from the national soil inventory. The latter information was related to ECa with R^2 0.01–0.71. The closest relationship was found at the more heterogeneous sites. On heterogeneous field sites, good correlation to yield could be found with R^2 up to 0.71. ECa measurements represent a fast technique to map soil heterogeneity and are useful to delineate different management zones.

Determination of soil texture and organic matter content by near-infrared spectroscopy

Spectra (1000–2840 nm) of dried and sieved (2 mm) soil samples were obtained by using a FT-NIR equipped with a PbS detector (Vector 22N, Bruker, Ettlingen, Germany). Multivariate calibrations were developed with PLSR and cross-validated using OPUS 4.0 (Bruker, Ettlingen, Germany).

Results of the cross-validation confirm the potential of NIRS models to accurately predict clay, silt, and organic matter content in soil. The corresponding regression coefficients were 0.91, 0.91, and 0.9 with prediction errors (RMSECV) of 11, 15, and 12% (Wagner et al., 2001). These results are in line with a more recent investigation (Sorensen and Dalsgaard, 2005) and indicate the feasibility of near-infrared spectroscopy for rapid non-destructive prediction of soil properties.

Remote sensing – Spatial detection of topsoil properties using hyper spectral sensing

The spatial variability of topsoil texture and organic matter across fields was studied using field spectroscopy and airborne hyper spectral imagery with the aim of improving soil-mapping procedures (Selige et al., 2006). Organic matter and clay content were correlated with spectral properties. Topsoil reflectance (330–2500 nm) was measured in the field using a GER 3700 field spectrometer and a Lambertian Spectralon reference panel of known reflectivity. The airborne HyMap sensor was used at an early flight campaign in May for recording hyper spectral images (420–2480 nm, 128 channels) of bare soil fields. PLSR was applied to develop and calibrate a model that establishes a quantitative relationship between the spectra and soil parameters.

Organic matter and clay content could be determined simultaneously from a single spectral signature since organic carbon largely responds to wavebands in the visible range and clay responds to wavebands in the near-infrared region (Selige and Schmidhalter, 2001). Complexity and auto-correlation between the soil parameters led to the use of multivariate calibration techniques, particularly PLSR. PLSR estimates of the organic matter content and the clay content of topsoils indicated R^2 values of 0.82–0.92 with prediction error values (RMSECV) of 0.4% for organic matter and 4–6% for clay content. It is shown that the clay and organic matter content can be predicted quantitatively using hyper spectral sensing (Selige et al., 2006).

Characterising soils for plant-available water capacity and yield potential using airborne remote sensing

Multi-spectral airborne remote sensing was used to improve the inventory of soil heterogeneity at the field level. Ground measurements of crop parameters were collected from representative soil sites. Spectral information at visible, infrared, and thermal wavebands was recorded from the airborne scanner Daedalus AADS 1268 at 11 spectral channels (Selige and Schmidhalter, 2001). The spectral information was transformed into soil information using bio-indicative transfer functions, based on cause and effect relationships of the soil–plant system. Soil properties, plant development, and crop stand conditions were measured on the ground at representative soil sites. The available water storage capacity and the rootability were derived from soil texture and texture changes within the soil profile. Grain yield and biomass of each soil site were determined. Relationships between the investigated parameters were established.

The variability of the plant-available water storage capacity of the rooting zone accounted for 93% of the variability of winter wheat biomass at the development stage BBCH 77 (milk ripeness) when the leaves started to become yellow (Selige and Schmidhalter, 2001). The biomass at this development stage also indicated the pattern of the later harvested grain yield. The crop stand condition at this development stage accounted for 96% of the grain yield variability of winter wheat. This result also suggests that the crop stand condition can be used to forecast yield and its pattern across fields. The correlation between plant-available water capacity and grain yield underlines the importance of soil water availability. The thermal emission and its relationship to the transpiration of crops were recognised as most suitable to detect quantitatively soil properties via crop stand conditions of winter wheat.

Detection of spatial crop heterogeneity – Yield data processing and yield mapping

During the harvesting process, readings from different sensors are permanently stored on a yield monitor. Apart from yield sensor and position information provided by GPS receivers, yield monitors also obtain data from moisture and speed sensors. Some yield monitoring systems use tilt sensor information in order to correct for tilt-induced errors in yield sensor readings.

The information contained in the yield data files varies depending on the yield monitoring system. Almost all sensors provide information on position, time and yield. Depending on the yield monitoring system, additional information on the quality of GPS, grain moisture, ground speed, tilt, cutting width, header status, and swath number is stored in the yield data files.

Data stored in yield monitors has been pre-processed and filtered to a different extent depending on the yield monitoring system. The calculation of yield at current position is a function of speed, cutting width and grain flow and as such a form of pre-processing. Results from investigations made with different yield monitoring systems installed on one combine harvester measuring the same grain flow indicate that the yield measurements from some systems have been filtered or averaged before storing (Steinmayr, 2002; Noack et al., 2003). The results suggest that rapid changes in yield grain flow cause different responses in the yield data recorded on the yield monitor.

However, on a load basis the absolute amounts of grain yield match very well when compared to the results obtained with scale weights (Auernhammer et al., 1993).

Yield data filtering

Yield data files logged with yield monitors contain erroneous measurements as the sensors are operating in harsh environments. Also, different factors such as unknown crop width entering the combine, the time the grain travels from the header to the sensor, and tilt of the combine affect the reliability of yield measurements. Errors occurring during the process of yield data collection have been very well described and classified by Blackmore and Marshall (1996).

The removal of potentially erroneous measurements from yield data files is a prerequisite for the creation of meaningful yield maps.

Simplistic approaches use upper and lower threshold values to filter yield datasets. The threshold values are either fixed or based on the standard deviation of the dataset. Global threshold filtering does not account for the local variance of yield and its spatial distribution and may therefore fail to remove erroneous measurements or may even remove reliable data.

Different authors have presented expert filters for filtering yield data files. Some of these filters rely on information that will not be available in all data file formats (Rands, 1995; Blackmore and Marshall, 1996; Beck et al., 2001; Taylor et al., 2000; Thylen et al., 2000; Steinmayr, 2002).

Noack et al. (2001) have developed a method that compares yield measurements in tracks with those in neighbouring tracks taking into account the standard deviation of yield within the tracks. It tries to use both the temporal and the spatial relations between yield measurements. This method has been tested on datasets collected with three yield-monitoring systems installed on one combine. By filtering with the H method (Noack et al., 2001) the comparability of the resulting yield maps from the different yield monitoring systems was notably increased.

Yield map creation

A yield map is the visual representation of yield variation within a field. One form of yield maps shows the GPS positions where the yield data was recorded (Figure 20, *right*). *Post maps* are simple to create, but they have several disadvantages when they are used as input for spatial analysis: Yield information is only available for discrete positions and is not related to neighbouring yield values. Post maps do not allow to distinguish areas with similar yield and does not allow to classify a field into higher and lower yielding zones.

Grid yield maps (Figure 20, *left*) are composed of tiled rectangles. A yield value is assigned to each rectangle (grid cell) so that yield information is available for any position within the field boundaries.

Grid yield maps can be created with different methods, inverse distance interpolation and Kriging interpolation being commonly used for yield mapping.

For the estimation of grid cell values, several other parameters apart from the grid cell size ('C' in Figure 21, *right*) have to be specified. The search radius determines the maximum distance that a data point may have from the centre of the grid cell in order to be included in the estimation of the grid cell value ('S' in Figure 21, *right*). The weight determines the weighting of each single data point in the estimation of the grid cell value. Generally, the weight is decreasing with increasing distance from the centre of the grid cell ('D' in Figure 21, *right*). The estimation of a grid cell value appreciates all data points within the search radius according to their weight.

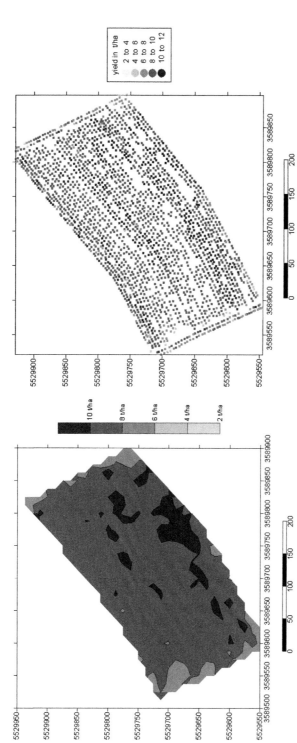

Figure 20 Different kinds of yield maps (*left*: grid yield map, *right*: post yield map).

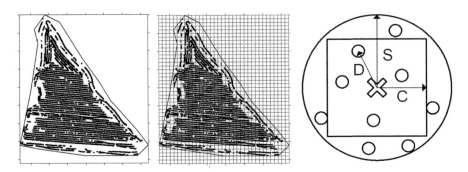

Figure 21 Steps during grid yield map generation (*left:* clipping, *middle:* grid creation, *right:* grid value calculation).

The inverse distance interpolation uses fixed values for the search radius and weight. The weight is expressed as the power of the inverse distance (e.g. weight 2 assigns the weight $1/D^2$ to each value).

The Kriging interpolation requires the calculation of a semivariogram. Semivariograms show how the variance of spatial data is related to the distance of data points. Semivariograms may be calculated either for the whole dataset (global semivariogram) or for a subset of data in the neighbourhood of the grid cell centre (local semivariogram). Fitting an appropriate model to the empirical curve is the most delicate step in Kriging interpolation: The model determines the search radius and the distance weighing for the grid cell value estimation.

The results of Kriging interpolation, especially Block Kriging (local semivariograms), are generally considered to be superior to the results from inverse distance interpolation. However, Kriging is very delicate due to its dependency on the choice of the correct model and can hardly be automated effectively. Minasny et al. (1999) have developed a very useful program for Kriging Interpolation of yield data.

Tractor-based non-destructive sensing of biomass and nitrogen status – Reflectance measurements

A tractor-based spectrometer was used to measure the reflectance in field-grown wheat and maize plants. The experimental design included different nitrogen applications (0, 90, 130, 170, and $210\,kg\,N\,ha^{-1}$). Individual plots were 15 m wide and 50–60 m long (Mistele, 2006). The spectrometer contained two units of Zeiss MMS1 silicon diode array with a spectral detection range from 400 to 1000 nm and a spectral resolution of 3.3 nm. One unit was linked with a diffuser and measured the sun radiation as a reference signal. Simultaneously the other unit measured the canopy reflectance with an oligo-view optic (Lammel et al., 2001). The spectrometer was connected with a four-in-one light fibre and the signal was optically averaged. The optical inputs were positioned with an azimuth angle of 80° between the front and rear side and 100° between the right and left side of the tractor. The zenith angle was set at 58 ± 6° to minimise the influence of the tractor (Reusch, 2003).

In front of the tractor the sensor system was mounted 1.90 m above the canopy. The field of view consisted of four ellipsoids with 1.23 m in length, together around 4.5 m^2. The reflectance was measured at five wavelengths, which were at 550, 670, 700, 740, and 780 nm. Various reflectance indices were calculated including the REIP, NDVI, IR/R, IR/G, and G/R.

Spectral indices were related to measurements of biomass and nitrogen content plants that were cut shortly after the spectral measurements. Small plots (plants) on both sides of the tractor were harvested, 1.5 m in width and around 8 m in length, matching exactly the measured area. A plot chopper equipped with a weighing unit was used for this purpose. A separate sub-sample was removed and dried after weighing to estimate the total dry matter. The dried samples were milled and analysed for total N content with an elemental analyser (Macro N, Varian).

The sensor was used to test the reliability of the 'YARA N-Sensor' to detect spatial differences in N status and biomass of crops in field experiments of 3-year duration. As judged against validations performed on harvested areas of 25 m^2, which varied in nitrogen supply, the results showed that strong correlations exist between reflectance indices and N uptake from the end of tillering to flowering ($R^2 = 0.90$) (Mistele et al., 2004; Mistele, 2006). Close relationships between spectral indices and biomass, nitrogen content, and particularly nitrogen uptake ($R^2 > 0.85$) were determined in four seasons from 2001 to 2004 for wheat (Schmidhalter et al., 2001a, 2003; Mistele et al., 2004). A good correlation between spectral indices and the final yield was observed as well. Reliable estimates could already be obtained at the 4-leaf stage of maize plants. Consistency in data normally requires that reflectance be measured only when the solar zenith angle provides sufficient irradiance, when sky conditions are uniform and bright, and when the sensor view angle is close to nadir (Major et al., 2003). The oligo-view optic tested outperformed existing techniques by enabling non-nadir measurements at solar zenith angles with reduced irradiance and non-uniform sky conditions. As such, the tractor-based passive sensor represents a fast and highly suitable means to measure the nitrogen status and biomass of wheat crops.

Laser-induced chlorophyll fluorescence measurements

The reliability of proximal remote-sensing measurements of the laser-induced chlorophyll fluorescence to determine chlorophyll and nitrogen content as well as biomass production in field-grown maize and wheat plants was determined with a newly developed sensor. A tractor-mounted fluorescence sensor developed by Planto GmbH company (Leipzig, Germany) was used that detects the fluorescence emitted at 690 and 730 nm. The sensor was mounted at the rear of the tractor at a height of around 3 m above the plant canopy. A laser beam stimulates the emission of fluorescence, which is detected at a distance of approximately 3.3 m between the canopy and the sensor. The canopy is scanned in a 0.5 m wide strip. Strips of approximately 15 m in length were measured and the total area sensed was around 6–7 m^2. The relationship between the ratio of laser-induced chlorophyll fluorescence intensities at 690 and 730 nm (F690/F730) and nitrogen supply in winter wheat was characterised (Bredemeier and Schmidhalter, 2005). The chlorophyll fluorescence ratio F690/F730 was then calculated. Destructive harvests for biomass and nitrogen content were done as described for the reflectance measurements by spatially matching sensor measurements and harvested area.

The fluorescence ratio F690/F730 and the biomass index were well correlated with shoot biomass and nitrogen uptake across different developmental stages. Similar relationships were found in wheat and maize. The fluorescence intensity at 690 and 730 nm increased as shoot biomass and SPAD values increased, while the ratio F690/F730 was inversely correlated with N uptake, shoot biomass, and SPAD values. The goodness of linear fits between nitrogen content, biomass, nitrogen uptake, and SPAD values to fluorescence ratio mean was as follows: 0.78, 0.87, 0.87, and 0.88 (Bredemeier and Schmidhalter, 2003). N fertilisation levels in the field could be differentiated by means of fluorescence ratio measurements. Shoot dry biomass could be determined by means of biomass index measurements independent of leaf chlorophyll content.

These results indicate that nitrogen uptake and biomass can be detected reliably through chlorophyll fluorescence measurements under field conditions (Bredemeier and Schmidhalter, 2005). In contrast to point data measurements, the establishment of scanning field fluorescence sensors opens new possibilities for N status and biomass measurements. Moreover, because the signal comes from green plant parts only, it has a very low background and is little affected by soil reflectance. Furthermore, the system allows the independent determination of biomass and chlorophyll density already at the seedling stage (Blesse and Schmidhalter, unpublished).

Nitrogen fixation of legumes and legume content of clover grass

Because N supply in organic farming mainly relies on symbiotic N_2 fixation, growth of the legumes is the Achilles' heel of this land use system. Therefore, a sufficient description of the reasons for variation in N_2 fixation does arouse much more interest than for the variation of crop yield in general. In organic farming, the main gateway for N to enter the crop rotation are legume–grass mixtures, because the N fixed by pulses is typically sold. Our research focused (i) on the development of a method to describe variation of N_2 fixation of legume–grass mixtures and (ii) to derive parameters that determine the observed variation in the field. By combining data about the variation of N_2 fixation and the variation of the yield of non-legumes, options for site-specific farming in organic farming can be lined out.

Nitrogen fixation and therefore its variation are basically determined by two parameters: the N yield of the legumes and the proportion of N derived from atmosphere (N_{dfA}).

In general, N_2 fixation can be calculated as follows (Eq. (1)):

$$N_2 \text{ fixed } [g/m^2] = \text{total yield } [g/m^2] \times \text{leg } [\%] \times N_{leg} [\%] \times N_{dfA} [\%] \qquad (1)$$

where total yield is the dry matter produced by the whole stand during the investigated time frame (i.e. the sum of shoot, roots, nodules, litterfall, and rhizodeposition); leg, proportion of the legumes within the dry matter production of the whole stand; N_{leg}, concentration of N in the total dry matter of the legume; and N_{dfA}, proportion of N derived from atmosphere within the total amount of N taken up by the legume during the investigated time frame.

Although the equation looks very simple, lots of pitfalls appear if N_2 fixation has to be calculated. In general, data of the actual shoot yield can be determined. The determination of N in litterfall, stubble and root residues, and rhizodeposition is difficult because these data typically remain in secrecy. On the basis of published data, Høgh-Jensen et al. (2004) derived an empirical model to calculate total N_2 fixation only based on shoot yield data of grass–clover mixtures. All the other data necessary to calculate total N_2 fixation are estimated via ratios to the shoot data. The model requires only data about shoot dry matter yield of the mixture, the proportion of legumes in the mixture, the concentration of N in the shoot dry matter of the legume, and N_{dfA} in the shoot-N. But, even then the determination of N_2 fixation is still challenging.

Total yield of legume–grass mixtures is rarely measured and on-the-go measurement systems are not yet available on the market (see above). Further on, the determination of the proportion of legumes within mixtures with non-legumes is simply laborious, because reliable data are only gained by hand sorting. For pulses, which are regularly grown in pure stands, the seed yield is usually measured. But this is not a sufficient indicator for N_2 fixation, because the ratio between seed and straw yield strongly varies (Beck et al., 1991), and in pure stands of legumes N_{dfA} may vary considerably at short distances (Mahler et al., 1979; Stevenson et al., 1995; Walley et al., 2001). So far there is no simple method available to determine N_{dfA} in field. However, at generally N-limited conditions and if legumes are grown in mixtures with non-legumes they will regularly reach high N_{dfA} values (>90%), because the non-legumes take up almost all plant-available soil N. Therefore, under such conditions the N yield of the legume will be closely correlated to their N_2 fixation (Boller, 1988;

Peoples et al., 1995). Such conditions were expected at the FAM research farm. Therefore, first of all, a method was needed to determine easily the proportion of legumes within a mixture.

During the past 20 years several papers have shown the potential of near-infrared reflectance spectroscopy (NIRS) to determine legume content in legume–grass mixtures. It is an easy-to-use technique, which can even be mounted onto harvesting machines (Dardenne and Féménias, 1999). However due to the dominant absorption of water within the spectral range of 0.9–2.5 µm, the determination of constituents other than water in fresh green plant samples is difficult. That is why determination of legume content in ground samples was so far tested with dried forage mixtures, binary (Petersen et al., 1987; Pitman et al., 1991; Shaffer et al., 1990; Wachendorf et al., 1999), and more complex mixtures of several legumes and grasses (Coleman et al., 1990; Pitman et al., 1991). However, there was still no study that showed the capability of NIRS in predicting legume content of multi-species legume–grass mixtures in widespread use in Western Europe, i.e. with white clover (*Trifolium repens* L.), red clover (*T. pratense* L.), and lucerne (*Medicago sativa* L.) as dominating legumes in varying proportions. Additionally, method development as published so far was mostly based on samples from well-defined plot experiments and the methods were not tested for their performance in real stands. But, an adequate validation is the most crucial aspect during the development of a NIRS method.

Therefore, three NIRS methods were developed based on samples taken from the fields of the FAM research farm (Locher et al., 2005a). Crucial aspects of calibrating a NIRS method (composition of the standards, measurement conditions, influence of the homogeneity in particle size within each sample, and the spectral range used for the estimation) were tested for their effect on precision and accuracy of the prediction of legume concentration in dried and ground mixtures. All of it proved to be of minor relevance presumably because of the overall diversity already inherent in the samples used for calibration. In contrast to the vegetation indices as introduced above, NIRS models that determine legume concentration require a broad range of the measured spectral information (in our case 75–90% of the recorded 1.0–2.86 µm range). However, comparing all six models only 70% of the selected spectral ranges were in common. Most likely this part of the spectral information describes the difference between the grasses and legumes in the samples (Figure 22). Reducing the models to this range slightly increased the standard error of prediction from less than 4% to less than 4.5% legume content (Locher et al., 2005a). Attempts to further restrict the spectral information failed, which is not surprising because the determination of complex parameters such as legume concentration will combine information about several constituents, which in principle are found in both grasses and legumes.

Finally, the applicability of the models to predict independent samples of legume–grass mixtures were broadly tested with samples from other

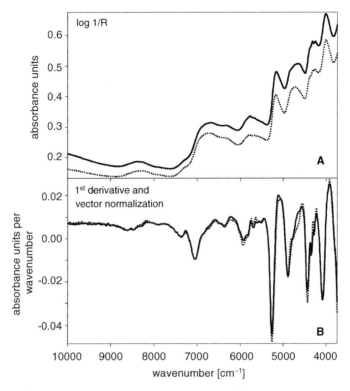

Figure 22 NIR spectra of a legume sample (solid line) and a grass sample (dotted line): original absorbance spectra (A), vector normalised first derivative spectra (B) (Locher et al., 2005a).

regions in Germany (Locher et al., 2005b) and at a small set of samples from Finland (Nykänen et al., 2005). With any of the test sets the calibrations proved to predict well legume content in dried ground mixtures. Instead of a strong linearity of the predicted to the real values (slope 0.93–1.09; $R^2 > 0.92$) standard error of prediction ranged between 3.3 and 12.5% legume concentration. Most of the errors were caused by a systematic bias of up to 10%, which was frequently observed but did not affect linearity. If estimates were bias corrected, the prediction error fell below 5% legume concentration, which is presumably well below any sampling error. It was concluded that without bias correction the NIRS values are already precise and can be compared within one field. For accurate values to compare fields or to calculate N_2 fixation, a bias correction is advised (Locher et al., 2005b).

The extensive validation done proved the capability of NIRS to precisely determine the proportion of legumes in dried legume–grass mixtures. However, the determination of legume content in fresh green plant samples remains an open task. A first test under lab conditions was promising, even by using the methods for dried samples a determination seemed to be possible (Locher, personal communication). However, for the purposes of precision farming a method is needed that already determines the proportion of legumes during harvesting. Nowadays several plot choppers use the NIRS technique (Dardenne and Féménias, 1999; Welle et al., 2003), but still analysis of data offers lots of pitfalls (Paul et al., 2002). The key problem is that the water status of the measured material strongly varies and therefore the spectral information representing water. If this problem can be solved it will be only a technical detail to install a NIRS system on a mower as used by Demmel et al. (2002) and Wild et al. (2004) (see above). Then on-the-go measurements of total yield of forages and several constituents will be possible.

At the FAM research station, the NIRS method was used to determine the in-field variability of legume proportion in multi-species legume grass, finally to determine the variation of N_2 fixation. As stated above, N_2 fixation of a legume–grass mixture is defined by four factors: dry matter yield of the mixture, the proportion of legumes therein, the N concentration of the legumes, and their fixing activity (N_{dfA}). Detailed investigations were carried out at the FAM research farm on different fields at different harvests and years (Locher, 2003). Results supported the expectation that only dry matter yield and the proportion of legume determined the variation of N_2 fixation of the investigated multi-species legume–grass. Measurements done in the fields confirmed earlier reports in the literature (Peoples et al., 1995; Weißbach, 1995; Boller, 1988; Lopotz, 1996) that N_{dfA} did not considerably vary under the conditions at the FAM research farm (Heuwinkel et al., unpublished). However, one should be aware that small increases in N availability may reduce N_{dfA} (Mallarino et al., 1990; Heuwinkel et al., 2005a). In field, variation of the N concentration of the legume dry matter was negligible, but not the variations between harvesting dates (Locher, 2003).

Strong in-field variation was observed for legume dry matter yield, which is the product of the total dry matter yield and legume concentration of the legume–grass mixture (Heuwinkel et al., 2005b). Additionally both parameters showed a strong seasonality, with the highest total dry matter yield at the first or second harvest. The proportion of legumes in the shoot dry matter yield steadily increased from moderate values in spring (30–70%) to high values in the following harvests (>70%) as shown for field A12 of the FAM research farm (Figure 23) (Locher, 2003; Heuwinkel et al., 2005b). Correlation analysis between total dry matter yield and proportion of legume done with the single data of all the 19 harvests revealed that both parameters were independent from each other in 15 of the 19 studied cases (Locher, 2003). It was concluded that in-field variation of N_2 fixation of legume–grass mixtures was determined by both the dry matter yield of the mixture and the proportion legumes therein (Heuwinkel et al., 2005b). This is visualised by one dataset of one harvest on field A09 of the FAM research station (Figure 24). Another striking

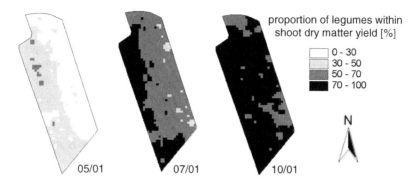

Figure 23 Variation of the proportion of legumes within the dry matter yield of a multi-species legume grass mixture grown at field A12 of the FAM research station. Data of all three harvests in May, July, and October 2001 are shown.

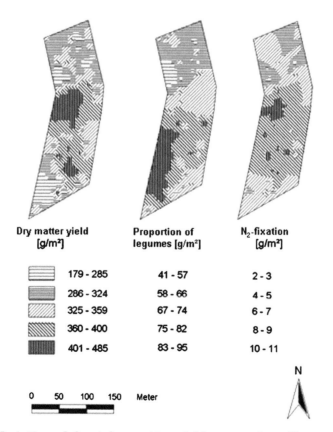

Figure 24 Variation of shoot dry matter yield, proportion of legumes, and N_2 fixation of the multi-species legume grass grown in field A09 of the FAM research station. Data are from the third harvest in 1999.

result is demonstrated by this figure: the strong variation of N_2 fixation within one field. Within all the 19 studied cases the coefficient of variation of shoot dry matter yield of the mixtures ranged from 14 to 55%, while this figure of the proportion of legumes ranges from 10 to 39%. The additive effect of both parameters on N_2 fixation resulted for 17 out of 19 datasets in a clearly higher coefficient of variation of this parameter (Locher, 2003).

2.3.3 Development of site-specific crop production strategies

As a consequence of the knowledge of the spatial variability of soils, crops and yield site-specific production strategies have to be developed and evaluated.

Principles of site-specific farming
Three major concepts based on mapping systems, on real-time sensor-actuator systems, and on a combination of both are known (Auernhammer and Schueller, 1999).

Map-based systems
Map-based systems are using satellite navigation systems such as GPS NAVSTAR (or other positioning systems) to establish a geographic basis for site-specific crop production (Figure 25).
Components of map-based systems include:
– Positioning systems to establish equipment location
– Sensors for yield detection and soil measurements
– Mapping software
– Controllers for map-based applications
– Actuators to perform the control

Map-based systems try to integrate information from different sources. Control maps have to be generated to execute field operations such as variable seeding, fertilisation, or pesticide application.

Real-time systems
Real-time systems use the actual information on a soil or plant parameter detected by a sensor system to deduce and take an appropriate action. They do not require positioning or mapping systems except for record of the action ('as applied map'). An example is the application of nitrogen fertiliser based upon sensing chlorophyll intensity and biomass quantity of the crop (Figure 26).

Real-time systems with maps
These systems combine the capabilities of map-based and real-time systems. Maps of yields, soil types, and nutrients can be used with real-time

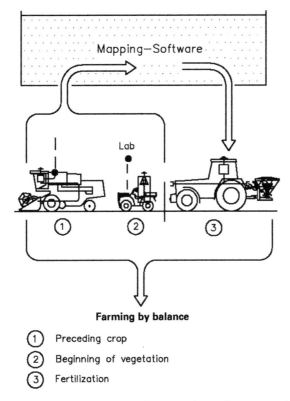

Farming by balance

① Preceding crop
② Beginning of vegetation
③ Fertilization

Figure 25 Components and structure for map-based systems (Auernhammer et al., 1999).

sensors of plant growth, soil moisture, and weed infestation to control field operations. Because such systems require all components listed previously, they are very complex. But they allow the optimisation of field operations, especially of fertiliser and pesticide application (Figure 27).

Problems coming up with real-time systems with maps come from the fact that more than one information source lead to a decision. Therefore, an information command structure or a sensor fusion model has to be applied to the system (Ostermeier et al., 2003).

Detection/determination of management zones – Causes of yield variability

Spatial variability of soil properties as a result of abiotic and biotic factors, as well as of considerable small-scale variations in topography and climate, has long been recognised. With the introduction of statistical analyses and global positioning systems, systematic recording and analysis of soil properties has become possible. The inherent high variability of

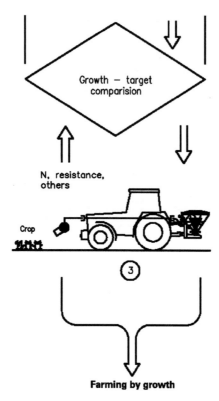

Figure 26 Components and structure for real-time systems (Auernhammer et al., 1999).

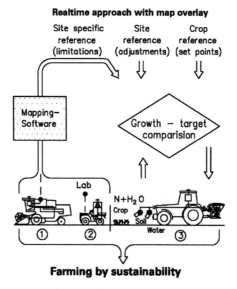

Figure 27 Components and structure for real-time systems with maps (Auernhammer et al., 1999).

inorganic and organic nitrogen in soils has been illustrated repeatedly (e.g. Van Meirvenne et al., 1990). As is the case for nitrogen, plant-available soil water varies frequently as a consequence of the variability of soil texture and is the norm rather than the exception in most fields. Precipitation variability is often no less important than that in soil, and its effect on the crop yield can often be even more considerable than that of spatial variability. In many areas of the world, available soil water and precipitation or irrigation amount and distribution are among the primary factors determining yields of crops.

To evaluate the interaction between plant-available soil water, precipitation (or irrigation), and nitrogen, fertilisation experiments were conducted over two consecutive years with winter wheat with two different N fertiliser treatments (180 and 120 N kg ha^{-1}) and three different water supply treatments (stress by rain sheltering, irrigation, and rain-fed only) on sites of different plant-available soil water.

Among the three factors (i.e. site, precipitation, and N fertilisation), site was the most important one influencing variability in grain yield, whereas precipitation, and in particular its distribution during the growing season, influenced the overall yield level in a given year. Increased N fertilisation generally increased yield, but its efficiency was low on lower yielding sites such as sandy soils if climatic conditions were unfavourable, thereby advocating reduced N fertilisation application on sites of lower plant-available soil water. This result also underscores the importance on such sites of sufficient water supply for efficient N use as already reported by others (Eck, 1988). For the maximum benefit, crop management should consider annual variability in yield in addition to soil conditions, and site-specific N fertilisation should be adapted to the actual plant growth.

At the FAM research farm the major sources of infield variability were soil depth and clay content, soil texture and shallowness, and soil texture and topography. Soil fertility levels except for nitrogen were adequate such that they were not yield limiting. Potassium and phosphorus removal by crops was regularly replaced. Soil physical and chemical properties of all sites were previously intensively characterised. This together with concomitant measurements of soil matrix potentials and previous more mechanistic investigations of the relationship between yield and varied water supply allow identifying water and nitrogen supplies as yield-limiting factors (Geesing et al., 2001).

In the organic farming the varied effect topography and N-mineralisation potential of the soil may have on different crops was investigated. Usually in hilly landscapes foothill positions are expected to be more fertile, i.e. their soil-N release will be higher. If legume–grass mixtures are grown in fields with a marked topography, one may expect changes in N$_2$ fixation as already shown for pulses (e.g. Stevenson et al., 1995). If legume–grass is frequently cut and removed, the amount of plant-available soil-N should decrease from harvest to harvest. In two of the four intensively investigated fields, the legume–grass strongly reflected this effect of N availability on species composition. At field A12 (Figure 23) foothill positions are along the east side of the field while the west represents hilltop positions. At each harvest the highest proportion of legumes was found in the west of the field and the lowest along its east border. Additionally, a strong increase over time was observed at any place in the field.

Field A09 (Figure 24) was taken for more detailed investigations. Because of a grassland history prior to the FAM study, its northern third had increased levels of organic matter (1.8% C as compared to 1.4% in the rest of the field), which can effect availability of soil-N. In this part of the field a much lower proportion of legumes was observed as compared to the remaining field at any sampling date (q.v. Figure 24). In this field, correlation analysis revealed a significant linear relationship of the proportion of legumes on the uptake of soil-N by the mixture at all seven harvesting dates (R^2 = 0.19–0.75; Heuwinkel et al., 2005b). This was found for the other fields as well, but the coefficient of determination was mostly low (0.3; Locher, 2003). In all datasets the regression had a negative slope and the intercept on the Y-axis was mostly close to the theoretically expected value of 100, i.e. at no N uptake from soil-N only legumes can grow.

In the year 2000, undisturbed soil samples were taken at six selected sites of field A09 and incubated in lab to determine the specific potential for N release. These data were correlated to field data about the uptake of N from soil by the legume–grass mixture and to the grain yield of the two following grains, wheat and rye (Figure 28). Surprisingly, legume–grass reflected much better the potential differences in N release from soil-N than the grains although their growth was clearly N limited. The most likely explanation for this are lateral fluxes of mineral N in field. With legume–grass these can be expected to be of minor importance because relevant amounts of mineral-N are seldom found with legume–grass. For the grains, higher concentrations of mineral-N will accumulate during winter, especially for winter wheat following the legume–grass. This N may move downward or lateral with the water. At the field A09 relatively high grain yields were found at two sites with a medium N release, but which are located right below a site with a relatively high N release in lab that was not reflected by grain yield in field. Other effects (i.e. water supply) are less likely, because the N uptake by legume–grass was high in both years. However, further research is needed to verify this explanation.

Strategies for precision farming in organic farming should be aware of any kind of varied interactions between crops and topography because they have to address much more effects on the crop rotations than for integrated farming.

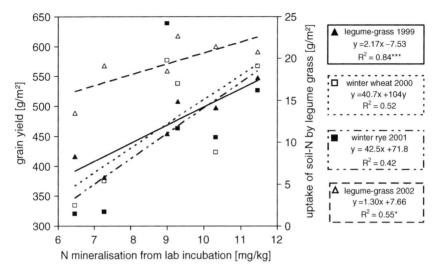

Figure 28 Correlation between N release from undisturbed soil cores incubated in lab and the uptake of soil N by legume–grass or the grain yield of winter wheat and winter rye grown at the sites in field A09 of the FAM research station (Heuwinkel et al., 2003).

Delimitation of management zones

Results from our studies indicate that zones of similar relative yield productivity can be precisely delineated using remote or proximal sensing information (Selige and Schmidhalter, 2001; Schmidhalter et al., 2001a). The information obtained from yield monitoring will reflect the yield more absolutely. The possibility for developing maps of classified management based on similar quality yield maps as obtained from farmers appears limited because of the high frequency of erroneous datasets, systematic errors in the recorded data, and their restricted yield predictive ability (Joernsgaard and Halmoe, 2003). More consistent information reflecting yield zones was obtained in this study and a previous one (Schmidhalter et al., 2001a) from the combination of several years of yield mapping and spectral information. Gaining information from proximal or remote sensing is considered to represent a powerful approach for the future for the delimitation of relative yield productivity areas. Remote sensing is especially appealing to identify management zones because it is non-invasive and low in cost. It also seems likely that further improvements can be obtained by combining mapping and sensor approaches. Combining the use of management zones with crop-based in-season remote sensing is suggested by others as well (Schepers et al., 2004).

Development of management strategies – Current N fertiliser recommendations

Different methods for deriving a N recommendation in winter wheat are available for farmers. Five strategies were tested in a 3-year field trial on several sites in Southern Bavaria (Hege et al., 2002). The strategies used various data for calculating N fertiliser doses. The strategies Expert-N and Hermes resulted from simulation models for estimating plant development and N release from soil. The strategy TUM relied on actual plant status. DSN as well as EUF used soil test data in order to adapt mineral N doses in winter wheat.

The strategy using actual plant information (TUM) proved to be valuable in achieving a high kernel yield as well as a pronounced quality of wheat kernels (Figure 29). The results were evident for the different trial sites. As only low N dressings were applied, N use efficiency was quite good. EUF reached the highest protein concentration of all strategies with 12.7%, but this was due to relatively high N dressings, leading to a reduced N use efficiency. Despite N use efficiency was highest for the strategies Expert-N and DSN, both fertilisation regimes were associated with a lower kernel yield and a lower kernel quality in relation to TUM.

For a strategy of site-specific N fertilisation, an adaptation of N doses to the heterogeneous conditions within agricultural fields is evident. Nevertheless, simulation models as well as soil test recommendations lose their importance in this context. Soil analysis in a small-scale resolution is cost-intensive, time-consuming, and not feasible. Input data for simulation models are often missing or not available in the required high-areal

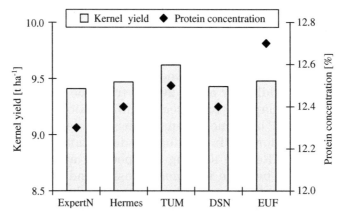

Figure 29 Kernel yield and protein concentration for different strategies of N fertilisation in winter wheat (adapted from Hege et al., 2002).

resolution. Nevertheless, plant information can be received on a small-scale mode, not via destructive biomass sampling and laboratory analysis, but by using sensors for online-detecting of N status of plants (see the subsection "Tractor-based non-destructive sensing of biomass and nitrogen status – Reflectance measurements").

Strategies for site-specific N fertilisation

The 'on-the-go' information about biomass and nitrogen status obtained using the sensing methods can be combined with a fertilising algorithm to control the amount of N fertiliser being applied. Unfortunately, however, nutrient recommendations corresponding to within-field site-specific characteristics are rarely available (Robert, 2001). This is true not only for the sensor-based approaches but also for mapping approaches that largely report results from short-term studies. Universal nitrogen fertiliser application strategies for heterogeneous fields do not exist. Furthermore, there is no current consensus as to how lower or higher yield productivity zones should be treated. Increasing nitrogen input to weaker crop stands would enhance yields, but is not particularly environmentally friendly. Alternatively, it has been variously argued whether higher yield productivity areas should receive higher nitrogen inputs. The situation becomes even more complicated in trying to generalise the strategies for regions that differ in climate and, even more so, in trying to account for any annual variation in climate, which might also interact differently at different locations.

In the FAM research project, site-specific management of agricultural fields was performed for the cultivars wheat and maize. In this chapter, some exemplary results are presented.

The heterogeneous fields of the research station Scheyern were divided into sub-parts according to their yielding potential:
- High Yielding Zone (HYZ): > 105% average yield
- Medium Yielding Zone (MYZ): 95–105% average yield
- Low Yielding Zone (LYZ) < 95% average yield

For winter wheat, strip trials were conducted in order to compare different strategies of site-specific N fertilisation. In this context, the fertilisation practise within several repeated strips with a width of 7.50 m was set to uniform (UA) or variable application (VA) of N fertiliser. Beside the homogeneous application of mineral N fertiliser for a whole strip, three strategies of variable application of mineral N fertiliser have been executed:
- Mapping approach
- Online approach
- Online with map-overlay.

The mapping approach usually relies on deriving an application map prior to the fertiliser application. Thus, mainly historic data is used, e.g. yield maps and maps of soil parameters such as texture and electric conductivity. In the FAM research project, the trial sites were divided into sub-parts with different yielding potential. For this purpose, 3-year data of combine harvester measurements (2 years with winter wheat and 1 year with maize) were used.

For each yielding zone, the potential yield and quality were assessed at the start of vegetation in spring. Mineral N fertiliser doses within the different yielding zones were calculated according to the expected N uptake at final harvest. Thus, N fertiliser doses were in accordance with the yielding potential of the sub-parts: High yielding zones received more mineral N fertiliser than middle yielding zones that exceeded the low yielding zones. Timing of mineral N doses was set to the three most sensitive growth stages during winter wheat development (Figure 30): start of vegetation in spring, EC 32, and EC 49. The definition of growth stages followed the scheme of Tottman (1987). Partitioning of N doses at the growth stages was 30, 40, and 30% for start of vegetation, EC 32, and EC 49, respectively.

In contrary to the mapping approach, the strategy Online used actual information about growth and nutrient conditions of the plants when applying mineral N fertiliser. Thus, sensors are necessary for determining biomass and nitrogen status of the canopy in an online mode. Up to now, different sensors are available as tractor-mounted tools (see the subsection "Tractor-based non-destructive sensing of biomass and nitrogen status – Reflectance measurements"). The active chlorophyll sensor uses a laser beam as excitation source for performing laser-induced chlorophyll

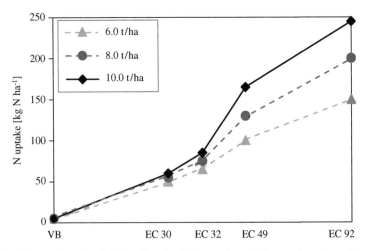

Figure 30 N uptake (kg N ha^{-1}) of winter wheat dependent on the kernel yield at final harvest (adapted from Diepolder, 1994).

measurements (Schächtl et al., 2005), whereas passive sensors rely on the reflection of sunlight. Such a reflectance sensor, the N-Sensor (YARA, Norway), is available as tractor-mounted mode (Link et al., 2002). We used this sensor for deriving our site-specific N recommendation in the online approach. For the first mineral N application at start of vegetation, N-Sensor values were influenced by soil reflectance and not reliable in deriving a N recommendation. Thus, a homogeneous and moderate amount of mineral N fertiliser was spread within the fields in order to enable a reaction with sensor measurements in EC 32.

For the second (EC 32) and third (EC 49) N application, wheat canopies were scanned with the N-Sensor (Co. YARA) and N fertiliser was applied according to these measurements. Core of the strategy was the establishment of homogeneous canopies. Thus, sub-parts of the field with an actually high biomass production and N supply received low mineral N doses in order to avoid lodging. In contrary, zones with an actually low biomass yield and N status of wheat plants received high dressings of mineral N doses.

The third examined site-specific N fertilisation regime 'online with map-overlay' combines the mapping approach with the online approach (Diepolder, 1994; Maidl et al., 2004). Our procedure in deriving an application map for this strategy was divided into two parts: First, the canopy was scanned with a reflectance sensor (2000: hand-held device, 2001: tractor-mounted N-Sensor), then mineral N doses were adjusted in relation to the site-specific yielding potential, i.e. core of the adaptation was the knowledge of optimal N supply of plants according to the yielding potential of the site (q.v. Figure 30).

An example for the derivation of mineral N application at EC 49 is given in Figure 31. The sensor value (REIP) indicates the N uptake of the canopy.

Effects of site-specific crop management in winter wheat – Influence on yield and quality

The fertiliser strategies lead to variable doses of N fertiliser applied to the wheat canopies (Table 10) at the three yielding zones. For uniform application of mineral N doses, all received 170 kg N ha^{-1} (2000) or 180 (2001) kg N ha^{-1}. The mapping approach was characterised by increased N dressings in the HYZ and reduced N doses in the LYZ. This effect becomes clear considering the high yield level and N uptake in the HYZ. Nevertheless, average values over yielding zones indicated a comparable amount of N fertiliser as with uniform application. For the online approach, N partitioning changed between the yielding zones and as expected, the LYZ received more N fertiliser than the HYZ.

For the strategy online with map-overlay, N partitioning was 195, 190, and 170 kg N ha^{-1} (2000) and 200, 205, and 170 kg N ha^{-1} (2001) at the

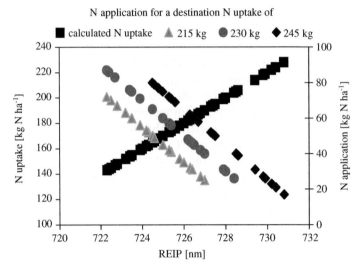

Figure 31 Online detection of N status at EC 49 with REIP values and corresponding N fertiliser application rates for the yielding zones HYZ, MYZ, and LYZ (Maidl et al., 2004).

Table 10 Amount of N fertiliser for the strategies uniform application (UA) and variable application (VA) (Maidl et al., 2004).

Year	Yielding zone	UA (kg N ha^{-1})	VA (kg N ha^{-1})		
			Mapping	Online	Online map
2000	HYZ[a]	170	205	–	195
	MYZ[b]	170	170	–	190
	LYZ[c]	170	135	–	170
	Average	170	170	–	187
2001	HYZ	180	200	185	200
	MYZ	180	180	210	205
	LYZ	180	160	220	170
	Average	180	180	203	191

[a] HYZ = High Yielding Zone.
[b] MYZ = Medium Yielding Zone.
[c] LYZ = Low Yielding Zone.

HYZ, MYZ, and LYZ, respectively. The average values exceeded the uniform application by 10% (2000) or 6% (2001) due to an adaptation of N fertiliser doses to actual plant status. The same was evident for the online approach. Reasons for this can be found in the calibration procedure of the

N-Sensor (YARA, Norway) as the respective amount of N fertiliser was adjusted at a sub-part of the field. Maybe this zone was not as representative for the whole field as expected.

Average kernel yield for uniform N application amounted to 10.0 and 9.9 t ha^{-1} in 2000 and 2001, respectively (Table 11). In 2000, site-specific differences between the yielding zones were negligible (0.4 t ha^{-1}). Thus, the strategies of variable N application reached comparable yield results of 9.8 and 9.9 t ha^{-1}. Nevertheless, the low amount of N fertiliser with the mapping approach in the LYZ led to a small reduction in kernel yield, whereas increased doses of mineral N in the HYZ had no effect on yield formation.

In 2001, the HYZ reached a kernel yield of 11.0 t ha^{-1}, 1.8 t ha^{-1} superior to the LYZ. As for the first year, the mapping approach increased these site-specific differences between the yielding zones to 2.4 t ha^{-1}. The reduced amount of N fertiliser in the LYZ impacted kernel yield, whereas increased N doses in HYZ had no effect.

The contrary was remarkable for the online approach where site-specific differences between the yielding zones diminished. This result is not surprising due to the aim of the online approach in gaining a more homogeneous plant stand within a heterogeneous field (Link et al., 2002).

The average yield of mapping approach was significantly increased when combining with sensor measurements in the strategy online with map-overlay. This is in accordance with Welsh et al. (2003) who showed that the actual situation of the plant is a better indicator of N demand

Table 11 Kernel yield for the strategies uniform application (UA) and variable application (VA) (Maidl et al., 2004).

Year	Yielding zone	UA (t ha^{-1})	VA (t ha^{-1})		
			Mapping	Online	Online map
2000	HYZ[a]	10.2 a	10.2 a	–	10.3 a
	MYZ[b]	9.7 a	9.7 a	–	9.6 a
	LYZ[c]	9.8 a	9.6 a	–	9.6 a
	Average	10.0 a	9.8 a	–	9.9 a
2001	HYZ	11.0 a	10.9 a	10.8 a	11.1 a
	MYZ	9.5 b	9.5 b	10.6 a	10.5 a
	LYZ	9.2 b	8.5 c	9.5 b	10.2 a
	Average	9.9 b	9.5 c	10.3 ab	10.6 a

Different letters in a line indicate significance at $p = 0.05$.
[a] HYZ = High Yielding Zone.
[b] MYZ = Medium Yielding Zone.
[c] LYZ = Low Yielding Zone.

than historic yield data. Regarding the results of both years, mainly the kernel yield in the LYZ was affected by the different strategies of variable N fertilisation.

The first year was characterised by a higher protein concentration in wheat kernels of 12.7% than the following growth period with 10.8% (Table 12). For uniform N application, lowest value of N concentration in kernels was achieved in the LYZ in both years. The mapping approach led to a further reduction of this quality parameter in the LYZ due to the low amount of N fertiliser. Nevertheless, an enhanced amount of mineral N in the HYZ enhanced protein concentration.

Protein concentration in the LYZ was increased when augmenting N doses in the online approach (Table 12). Thus, this strategy leads to a higher average protein concentration for the whole field as well as to a more homogeneous wheat quality. A comparable effect was observed for the strategy online with map-overlay. These results are not surprising as Reckleben and Rademacher (2004) point at the importance of giving high doses of mineral N to wheat canopies with a high yielding potential in order to obtain an increase in wheat quality.

Influence on the environment – N use efficiency

Regarding the implications of N fertiliser applications, not only the effects on kernel yield and kernel quality are of importance. N use efficiency (NUE) indicates the amount of N uptake of the kernels in relation to the amount of mineral N: $NUE = N \text{ uptake} \times (N \text{ rate})^{-1}$.

Table 12 Protein concentration of wheat kernels for the strategies uniform application (UA) and variable application (VA) (Maidl et al., 2004).

Year	Yielding zone	UA (%)	VA (%)		
			Mapping	Online	Online map
2000	HYZ[a]	12.6 b	12.7 b	–	13.4 a
	MYZ[b]	12.8 b	12.8 b	–	13.1 a
	LYZ[c]	12.4 a	11.8 b	–	12.5 a
	Average	12.7 ab	12.4 b	–	13.1 a
2001	HYZ	11.6 b	12.3 a	12.2 a	11.7 b
	MYZ	10.2 b	10.2 b	11.4 a	11.7 a
	LYZ	10.4 b	9.7 c	11.4 a	10.9 b
	Average	10.8 b	10.8 b	11.7 a	11.4 a

Different letters in a line indicate significance at $p = 0.05$.
[a] HYZ = High Yielding Zone.
[b] MYZ = Medium Yielding Zone.
[c] LYZ = Low Yielding Zone.

First of all in 2000 the average NUE was always higher than 100% and always below in 2001. The amount of N fertiliser corresponded with N use efficiency for the mapping approach in 2000 (Table 13). High N dressings led to a reduction in N use efficiency in the HYZ, whereas low N doses increased N use efficiency in the LYZ.

The strategy online with map-overlay resulted in both years an average N use efficiency close to 100%. With the other strategies, average values of N use efficiency were lower or higher especially in the LYZ.

Static field trials of a long-term or multi-year character were further established to test whether targeted, site-specific nitrogen fertiliser application can enhance nitrogen use efficiency as compared to optimal uniform nitrogen application while still maintaining yields. Mapping and online (sensor) variable rate nitrogen fertiliser application strategies at several locations (Ebertseder et al., 2005; Schmidhalter et al., 2006). In general, high yields were found on field sites representing moderate in-field variability. Despite highly contrasting weather conditions between years, similar responses of lower and higher yield zones were observed, and the effects of the different strategies were found to be relatively consistent. The results indicate considerable potential to increase nitrogen use efficiency while simultaneously maintaining yields. Similar to the results reported above, the mapping approach, which considers the long-term yield potential, indicated substantial gains for the environment in areas of lower yield productivity and fertile colluvial deposit zones, whereas the sensor approach allowed for nitrogen use efficiency to be optimised in areas of higher yield productivity.

Table 13 N use efficiency for the strategies uniform application (UA) and variable application (VA) (Maidl et al., 2004).

Year	Yielding zone	UA (%)	VA (%)		
			Mapping	Online	Online map
2000	HYZ[a]	114 a	84 b	–	107 c
	MYZ[b]	110 a	110 a	–	100 c
	LYZ[c]	108 a	126 b	–	107 a
	Average	113 a	102 b	–	105 b
2001	HYZ	107 a	101 b	108 a	98 b
	MYZ	82 a	82 a	86 b	90 c
	LYZ	82 a	79 b	76 c	99 d
	Average	93 a	87 b	92 a	96 c

Different letters in a line indicate significance at $p = 0.05$.
[a] HYZ = High Yielding Zone.
[b] MYZ = Medium Yielding Zone.
[c] LYZ = Low Yielding Zone.

N balance

N balance was calculated as difference between N fertiliser rates and N uptake of wheat kernels at final harvest. Negative values of this parameter indicate that more N was removed from the field than the amount applied via N fertiliser. On the other hand, positive values show that N rates exceeded N transfer via wheat kernels, resulting in an accumulation of nitrogen in the field. This amount of nitrogen is potentially available for leaching hazard.

In 2000, positive values of N balance were only found for the mapping approach in the HYZ (Table 14). The high amount of N fertiliser in these parts of the field could not be removed with kernel harvest. The strategy online with map-overlay enabled values of –1 and –13 kg N ha^{-1}. Applying a homogeneous amount of N fertiliser within the whole field resulted in a N balance of -16 kg N ha^{-1}. Nevertheless, repeatedly high negative N balances may induce a decrease in soil fertility.

In 2001, all fertilisation strategies led to positive N balances regarding average values for the whole field. Nevertheless, the strategy online with map-overlay allowed the slightest value of 9 kg N ha^{-1}. Regarding the separate yielding zones, N balance was well adjusted using this strategy. The other regimes of site-specific mineral N fertilisation resulted in great differences of N balance within the yielding zones, ranging from –12 to +36 kg N ha^{-1}, for the uniform application, +1 to +36 kg N ha^{-1} for the mapping approach, and –14 kg N ha^{-1} up to +57 kg N ha^{-1}for the online approach. Whereas the HYZ was characterised by negative N balances,

Table 14 N balance for the strategies uniform application (UA) and variable application (VA).

Year	Yielding zone	UA (kg N ha^{-1})	VA (kg N ha^{-1})		
			Mapping	Online	Online map
2000	HYZ[a]	-23 a	$+10$ b	–	-13 b
	MYZ[b]	-17 a	-17 a	–	-1 b
	LYZ[c]	-13 a	-35 b	–	-11 a
	Average	-21 a	-13 b	–	-8 b
2001	HYZ	-12 a	$+1$ b	-14 a	$+4$ b
	MYZ	$+34$ a	$+34$ a	$+28$ b	$+20$ b
	LYZ	$+36$ a	$+36$ a	$+57$ b	$+2$ c
	Average	$+19$ a	$+25$ b	$+22$ b	$+9$ c

Different letters in a line indicate significance at $p = 0.05$.
[a] HYZ = High Yielding Zone.
[b] MYZ = Medium Yielding Zone.
[c] LYZ = Low Yielding Zone.

high positive values in the LYZ display a great problem regarding the potential of N leaching hazard (Schächtl, 2004).

Transborder farming

In small-structured agricultural regions, many farmers have competitive disadvantages due to small-sized fields and a multitude of single plots resulting in extensive operation times and use of resources per area. An alternative for the time consuming and expensive restructuring by regular land consolidation is the virtual land consolidation in terms of a transborder farming system. A transborder field is a number of adjoined plots, surrounded by natural borders or farm roads. Several single plots, cultivated by several farmers, can be farmed as one big field across the property borders. This requires cooperation of the participating farmers. Transborder farming means an enlargement of cultivation units and thus declining operation time and costs. At the same time, production methods can be better adjusted to ecological requirements.

In a transborder farming system, small fields situated side by side, are farmed together as one large field (Figure 32). Because of bigger cultivation units, farmers can realise increased gross margins through decreased labour and resource costs (Auernhammer et al., 2000a). Farmers have to agree on a common crop rotation on the transborder field and should use the best farming techniques and technologies that are available for each production process.

For evaluating the transborder farming system, as well as for accounting each individual farmer, a system for exact documentation of the field work for tractors and harvesters is essential (Rothmund et al., 2003a; 2002a). Using these systems, yields can be allocated site-specifically by DGPS data. All process data, containing position information, can be allocated to known field boundaries of each farmer's property within the transborder field (Figure 33).

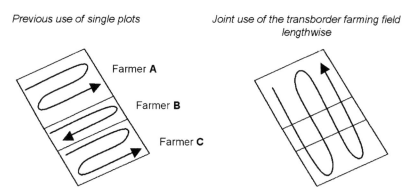

Figure 32 Changing of cultivation due to the formation of a transborder field.

online-acquisition of operating-   allocation of times and   property based validation
times, application- and yield-   volumes within the outlines of   for each participating
volumes   parcels   farmer

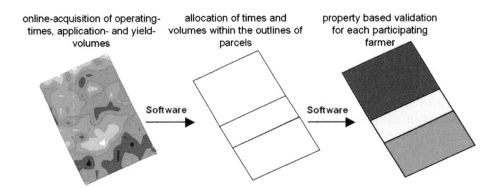

Figure 33 Specific accounting of mapped process data by boundaries within the transborder field.

The result of combining the automatic process data acquisition with an adjusted data management system is an 'Automated Documentation and Management System' for transborder farming. It allows field work on different plots, for instance in transborder fields, without paying attention to manual documentation. The high spatio-temporal resolution of GPS and attribute data from machines could never be realised by manual or just PC-supported systems. Using local or web-based software for allocating and validating field work data to each farmer's plots by a few mouse clicks, considerably reduces the time required for managing transborder farming.

Investigations in simulating the production on a virtual 'transborder farm' on 5–7 ha sized transborder fields instead of 1–2 ha sized fields resulted in a reduction of labour costs of about 30% and about 25% of machinery costs. There are some reductions of costs by sales discounts and there is a 'know-how effect', which can be from € 0 up to € 150 per hectare (Rothmund et al., 2002b). The results of a survey showed that the possible effects caused by specialised knowledge of single farmers in a transborder farming group could be more than 100% yield difference under equal soil conditions. Consequently, this 'know-how effect' can cause a big increase in yield for farmers with previous low yields and there will be no such effect for farmers with the previous highest yields. All in all, increases of gross margin from about € 100 up to more than € 300 per hectare and year are possible (Rothmund, 2006).

Both, in practical trials and in various simulations, the excellence of creating transborder farming systems in small-structured agricultural regions has been proved. Often this virtual land consolidation can be simply realised by working across the previous field boundaries without changing landscape or boundary ridges. Indeed, the realisation of a transborder farming system mostly fails due to another reason: the missing will for cooperation between the farmers. Even cognition of the economical facts cannot persuade, because farmers are afraid of losing their freedom of decision. Only economic forces combined with an increase in corporate thinking among the farmers can change this situation.

Documentation and traceability

Today site-specific acquisition of process data is enabled by a lot of sensors commonly integrated in modern tractors and farm tools combined with a satellite-based positioning system such as GPS. The automated and geo-referenced acquisition of process data on agricultural machinery can be implemented easily, presupposed that a standardised electronic communication BUS system is available (Auernhammer et al., 2000a). By introducing the ISO 11783 standard (ISOBUS) for tractor-implement systems a widespread use of automated data acquisition in practice can be expected. Combined with an adequate data and information management it enables an automated documentation system. This is the base for a number of internal, inter-company, and external farm applications such as optimised farm management, site-specific crop management, machinery management for joint machinery use and contractors as well as the traceability of food and animal feed (Rothmund et al., 2003b). For these purposes, exclusively automated systems guarantee high quality of acquired information in a sufficient temporal and spatial resolution as well as adequate safety from falsification.

Some important components for an automated documentation system are already available, some however not yet. At the Technische Universität München an automatic process data acquisition system based on Agricultural BUS System (DIN 9684) has been developed (Auernhammer et al., 2000b). The new international standard for electronic communication systems ISOBUS (ISO 11783) is nearly accomplished (Böttinger and Autermann, 2003). The discussion on data interfaces and exchange formats based on the XML standard is on the way. To get a practical documentation system that produces additional benefits by using acquired information, missing components (i.e. data processing methods) have to be conceived and linked together with existing components (Demmel et al., 2001). The project's aim was to design an information management system, which starts with data acquisition and ends with the use of process information in different applications. The concepts have to include data storing, data transfer, and data processing up to providing information and diverse data interfaces.

Different information system models, which could be denoted as offline and online dataflow should be investigated and implemented in prototypes. For these models, advantages and disadvantages should be worked out and analysed. Especially data security and faultless data handling and processing a huge amount of data are points of interest. Accordingly, different types of use of information from automated documentation systems should be investigated. It was assumed that farmers, cooperatives, and contractors on the one hand and consumers or authorities on the other hand do not concern in the same information contents about agricultural production processes.

The following description is divided into three steps: first, data acquisition and transfer, then data processing and information providing and last but not least the information use.

Automatic process data acquisition within standardised communication networks

First of all, automatic data acquisition means that data recording is not affected by the driver. Starting the engine of a machine means starting data recording automatically. This approach differs from previous recording systems, which were operator or job orientated and data logging had to be started and stopped by pressing a button. In an automatic system the centre is a CAN-BUS that builds up a 'data transfer highway' for a number of connected electronic control units (ECUs), which are distributed on tractor and farm tools. These ECUs can deliver process data, like speed and hitch parameters of the tractor, working and application parameters of working implements, measurement values of crop sensors, and many more. An additional module called data recorder acts as a file server and inquires and stores the process data from different sources as shown in Figure 34. All the recorded data can be allocated to an exact position in landscape, assuming that a GPS receiver is connected to the system. The interoperability between different ECUs and the communication by the CAN-BUS is enabled by standardised data transfer protocols. Until today the German standard for agricultural BUS systems (LBS DIN 9684) had been used. It is now followed by the international standard ISOBUS (ISO 11783).

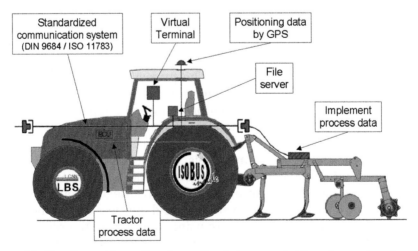

Figure 34 Configuration of automatic data acquisition in a standardised communication network (from Rothmund, 2006).

In the prototype conceived and realised at the Technische Universität München, all position and process data were stored once per second in a relational data model. For data storing and transfer PCMCIA memory cards and an ASCII format were used. In the current prototypes, data have to be transferred by reading out the memory cards to a local PC. For data transfer more advanced methods such as data synchronising using Bluetooth® standard or connecting machines directly to the Internet by WLAN may be conceived. Transferring all process data via GSM (mobile phone standard) seems to be impossible at the moment because of the huge amount of data. As it is just being worked out in the ISOBUS standardisation process, the raw data transfer format will change to an XML-based data interface in the future.

Farm-based and web-based data processing methods

The new approach of fully automated data recording on agricultural machinery offers a new quality of data with a very high spatial and temporal resolution of information. But the huge amount of raw data requires sufficient data processing systems, database, and information management. Otherwise 'data graveyards' would be produced and the effort for data collecting would not be profitable. In principle there are two ways of data processing. The first way is farm-based data processing and storing using local software and database systems. This way could be called offline dataflow. The second way is a centralised data processing and data management for several or many farmers by external service providers. This also could be called online dataflow or web-based information system (Figure 35).

By conceiving, programming, and testing a prototype information system in each case – the offline and the online way – experiences for analysing problems and advantages of those methods could be gained. As a tool for local use on a farmer's PC, a Microsoft Access-based data evaluation tool was built. The complete functionality and the user interfaces for data processing, data keeping, and information providing were integrated within

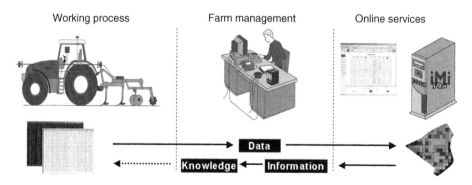

Figure 35 Scheme of online data and information flow in a web-based information system based on automated data acquisition systems.

the Microsoft Access database environment. For the online data processing a web-based information system using exclusively open source tools has been developed alternatively. The main components are a powerful database system and web developing tools such as mark-up and scripting languages. Data upload and information download by users occur by browser interfaces via Internet.

Demands on information contents for different user groups

The raw data from automated process data acquisition allows a big variety of different evaluation targets. It could be the base for classic farm software tools such as field books, machinery management, or invoicing software (Rothmund et al., 2002b). Furthermore, it can be the main data source for extended applications, which need detailed geographical information, like site-specific farming or inter-company farming (e.g. trans-border farming systems in a virtual land consolidation). Regarding the target of information use, the requirements on processing algorithms and information structure can be very different, starting with simple aggregated schedular information or diagrams up to detailed maps and geo-statistical analyses. Beyond those mentioned internal or inter-company farm applications there is an important external use of information. Complete datasets of production processes enable documentation for the trace of products. This requirement of dealers, commercial processors, and consumers has to be fulfilled more and more by the farmers (Auernhammer, 2002).

To realise different information structures for different user groups considering data protection and privacy of information in a centralised database, variable aggregation levels and access authorisation structures have been modelled. Indeed those complex structures have not been fully integrated in the first prototype implementation yet.

2.3.4 Conclusions

Existing methods of soil and plant analysis are costly and time-consuming in delivering information on the actual and spatially resolved site-specific soil properties as well as the biomass and nutritional status of crops. Non-destructive techniques to sense soil and plant properties can contribute to improvements.

New developments of non-destructive techniques to sense soil and plant properties have been validated in the FAM project. Spatially resolved soil information can be gained by electromagnetic induction, near-infrared spectroscopy, and indirectly by correlating plant stands to soil properties. With such methods, soil texture, soil carbon, and plant-available water in the soil can be characterised. Derivation of relevant soil properties by non-contacting sensor techniques is highly effective and will provide long-term information for optimised management.

Remote and proximal sensing allows determining plant biomass, nitrogen content, and nitrogen uptake. The methods tested and further developed included aerial and ground-based reflectance-based measurements and tractor-based laser-induced chlorophyll measurements. Assessment of the nitrogen content, biomass, and nitrogen uptake of plants by contact less optical measurements is seen to be a promising technique for management decisions of farmers. These techniques have the capability of sampling a high number of plants in a short time rather than a single leaf point and allow a fast assessment of the spatial and temporal variability of plant growth.

Together such methods allow for optimised management to better adapt plants and inputs to heterogeneous field sites.

Site-specific agriculture aims at optimising inputs on the field and farm levels, and thus can benefit both the farmer (in terms of net return) and the environment (through lower emission levels). The results, derived from multi-year and multi-location field studies, recommend the adoption of variable rate nitrogen fertiliser application. A sensible increase in the already high yields by using site-specific farming is not possible. Site-specific farming can increase nitrogen efficiency and reduce environmental impacts. In general, it appears that potential benefits to the environment of site-specific nitrogen fertiliser application increase the higher the yield differences on a field are and the less favourable the weather conditions are. A combination of mapping and online approaches will further contribute to improvements. Techniques applied and newly developed within the FAM study offer a great potential for further applications. Promising fields are seen in site-specific soil cultivation, seeding, and plant protection. Both integrated and organic farming will benefit from such achievements.

Acknowledgements

The scientific activities of the *FAM Munich Research Network on Agroecosystems* were financially supported by the German Federal Ministry of Education and Research (BMBF 0339370). Overhead costs of the Research Station Scheyern were funded by the Bavarian State Ministry for Science, Research and the Arts.

References

Agati G, Fusi F, Mazzinghi P, Lipucci di Paola M, 1993. A simple approach to the evaluation of the reabsorption of chlorophyll fluorescence spectra in intact leaves. Journal of Photochemistry and Photobiology B: Biology 17, 163–171.

Auernhammer H, 1997. Elektronische Traktor-Gerätekommunikation LBS. In: Proceedings of the International Symposium on Mobile Agricultural

Bus-System LBS and PA for the Large-Scale Farm Mechanization. Sapporo: Hokkoaido Branch of the Japanese Society of Agricultural Machinery, 1–22 (deutsch); 23–38 (japanisch).

Auernhammer H, 2002. The role of mechatronics in crop product traceability. CIGR-Ejournal 4, 21 (Invited Overview Articles no. 15).

Auernhammer H, Demmel M, Maidl FX, Schmidhalter U, 1999. An on-farm communication system for precision farming with nitrogen real-time application. ASAE St. Joseph, Paper No. 99 11 50.

Auernhammer H, Demmel M, Muhr T, 1994. GPS and DGPS as a challenge for environment friendly agriculture. In: Proceeding of EURNAV 94, 3rd International Conference on Land Vehicle Navigation 14–16 June 1994, Dresden, Editor: Deutsche Gesellschaft für Ortung und Navigation.

Auernhammer H, Demmel M, Muhr T, Rottmeier J, 1991. Future developments for fertilizing in Germany. Paper No. 91 1040. ASAE, St. Joseph, MI, USA.

Auernhammer H, Demmel M, Muhr T, Rottmeier J, Wild K, 1993. Yield measurement on combine harvesters. ASAE Paper No. 93 1506, ASAE, St. Joseph, MI, USA.

Auernhammer H, Demmel M, Pirro PJM, 1997. Durchsatz- und Ertragsermittlung im selbstfahrenden Feldhäcksler. VDI Bericht 1356, 135–138.

Auernhammer H, Mayer M, Demmel M, 2000a. Transborder farming in small-scale land use systems. Abstracts of the XIV Memorial CIGR World Congress 2000, CIGR, Tsukuba, Japan, pp. 189.

Auernhammer H, Rottmeier J, 1990. Weight determination in transport vehicles – Exemplary shown on selfloading trailers. In: Technical Papers and Posters Abstracts of AgEng'90, 24–26. October 1990, Berlin, 100–101.

Auernhammer H, Schueller JK, 1999. Precision agriculture. In: CIGR-Handbook of Agricultural Engineering. Vol. III: Plant Production Engineering. ASAE, St. Joseph, pp. 598–616.

Auernhammer H, Spangler A, Demmel, M, 2000b. Automatic process data acquisition with GPS and LBS. In: Proceedings of the AgEng Conference 2000, Paper No. 00-IT-005, EurAgEng, Silsoe, UK.

Barnet NG, Shinners KJ, 1998. Analysis of systems to measure mass-flow-rate and moisture on forage harvesters. ASAE Paper No. 981118, ASAE, St. Joseph, MI, USA.

Baumgärtel G, 2001. Vier Stickstoff-Düngesysteme im Vergleich. Getreidemagazin 7(1), 44–47.

Beck AD, Searcy SW, Roades JP, 2001. Yield data filtering techniques for improved map accuracy. Applied Engineering in Agriculture. ASAE, St. Joseph, MI, USA. 17(4), 423–431.

Beck DP, Wery J, Saxena MC, Ayadi A, 1991. Dinitrogen fixation and nitrogen-balance in cool-season food legumes. Agronomy Journal 83, 334–341.

Behme JA, Schinstock JL, Bashford LL, Leviticus LI, 1997. Site specific yield for forages. Paper No. 97 1054, ASAE, St. Joseph, MI, USA.

Ben-Dor E, Banin A, 1995. Near-infrared analysis as a rapid method to simultaneously evaluate several soil properties. Soil Science Society of America Journal 59, 364–372.

Benjamin CE, Price RR, Mailander MP, 2001. Sugar cane monitoring system. Paper No. 01 1189, ASAE, St. Joseph, MI, USA.

Blackmore S, Marshall C, 1996. Yield mapping; errors and algorithms. In: Robert PC, Rust AH, Larson WE (Eds.), Proceedings of the 3rd International Conference on Precision Agriculture, June 23–26, 1996 in Minneapolis. ASA; CSSA; SSSA. Madison, WI, USA, pp. 403–415.

Boller BC, 1988. Biologische Stickstoff-Fixierung von Weiß- und Rotklee unter Feldbedingungen. Landwirtsch. Schweiz 1, 251–253.

Bongiovanni R, Lowenberg-Deboer J, 2004. Precision agriculture and sustainability. Precision Agriculture 5, 359–387.

Böttinger S, Autermann L, 2003. ISOBUS (ISO 11783) – Technische Umsetzung und Erinführung im Markt. In: VDI-MEG Tagung Landtechnik 2003, VDI-Verlag Düsseldorf, Germany, S.109–S.113.

Bredemeier C, Schmidhalter U, 2001. Laser-induced chlorophyll fluorescence to determine the nitrogen status of plants. In: 14th International Plant Nutrition Colloquium, Hannover, 27.07.-03.08.01. Kluwer Academic Publishers, Dordrecht, Developments in Plant and Soil Sciences, Vol. 92, 726–727.

Bredemeier C, Schmidhalter U, 2003. Non-contacting chlorophyll fluorescence sensing for site-specific nitrogen fertilization in wheat and maize. Proceedings of the 4th European Conference of Precision Agriculture, Berlin, 103–108.

Bredemeier C, Schmidhalter U, 2005. Laser-induced chlorophyll fluorescence sensing to determine biomass and nitrogen uptake of winter wheat under controlled environment and field conditions. In: Stafford JV (Ed.), Precision Agriculture '05, Wageningen Academic Publishers, The Netherlands, pp. 273–280.

Broos B, Missotten B, Reybrouck W, De Baerdemaker J, 1998. Mapping and interpretation of sugar beet yield differences. Proceedings of the 4th International Confernce on Precision Agriculture, Robert PC, Rust RH, Larson WE (Eds.), ASA, Madison, WI, USA.

Brunner R, 1998. Untersuchungen zu den Ursachen kleinräumiger Ertragsschwankungen auf einem Standort des Tertiärhügellandes (Scheyern). PhD thesis, TU München, Shaker Verlag, Aachen, ISBN 3-8265-3275-9, FAM-Bericht; 21, 207 pp.

Campbell RH, Rawlins SL, Han S, 1994. Monitoring methods for potato yield mapping. ASE Paper No. 94 3184, ASAE, St. Joseph, MI, USA.

Coleman SW, Christiansen S, Shenk JS, 1990. Prediction of botanical composition using NIRS calibrations developed from botanically pure samples. Crop Science 30, 202–207.

Cox GJ, Harris HD, Cox DR, 1998. Application of precision agriculture to sugar cane. Proceedings of the Fourth International Conference on Precision Agriculture. ASA-CSSA-SSSA, St. Paul, MN, USA.

Dammer KH, Wartenberg G, Ehlert D, Hammen V, Reh A, Wagner U, Dohmen B, 2001. Recording of present plant parameters by pendulum sensor, remote sensing, and ground measurements as fundamentals for site specific fungicide application in winter wheat. In: Proceedings of 3rd European Conference on Precision Farming in Agriculture, June 18–20, 2001. Montpellier, France, pp. 647–652.

Dardenne P, Féménias N, 1999. Diode array near infrared instrument to analyse fresh forages on a harvest machine. In: Davies AMC, Giangiacomo R (Eds.), Proceedings of NIR-99, Verona, Italy. NIR Publications, Chichester, UK, pp. 121–124.

De Jong E, Ballantyne AK, Cameron DR, Read DWL, 1979. Measurement of apparent electrical conductivity of soils by an electromagnetic induction probe to aid salinity surveys. Soil Science Society of America Journal 43, 810–812.

Demmel M, 2001. Ertragsermittlung im Mähdrescher – Ertragsmessgeräte für die lokale Ertragsermittlung. DLG Merkblatt 303. Hrsg: Deutsche Landwirtschafts-Gesellschaft, Fachbereich Landtechnik, Ausschuss für Arbeitswirtschaft und Prozesstechnik, Deutsche Landwirtschafts-Gesellschaft, 20 pp.

Demmel M, 2002. Wann kommen die Feldroboter – Autonome Führung landwirtschaftlicher Fahrzeuge machen Fortschritt. In: Neue Landwirtschaft 12/2002, 13. Jhrg., Berlin 2002, 40–43.

Demmel M, Auernhammer H, 1999. Local yield measurement in a potato harvester and overall yield pattern in a cereal – Potato crop rotation. ASAE Paper No. 99 1149, ASAE, St. Joseph, MI, USA.

Demmel M, Auernhammer H, Rottmeier J, 1998. Georeferenced data collection and yield measurement on a self propelled six row sugar beet harvester. ASAE Paper No. 98 3103, ASAE, St. Joseph, MI, USA.

Demmel M, Rothmund M, Spangler A, Auernhammer H, 2001. Algorithms for data analysis and first results of automatic data acquisition with GPS and LBS on tractor-implement combinations. In: Proceedings of 3rd European Conference on Precision Farming in Agriculture, June 18–20, 2001. Montpellier, France.

Demmel M, Schwenke T, Böck J, Heuwinkel H, Locher F, Rottmeier J, 2002. Development and field test of a yield measurement system in a mower conditioner. EurAgEng Paper Number 02-PA-032, AgEng Budapest.

Diepolder M, 1994. Untersuchungen zur Ableitung von Richtlinien für die Optimierung der N-Düngung zu Winterweizen (Investigations on the derivation of guidelines for an optimization of N fertilization in winter wheat). PhD thesis, TU München, 184 pp.

Durrance JS, Perry CD, Vellidis G, Thomas DL, Kvien CK, 1998. Commercially available cotton yield monitors in Georgia field conditions. Paper No. 98 3106, ASAE, St. Joseph, MI, USA.

Ebertseder T, Schmidhalter U, Gutser R, Hege U, Jungert S, 2005. Evaluation of mapping and one-line nitrogen fertilizer application strategies in multi-year and multi-location static field trials for increasing nitrogen use efficiency of cereals. In: Stafford JV (Ed.), Precision Agriculture '05, Wageningen Academic Publishers, The Netherlands, pp. 327–335.

Eck HV, 1988. Winter wheat response to nitrogen and irrigation. Agronomy Journal 80, 902–908.

Ehlert D, 1999. Durchsatzermittlung zur Ertragskartierung im Feldhäcksler. Agrartechnische Forschung 5/1, 19–25.

Ehlert D, 2004. Stickstoff-Einsatz auspendeln? dlz agrarmagazin 7, 48–51.

Ehlert D, Dammer KH, Völker U, 2004a. Applikation nach Pflanzenmasse. Agrartechnische Forschung 10/2, 28–32.

Ehlert D, Schmerler J, Voelker U, 2004b. Variable rate nitrogen fertilisation of winter wheat based on a crop density sensor. Precision Agriculture 5, 263–273.

Ehrl M, 2005. ISO 11783 Communication Standard. Yanmar Corp., Maibara/Japan, 21.02.2005.

Ehrl M, Stempfhuber W, Demmel M, Auernhammer H, 2003. Quality assessment of agricultural positioning and communication systems. In: Stafford J, Werner A (Eds.), Precision Agriculture – Proceedings of the European Conference on Precision Agriculture 2003. Wageningen Academic Publishers, The Netherlands 2003, 205–211.

Ercoli L, Mariotti M, Masoni A, Massantini F, 1993. Relationships between nitrogen and chlorophyll content and spectral properties in maize leaves. European Journal of Agronomy 2, 113–117.

Funk R, Maidl, FX, 1997. Heterogenität der Ertragsbildung von Winterweizen auf Praxisschlägen des oberbayerischen Tertiärhügellandes im Hinblick auf eine teilschlagspezifische Bestandesführung. In: Pflanzenbauwissenschaften, 1(3), S. 117–126, (1997) ISSN 1431-8857, Verlag Eugen Ulmer GmbH & Co., Stuttgart.

Geesing D, Gutser R, Schmidhalter U, 2001. Importance of spatial and temporal soil water variability for nitrogen management decisions. In: Third European Conference on Precision Agriculture, Montpellier, 659–664.

Glancey JL, Kee WE, Lynch M, 1997. A preliminary evaluation of yield monitoring techniques for mechanically harvested processed vegetables. Paper No. 97 1060, ASAE, St. Joseph, MI, USA.

Godwin RJ, Wheeler PN, 1997. Yield mapping by mass accumulation. Paper No. 97 1061, ASAE, St. Joseph, MI, USA.

Hall TL, Backer LL, Hofmann VL, Smith LJ, 1997. Monitoring sugar beet yield on a harvester. ASAE Paper No. 97 3139, ASAE, St. Joseph, MI, USA.

Hege U, Maidl FX, Dennert J, Liebler J, Offenberger K, 2002. Düngestrategien für Stickstoff zu Winterweizen: Ein Vergleich von Simulationsmodellen und Düngeberatungssystemen (Fertilization strategies for nitrogen in winter wheat: a comparison of simulation models and fertilization recommendation systems). Pflanzenbauwiss (German Journal of Agronomy) 6, 25–35.

Heil K, Schmidhalter U, 2003. Derivation of soil texture and soil water content from electromagnetic induction measurements. Program book of the joint conference ECPA-ECPLF, Berlin, 429–430.

Hennes D, Baert J, De Baerdemaeker J, Ramon H, 2002. Yield mapping of sugar beets with a momentum type flow rate sensor. In: Proceedings of the Conference on Agricultural Engineering, Halle 2002. VDI, Düsseldorf, 247–252.

Heuwinkel H, Gutser R, Schmidhalter U, 2005a. Does N-cycling impair the N_2-fixing activity of mulched legume–grass in field? In: Frankow-Lindberg BE, Collins RP, Lüscher A, Sèbastia T, Helgadóttir Á (Eds.), Adaptation and management of forage legumes- strategies for improved reliability in mixed swards. Proceedings of the 1st COST 852 Workshop, Ystad, Schweden 20-23.09.2004, 141–144.

Heuwinkel H, Locher F, Gutser R, Schmidhalter U, 2003. Ausmaß und Ursachen der schlaginternen Variabilität der N_2-Fixierung von Kleegras. In: Arbeitsgemeinschaft Grünland und Futterbau (AGGF) in der Gesellschaft für Pflanzenbauwissenschaften (Hrsg.), Jahrestagung 2003 in Braunschweig, 75–78.

Heuwinkel H, Locher F, Gutser R, Schmidhalter U, 2005b. How and why does legume content of multispecies legume–grass vary in field? In: Frankow-Lindberg BE, Collins RP, Lüscher A, Sèbastia T, Helgadóttir Á (Eds.), Adaptation and management of forage legumes- strategies for improved reliability in mixed swards. Proceedings of the 1st COST 852 Workshop, Ystad, Schweden 20-23.09.2004, 262–265.

Høgh-Jensen H, Loges R, Jorgensen FV, Vinther FP, Jensen ES, 2004. An empirical model for quantification of symbiotic nitrogen fixation in grass-clover mixtures. Agricultural Systems 82, 181–194.

Joernsgaard B, Halmoe S, 2003. Intra-field yield variation over crops and years. European Journal of Agronomy 19, 23–33.

Kachanoski RG, Gregorich EG, Van Wesenbeck J, 1988. Estimating spatial variations of soil water content using noncontacting electromagnetic induction methods. Canadian Journal of Soil Science 68, 715–722.

Kormann G, Demmel M, Auernhammer H, 1998. Testing stand for yield measurement systems in combine harvesters. International AgEng Conference 98, Oslo, Paper No.: 98-A-054.

Kromer KH, Degen P, 1998. Volume and scale based measuring machine capacity and yield and soil tare of sugar beet. Paper No. 98 3107, ASAE, St. Joseph, MI, USA.

Kumhala F, Kroulik M, Hermanek P, Prosek V, 2001. Yield mapping of forage harvested by mowing machines. VDI Berichte 1636, Düsseldorf, 267–272.

Lammel J, Wollring J, Reusch S, 2001. Tractor based remote sensing for variable nitrogen fertilizer application. In: Horst WJ, Schenk MK, Bürkert A, Claassen N, Flessa H, Frommer WB, Goldbach H, Olfs HW, Römheld V, Sattelmacher B, Schmidhalter U, Schubert S, v. Wirén N, Wittenmayer L. (Eds.), Plant Nutrition – Food Security and Sustainability. Springer, New York, USA, Dordrecht Boston, London, pp. 694–695.

Liebler J, 2003. Feldspektroskopische Messungen zur Ermittlung des Stickstoffstatus von Winterweizen und Mais auf heterogenen Schlägen. PhD thesis, TU München, Utz Verlag, München, ISBN 3-8316-0294-8, 192 pp.

Liebler J, Sticksel E, Maidl FX, 2001. Field spectroscopic mesurements to characterise nitrogen status and dry matter production of Winter Wheat. 3. European Conference on Precision Agriculture, 935–940.

Link A, Panitzki M, Reusch S, 2002. HYDRO-N-Sensor: Tractor-mounted remote sensing for variable nitrogen fertilization. In: Robert PC (Ed.), Proceedings of the 6th International Conference on Precision Agriculture and Other Precision Resources Management. July 14–17, Minneapolis, MN, USA. Published as CD-ROM.

Locher F, 2003. Near infrared reflectance spectroscopy to predict legume content in legume–grass mixtures as a key parameter in N_2-fixation-method development, validation and application. PhD thesis, TU München, Shaker Verlag, Aachen, ISBN 3-8322-2071-2, FAM-Bericht 58, 68 pp.

Locher F, Heuwinkel H, Gutser R, Schmidhalter U, 2005a. Development of a NIRS calibration to estimate legume content of multispecies legume–grass mixtures. Agronomy Journal 97, 11–17.

Locher F, Heuwinkel H, Gutser R, Schmidhalter U, 2005b. The legume content in multispecies mixtures as estimated with near infrared reflectance spectroscopy: Method validation. Agronomy Journal 97, 18–25.

Lopotz HW, 1996. Biologische N_2-Fixierung von Klee-Reinbeständen und Klee-Gras-Gemengen unter besonderer Berücksichtigung des Einflusses der N-Nachlieferung des Bodens. PhD thesis, Rheinische Friedrich-Wilhelm-Universität Bonn, Germany.

MacKerron DKL, Young MW, Davies HV, 1995. A critical assessment of the value of petiole sap analysis in optimizing the nitrogen nutrition of the potato crop. Plant and Soil 172, 247–260.

Mahler RL, Bezdicek DF, Witters RE, 1979. Influence of slope position on nitrogen fixation and yield of dry peas. Agronomy Journal 71, 348–351.

Maidl FX, Huber G, Schächtl J, 2004. Strategies for site specific nitrogen fertilisation in winter wheat. Seventh International Conference on Precision Agriculture, Minnesota/USA, 1938–1948.

Maidl FX, Liebler J, Sticksel E, 2001. Bestimmung von Biomasseaufwuchs und Stickstoffstatus von Winterweizen mittels Feldspektroskopie. Mitt. Ges. Pflanzenbauw., Bd. 13, 121–122.

Major DL, Baumeister R, Touré A, Zhao S, 2003. Methods of measuring and characterizing the effects of stresses on leaf and canopy signatures. In: Van Toai T, Major D, McDonald M, Schepers J, Tarpley L. (Eds.), Digital Imaging and Spectral Techniques: Applications to Precision Agriculture and Crop Physiology. ASA, Special Publication Number 66.

Mallarino AP, Wedin WF, Goyenola RS, Perdomo CH, West CP, 1990. Legume species and proportion effects on symbiontic dinitrogen fixation in legume–grass mixtures. Agronomy Journal 82, 785–789.

Martinez DE, Guiamet JJ, 2004. Distortion of the SPAD 502 chlorophyll meter readings by changes in irradiance and leaf water status. Agronomie 24, 41–46.

Minasny B, McBratney AB, Whelan BM, 1999. VESPER version 1.0. Australian Centre for Precision Agriculture, McMillan Building A05, The University of Sydney, NSW 2006 (http://www.usyd.edu.au/su/agric/acpa).

Missotten B, Broos B, Strubbe G, De Baerdemaeker J, 1997. A yield sensor for forage harvesters. In: Stafford JV (Ed.), Precision Agriculture '97. Bios Scientific Publishers, Oxford, 529–536.

Mistele B, 2006. Tractor based spectral reflectance measurements using an oligo view optic to detect biomass, nitrogen content and nitrogen uptake of wheat and maize and the nitrogen nutrition index. Ph.D. thesis TU München, http://mediatum.ub.tum.de, 72 pp.

Mistele B, Gutser R, Schmidhalter U, 2004. Validation of field-scaled spectral measurements of the nitrogen status in winter wheat. Program book of the Joint Conference ICPA, Minneapolis, 1187–1195.

Neudecker E, Schmidhalter U, Sperl C, Selige T, 2001. Site-specific soil mapping by electromagnetic induction. In: Grenier G, Blackmore S (Eds.), Proceedings of the Third European Conference on Precision Agriculture, Montpellier, 271–276.

Noack PO, Muhr T, Demmel M, 2001. Long term studies on determination and elimination of errors occurring during the process of georeferenced yield data collection on combine harvesters. In: Grenier G, Blackmore S (Eds.), Proceedings of the Third European Conference on Precision Agriculture, Montpellier 2001, 2, 833–837.

Noack PO, Muhr T, Demmel M, 2003. Relative accuracy of different yield mapping systems installed on a single combine harvester. In: Stafford J, Werner A (Eds.), Precision Agriculture – Proceedings of the Conference of the European Conference on Precision Agriculture 2003. Wageningen Academic Publishers, The Netherlands, 451–457.

Nykänen A, Heuwinkel H, Loges R, Locher F, 2005. Comparison of NIRS based methods to determine legume content of mixed swards. In: Frankow-Lindberg BE, Collins RP, Lüscher A, Sèbastia T, Helgadóttir Á (Eds), Adaptation and management of forage legumes- strategies for improved reliability in mixed swards. Proceedings of the 1st COST 852 Workshop, Ystad, Schweden 20-23.09.2004, 282–285.

Ostermeier R, Auernhammer H, Demmel M, 2003. Development of an in-field controller for an agricultural bus-system based on open source program library lbs-lib. In: Stafford J, Werner A (Eds.), Precision Agriculture. Wageningen Academic Publishers Wageningen, 515–520.

Paul C, Dietrich F, Rode M, 2002. Influence of sample temperature and the assessment of quality characteristics in undried forages by near infrared reflectance spectroscopy (NIRS). Landbauforschung Völkenrode 4(52), 229–237.

Pelletier MG, Upadhyaya SK, 1998. Development of a tomato yield monitor. In: Robert PC et al. (Eds.), Proceedings of the Fourth International Conference on Precision Agriculture, American Society of Agronomy, Madison, WI, 1119–1129.

Peoples MB, Ladha JK, Herridge DF, 1995. Enhancing legume N_2 fixation through plant and soil management. Plant Soil 174, 83–101.

Perry CD, Durrence JS, Vellidis G, Thomas DL, Hill RW, Kvien CS, 1998. Experiences with a prototype peanut yield monitor. Paper No. 98 3095, ASAE, St. Joseph, MI, USA.

Petersen JC, Barton II FE, Windham WR, Hoveland CS, 1987. Botanical composition definition of tall fescue-white clover mixtures by near infrared reflectance spectroscopy. Crop Science 27, 1077–1080.

Pitman WD, Piacitelli CK, Aiken GE, and Barton II FE, 1991. Botanical composition of tropical grass-legume pastures estimated with near-infrared reflectance spectroscopy. Agronomy Journal 83, 103–107.

Rands M, 1995. The development of an expert filter to improve the quality of yield data. Unpublished Masters thesis. Silsoe College, Department of Agricultural and Environmental Engineering.

Rawlins SL, Campbell GS, Campbell RH, Hess JR, 1995. Yield mapping of potatoes. In: Robert PC, Rust RH, Larson WE (Eds.), Proceedings of the Second International Conference on Site-Specific Management for Agricultural Systems, ASA, Madison, WI, USA, 59–69.

Reckleben Y, Rademacher J, 2004. Sensorgestützte Düngestrategien zur Qualitätssteigerung (Sensor-based strategies for improvement of quality production) In: Maidl FX, Schmid A, Schmidhalter U, Mistele B (Eds.),

Workshop Precision Farming, June 02–03, Weihenstephan, Germany, 15–16.

Reusch S, 1997. Entwicklung eines reflexionsoptischen Sensors zur Erfassung der Stickstoffversorgung landwirtschaftlicher Kulturpflanzen. PhD thesis, Christian-Albrecht-Universität Kiel.

Reusch S, 2003. Optimisation of oblique-view remote measurement of crop N-uptake under changing irradiance conditions. In: Stafford J, Werner A (Eds.), Precision agriculture: Papers from the 4th European Conference on Precision Agriculture. Berlin, Germany. Wageningen Academic Publishers, 573–578.

Robert PC, 2001. Precision Agriculture: A challenge for crop nutrition management. In: Horst WJ, Schenk MK, Bürkert A, Claassen N, Flessa H, Frommer WB, Goldbach H, Olfs HW, Römheld V, Sattelmacher B, Schmidhalter U, Schubert S, v. Wirén N, Wittenmayer L. (Eds.), Plant Nutrition – Food Security and Sustainability. Springer, New York, USA, pp. 692–693.

Rothmund M, 2006. Technische Umsetzung einer Gewannebewirtschaftung als "Virtuelle Flurbereinigung" mit ihren ökonomischen und ökologischen Potenzialen. PhD thesis. Forschungsbericht Agrartechnik 441, VDI-MEG, ISSN 0931–6264.

Rothmund M, Auernhammer H, Demmel M, 2002a. First results of Transborder-Farming in Zeilitzheim (Bavaria). In: Proceedings of the EurAgEng Conference 2002, Budapest (Hungary), Paper Number 02-RD-006.

Rothmund M, Demmel M, Auernhammer H, 2002b. Nutzung von Informationen aus der automatischen Prozessdatenerfassung (Use of information from automatic process data acquisition.) Landtechnik 57/3, 48–149, Münster, Germany.

Rothmund M, Demmel M, Auernhammer H, 2003a. Data management for transborder farming. In: Precision Agriculture, European Conference on Precision Agriculture, ECPA 2003, Berlin, Germany, ISBN 9076998213, 597–602.

Rothmund M, Demmel M, Auernhammer H, 2003b. Methods and services of data processing for data logged by automatic process data acquisition systems. In: Implications of N Fertiliser Applications, Not Only the Effect on Kernel. XXX CIOSTA–CIGR V Congress Proceedings, Vol. 2, Turin, Italy, 713–721.

Schächtl J, 2004. Sensorgestützte Bonitur von Aufwuchs und Stickstoffversorgung bei Weizen- und Kartoffelbeständen. PhD thesis, TU München, Shaker Verlag, ISBN 3-8322-3185-4.

Schächtl J, Maidl FX, 2002. Spectroscopic measurements of potato canopies. Proceedings European Society for Agronomy, 711–712.

Schächtl J, Maidl FX, Huber G, Sticksel E, Schulz J, Haschberger P, 2005. Laser-induced fluorescence measurements for detecting the nitrogen status of wheat canopies (*Triticum aestivum* L.). Precision Agriculture 6, 143–156.

Schepers AR, Shanahan JF, Liebig MA, Schepers JS, Johnson SH, Luchiari A, 2004. Appropriateness of management zones for characterizing spatial variability of soil properties and irrigated corn yields across years. Agronomy Journal 96, 195–203.

Schmidhalter U, 2001. Geophysikalische Kartierung von Bodeneigenschaften für die teilflächenspezifische Bewirtschaftung. Landtechnik 6, 417.

Schmidhalter U, Bredemeier C, Geesing D, Mistele B, Selige T, Jungert S, 2006. Precision agriculture: Spatial and temporal variability of soil water, soil nitrogen and plant crop response. Fragmenta Agronomica 11, Part III, 97–106.

Schmidhalter U, Bredemeier C, Jungert S, Blesse D, 2004. Lasersensorik zur Erfassung des Stickstoffstatus und der Biomasse von Pflanzen ATB, Bornimer Agrartechnische Berichte 36, 29–33.

Schmidhalter U, Glas J, Heigl R, Manhart R, Wiesent S, Gutser R, Neudecker E, 2001a. Application and testing of a crop scanning instrument – Field experiments with reduced crop width, tall maize plants and monitoring of cereal yield. Third European Conference on Precision Agriculture, Montpellier, 953–958.

Schmidhalter U, Jungert S, Bredemeier C, Gutser R, Manhart R, Mistele B, Gerl G, 2003. Field-scale validation of a tractor based multispectral crop scanner to determine biomass and nitrogen uptake of winter wheat. Proceedings of the Fourth European Conference of Precision Agriculture, Berlin, 615–619.

Schmidhalter U, Oertli JJ, 1991. Transpiration/biomass ratio for carrots as affected by salinity, nutrient supply and soil aeration. Plant and Soil 135, 125–132.

Schmidhalter U, Zintel A, Neudecker E, 2001b. Calibration of electromagnetic induction measurements to survey the spatial variability of soils. Third European Conference on Precision Agriculture, Montpellier, 479–484.

Schmittmann O, Osman AM, Kromer KH, 2000. Durchsatzmessung bei Feldhäckslern. Landtechnik 55(4), 286–287.

Searcy SW, Schueller JK, Bae YH, Borgelt SC, Stout BA, 1989. Mapping of spatially-variable yield during grain combining. Transactions of the ASAE 32(3), 826–829.

Selige T, Böhner J, Schmidhalter U, 2006. High resolution topsoil mapping using hyper spectral image and field data in multivariate regression modelling procedures. Geoderma 136, 235–244.

Selige T, Schmidhalter U, 2001. Characterizing soils for plant available water capacity and yield potential uring airborne remote sensing. In: Tupper G (Ed.), Proceedings of the Australian Geospatial Information and Agriculture Conference Incorporating Precision Agriculture in

Australasia 5th Annual Symposium NSW Agriculture, Orange, NSW, Australia, 308–314.

Shaffer JA, Jung GA, Shenk JS, Abrams SM, 1990. Estimation of botanical composition in Alfalfa/Ryegrass mixtures by Near Infrared Spectroscopy. Agronomy Journal, 82, 669–673.

Sommer M, Wehrhan M, Zipprich M, Weller U, 2007. Assessment of soil landscape variability. In: Schröder P, Pfadenhauer J, Munch JC (Eds.), Perspectives for agroecosystem management – balancing environmental and socioeconomic demands, pp. 351–373.

Sorensen LK, Dalsgaard S, 2005. Determination of clay and other soil properties by near infrared spectroscopy. Soil Science Society of America Journal 69, 159–167.

Steinmayr T, 2002. Fehleranalyse und Fehlerkorrektur bei der lokalen Ertragsermittlung im Mähdrescher zur Ableitung eines standardisierten Algorithmus für die Ertragskartierung. PhD thesis. TU München, http://mediatum.ub.tum.de, 227 pp.

Steinmayr T, Auernhammer H, Mauer W, 2000. First evaluation of the applicability of infra-red-tracking-systems for examining the accuracy of DGPS in field works. In: Proceedings of 28th International Symposium on Agricultural Engineering. Opatija (Croatia), 01.-04.02.2000, 53–61.

Stempfhuber W, 2001. The integration of kinematic measuring sensors for precision farming calibration. Third International Symposium on Mobile Mapping Technology, FIG Proceedings.

Stenberg B, Nordkvist E, Salomonsson L, 1995. Use of near infrared reflectance spectra of soils for objective selection of samples. Soil Science 159, 109–114.

Stevenson FC, Knight JD, Kessel C, 1995. Dinitrogen fixation in pea – Controls at the landscape-scale and microscale. Soil Science Society of America Journal 59, 1601–1611.

Sticksel E, Maidl FX, Schächtl J, Huber G, Schulz J, 2001. Laser-induced chlorophyll-fluorescence – A tool for online detecting nitrogen status in crop stands. 3. European Conference on Precision Agriculture, 959–964.

Sticksel E, Schächtl J, Huber G, Liebler J, Maidl FX, 2004. Diurnal variation in hyperspectral vegetation indices related to winter wheat biomass formation. Precision Agriculture 5, 509–520.

Taylor RK, Kastens DL, Kastens TL, 2000. Creating yield maps from yield monitor data using multi-purpose grid mapping (MPGM). In: Robert PC (Ed.), Proceedings of the Fifth International Conference on Precision Agriculture and Other Precision Resources Management, Bloomington/Minneapolis, MN, July 16–20.

Thylen L, Algerbo PA, Giebel A, 2000. An expert filter removing erroneous yield data. In: Robert PC (Ed.), Proceedings of the Fifth International Conference on Precision Agriculture and Other Precision Resources Management. Bloomington/Minneapolis, MN, July 16–20.

Tisseyre B, Mazzoni C, Ardoin N, Clipet C, 2001. Yield and harvest quality measurement in precision viticulture – Application for selective vintage. In: Grenier G, Blackmore S (Eds.), Third European Conference on Precision Agriculture, Montpelier, 133–138.

Tottman DR, 1987. The decimal code for the growth stage of cereals, with illustrations. Annals Applied Biology 110, 441–454.

Van Meirvenne M, Hofman G, Demyttenaere P, 1990. Spatial variability of N fertilizer application and wheat yield. Fertilizer Research 23, 15–23.

Vansichen R, De Baerdemaeker J, 1993. A measurement technique for yield mapping of corn silage. Journal of Agricultural Engineering Research 55(1), 1–10.

Wachendorf M, Ingwersen B, Taube F, 1999. Prediction of the clover content of red clover- and white clover-grass mixtures by near-infrared reflectance spectroscopy. Grass Forage Science 54, 87–90.

Wagner B, Gutser R, Schmidhalter U, 2001. NIR-spectroscopy to estimate soil nitrogen supply. Fourteenth International Plant Nutrition Colloquium, Hannover. In: Horst WJ, Schenk MK, Bürkert A, Claassen N, Flessa H, Frommer WB, Goldbach H, Olfs HW, Römheld V, Sattelmacher B, Schmidhalter U, Schubert S, v. Wirén N, Wittenmayer L. (Eds.), Kluwer Academic Publishers, Dordrecht, Developments in Plant and Soil Sciences, Vol. 92, pp. 752–753.

Walley F, Fu G, van Groenigen JW, van Kessel C, 2001. Short-range spatial variability of nitrogen fixation by field-grown chickpea. Soil Science Society of America Journal 65, 1717–1722.

Walter JD, Hofmann VL, Backer LF, 1996. Site specific sugar beet yield monitoring. Proceedings of the Third International Conference on Precision Agriculture. ASA-CSSA-SSSA, Madison, WI, USA.

Weißbach F, 1995. Über die Schätzung des Beitrags der symbiontischen N_2-Fixierung durch Weißklee zur Stickstoffbilanz von Grünlandflächen. Landbauforschung Völkenrode 45, 67–74.

Welle R, Greten W, Rietmann B, Alley S, Sinnaeve G, Dardenne P, 2003. Near-infrared spectroscopy on chopper to measure maize forage quality parameters online. Crop Science 43, 1407–1413.

Welsh JP, Wood GA, Godwin RJ, Taylor JC, Earl R, Blackmore S, Knight SM, 2003. Developing strategies for spatially variable nitrogen application in cereals, Part II: Wheat. Biosystems Engineering 84, 495–511.

Wild K, Ruhland S, Haedicke S, 2004. A conveyor belt based system for local yield measurement in a mower conditioner. Proceedings of AgEng '94, Leuven.

Part III

Influence of Land Use Changes on the
Biotic Environment

Chapter 3.1

Effects of Land Use Changes on the Plant Species Diversity in Agricultural Ecosystems

H. Albrecht, G. Anderlik-Wesinger, N. Kühn, A. Mattheis and J. Pfadenhauer

3.1.1 Introduction

During recent decades many plant species characteristic for agricultural landscapes have suffered a severe decline. Reports on the decrease of populations and habitats include arable land (e.g. Albrecht, 1995; Andreasen et al., 1996; Sutcliffe and Kay, 2000; Robinson and Southerland, 2002), grassland (e.g. Meisel and Hübschmann, 1976; Dierschke and Wittig, 1991; Willems, 1990; Schrautzer and Wiebe, 1993; Hentschel, 2001) and boundary structures (e.g. Baudry and Burel, 1984; Dowdeswell, 1987; Chapman and Sheail, 1994; Knauer, 1995; Steidl and Ringler, 1997; Boutin and Jobin, 1998). In Germany, intensification of agricultural exploitation and abandoning farmland are considered the main reasons why 455 out of 819 vascular plants are listed in the national

Perspectives for Agroecosystem Management
Edited by P. Schröder, J. Pfadenhauer and J.C. Munch

Red Data Book of endangered species (Korneck et al., 1998). To reduce these losses and to maintain species diversity in agricultural landscapes, several programmes such as a field margin strip programme, a programme to reduce the intensive use of grassland, and another one to support the establishment of boundary structures have been established. Such programmes are – in all European countries – based on EC decree 2078/92 (European Communities, 1992). Experience gained up to now suggests that these programmes can be quite successful in protecting biotic resources (Marggraf, 2003). However, if the corresponding guidelines directly affect management practices, efficiency seems to depend strongly on a regular control and a continuous payment of compensations (Pfadenhauer and Ganzert, 1992; Mattheis and Otte, 1994; Frieben, 1995; Wicke, 1998).

Therefore, the *FAM Munich Research Network on Agroecosystems* was founded as a model for sustainable land use which integrates resource protection and nature conservation with agricultural land use. Results analysed up to now have shown that the measures taken in the FAM research area have been very successful in protecting non-biotic resources (Auerswald et al., 1996) and in conserving soil and aboveground animals (Agricola et al., 1996; Lang and Barthel, 2008; Schröder, 2008). The objective of this study is to evaluate to what extent changes in landscape feature and management affect plant species diversity.

3.1.2 Land use systems practised in the research area

The FAM project investigates changes in agroecosystems induced by redesigning landscape and introducing new management systems on the 110 ha of farmland in the Tertiary Hills in southern Bavaria (Schröder et al., 2008). The characteristic features of this landscape are a marked difference in relief and considerable gradients in soil properties (Sinowski, 1995).

Before FAM took over, most of the investigated area was managed by the Scheyern Benedictine Abbey agricultural administration. Their main concern was arable farming. Cash crops such as wheat, barley and oil seed rape were the crop species cultivated most frequently. In addition, several hop fields were grown. Grassland was restricted to sites where arable use is impossible. It was leased or stocked with cattle belonging to other farmers. During the initial phase of the FAM project, conventional cropping and the old field patterns were maintained and the conditions of the different ecosystem components were recorded by the FAM scientists. Arable fields were all cultivated in the same way to ascertain comparability between the different sites. That means, in 1991 all plots were sown with winter wheat and in 1992 with summer barley. After harvest in autumn 1992, the second and main phase of the project was started. In this phase, the protection of biotic and non-biotic resources should be integrated into the land use by minimising soil erosion, leaching of xenobiotica, and nutrients as well as by increasing structural and species diversity. For this purpose the area was re-designed (Schröder et al., 2008; Figures 3 and 4 of Chapter 1.1). Large fields were divided by different types of boundary structures: fallow stretches as well as planted and dead wood hedges. Existing but very narrow structures were enlarged. Arable fields with a slope of more than 20% as well as erosion troughs and creek flood plains were converted into set-aside or grassland. Set-aside land was left to lie fallow, and boundary structures were mulched in August 1996 due to significant increase in noxious weeds such as *Cirsium arvense* and *Elymus repens* (Anderlik-Wesinger et al., 1998).

By establishment of both farming systems the percentage of unused land was extended to 28.5% in the integrated and to 13.5% in the organic management area. In addition, the number of fields increased and their average size was reduced. More detailed information on the changes in field cultivation and grassland management is given by Reents et al. (2008).

3.1.3 Vegetation analysis

To record the spatial distribution of plant species a 50×50 m grid of reference points was fixed throughout the experimental area of 110 ha. At each of the 412 grid points, a 10×10 m area was established to monitor the development of vegetation. In boundary structures 20 permanent plots ranging from 20 to 100 m^2, varying in proportion to the width of the structure, were established independently from the grid. In this type of ecosystem, only sites with herbaceous vegetation were recorded.

In a time series of 6 years from 1992 to 1997, two relevés were carried out per year in accordance with the Zurich-Montpellier School method (Braun-Blanquet, 1964) estimating the cover abundance of each individual species. Relevés in areas regularly used were performed before important agricultural measures like weed control or harvest took place. In fallow land and boundary structures these investigations were carried out in May and August. For data analysis, the higher of the two cover abundances was used. In fallow land and in the arable fields, vegetation analysis was continued until 2000 and 2002, respectively.

Vegetation on the farm includes species characteristic of a certain habitat and the ubiquitous generalists. As the characteristic species are much more affected by changes in land use and intensification, they must be considered to be more important for nature conservation impact. Thus, the development of typical arable weeds (order of *Stellarietea mediae*; Hofmeister and Garve, 1998), grassland species (*Molinio-Arrhenatheretea;* Dierschke and Briemle, 2002) and plants of herbaceous hedgerowsand forest fringes (*Trifolio-Geranietea sanguinei*; Müller, 1977) was analysed separately from species occurring more widely. Nomenclature of species follows Wisskirchen and Haeupler (1998).

As the background populations from which the samples were drawn cannot be assumed to be normally distributed, the median, the 95% interval of confidence, the Wilcoxon matched pairs signed rank test, the sign test by Dixon and Mood and the Friedman test, which all are non-parametric methods, were used to describe vegetation changes (Kent and Coker, 1992; Sokal and Rohlf, 1998). These analyses were computed using SPSS 11.5 statistical package for Windows (Anonymous, 2002). Changes in species composition were described with the detrended correspondence analysis (DCA) ordination technique (Hill and Gauch, 1980) using the PC-ORD4 computer program (McCune and Mefford, 1999).

3.1.4 Results

Numbers of plant species

Table 1 shows the change in the total number of plant species per grid point during the first 5 years after land use change. It significantly increased in pastures and meadows, in newly seeded grassland, in fields of organic farming, and in fallows adjacent to or surrounded by arable fields. Only in fallow grassland, in boundary structures, and in fields under integrated farming it remained unchanged.

The 'characteristic' species which are typical for certain ecosystems significantly increased in pastures and meadows, in newly seeded grassland, in arable fields under organic farming and in boundary structures which were enlarged. In the newly established boundary strips, increases remained just below the significance level. Only in unchanged boundary structures and in fields of the integrated system, where herbicides continued to be applied after the change in land use, did the number of characteristic species remain constant or decrease. It was not possible to analyse changes of 'characteristic' species on fallow land where the spectrum of species distinctly alters when arable fields or grassland are abandoned.

In the fields under organic farming the development of species abundance was observed over a 10 years period following the conversion (Figure 1). Changes could not be tested statistically because a varying percentage of grid points was sown with cover crops each year. There, the closed canopy made it impossible to record accurately the few weed plants growing in the understorey. In other crops, however, the conversion clearly increased the number of species above the initial level of 16 per 100 m². The highest number of species was 34.5 which occurred in the 5th year of organic management. Thereafter, the values dropped back and ranged between 23 and 28 from the 7th to the 10th year. In the fields with integrated exploitation and

Table 1 Development in the median number of plant species per grid point 5 years after establishing new land use (1992–1997).

	Arable fields, organic	Arable fields, integrated	Set-aside fields	Set-aside grassland	Seeded grassland	Pastures, organic	Meadows, organic	Boundary structures, new	Boundary structures, enlarged	Boundary structures, unchanged	Total
Number of samples	129	129	26	12	29	42	25	6	8	6	412
Number of species 92 (Median)	17	19	17	20	19	25.5	28	20	24	35	20
Number of species 97 (Median)	32	19	39.5	23.5	28	32	40	30	31	32	30
Significance of change	+***	n.s.	+***	n.s.	+***	+***	+***	n.s.	n.s.	n.s.	+***
Characteristic species 92 (Median)	11	13			1	13.5	16	0	1	3	
Characteristic species 97 (Median)	18	10			16	18	26	1	2	2	
Significance of change	+***	–***			+***	+***	+***	n.s.	+=	n.s.	
Characteristic species 92 (total)	56	48				49	73				
Characteristic species 97 (total)	58	46				51	74				
Other species 92 (total)	49	57				34	58				
Other species 97 (total)	94	95				83	83				

Significance according to the sign test by Dixon and Mood.

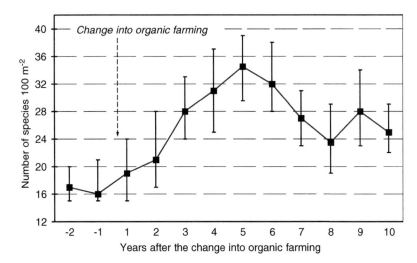

Figure 1 Development of the number of plant species per grid point in the course of 10 years after conversion from conventional into organic farming (median values and quartiles). Grid points sown with cover crops are not included in the calculation (Redrawn from Belde et al., 2003).

reduced tillage, species numbers increased from 17 to 22.5 per 100 m^2 within the first 5 years. Until the 10th year, however, it fell below the initial level to only 13 species per 100 m^2.

In fallow land, the number of species significantly increased from 17 under arable farming to a level of 35 to 40 species from the 2nd to the 6th year of abandonment (Figure 2). Until the 8th year, however, it declined again to 25 species per 100 m^2. A complete vegetation cover ($\geq$95%) was attained in the 4th year of succession and in the 5th year the cover of litter increased from 20 to 50%.

During the first 4 years after the change all linear structures were left to an undisturbed development. In the initial phase of this time period, the number of species significantly increased in both the newly established and the enlarged structures (Figure 3). In structures which still existed before the landscape was re-designed it remained constant. During the 3rd and 4th year after abandoning management, however, the numbers of species re-declined in all three types of boundary structures. This development went along with an increasing percentage cover of geophyte plants such as *Cirsium arvense* and *Elymus repens*. To stop this tendency, mulching was started at the end of the 4th year. Already in the 5th year first effects were observed in increasing numbers of species and a slightly decreasing geophyte cover.

In Figure 4 the change in the median number of plant species per grid point is compared to the development of the sum of plant species found at

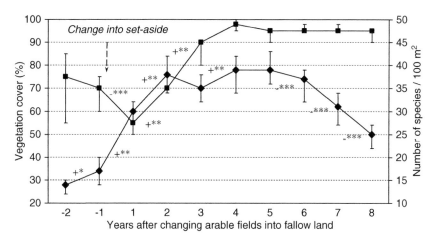

Figure 2 Development of vegetation cover (■) and number of plant species
(♦) per grid point in the course of 8 years after establishing fallow land (median values with 95% intervals of confidence). The
Friedman test showed that species numbers generally differed
($p < 0.001$), the Wilcoxon test was used to test changes from year
to year.

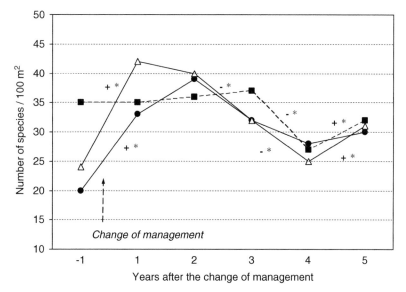

Figure 3 Development of a number of plant species in the course of 5
years after establishing (Δ), enlarging (●) as well as unchanged
(■) boundary structures (medians with 95% intervals of confidence). The Friedman test showed that values generally differed
($p < 0.001$); the Wilcoxon test was used to test the changes from
year to year.

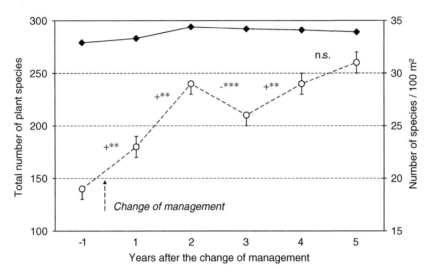

Figure 4 Development of the median number of species per grid point (○)
and the total number of species at 412 grid points in the whole
research area (◆) in the course of 5 years after management conversion. $I = 5\%$ confidence interval of the medians. The
Friedman test showed that grid point values generally differed
($p < 0.001$); the Wilcoxon test was used to test the changes from
year to year.

the total number of relevé points in all vegetation types. On grid point
scale, the number of species increased continuously from initially 19 to 31
in the 5th year after the change of land use. Only in the 3rd year (1995)
which had an unusually dry vegetation period, a decrease was observed.
In contrast, the increase of the total number of plant species from 279 in
the preliminary phase to 289 in 1997 is marginal.

Marginal changes occurred in the number of species characteristic for
different types of agro-ecosystems if they are regarded not on the grid
point but on the farm scale. Thus, the total number of typical grassland
species found at all 42 sampling points in the organically managed pastures increased only slightly from 49 to 51 (data not shown). The corresponding values for the meadows were 71 and 76, respectively. In contrast,
generalists which are not restricted to the specific living conditions in
grassland and which can also survive outside this type of ecosystem
increased from 34 to 83 in the pastures and from 58 to 83 in the meadows,
respectively. Similar results were also obtained for the arable fields.
Under organic farming, the total number of the obligatory arable weeds
changed from 56 to 58 while the number of ubiquitous species almost dou-
bled from 49 to 94. In integrated farming the characteristic weeds even
decreased from 48 to 46 while the 'other' species increased from 57 to 95.

Species composition

The changes in species composition are demonstrated by a DCA ordination of all 2783 vegetation relevés. The mean values of the relevé coordinates for each year and each ecosystem type are given in Figure 5. It can be seen that these ecosystem types clearly differ in their degree of change. Also, where land use altered only slightly the changes were moderate. This applies to meadows, pastures, established boundary structures and arable fields. In the ecosystem types where changes in management practice were more substantial, species composition rapidly began to shift to a new equilibrium. This was the case for fallow fields, for abandoned and newly sown grassland as well as for new and enlarged boundary structures.

Individual species

Changes in the cover abundance of the more frequent species during the study period are shown in Table 2. In fields under integrated and organic farming, in fallow fields, and in sown grassland the number of

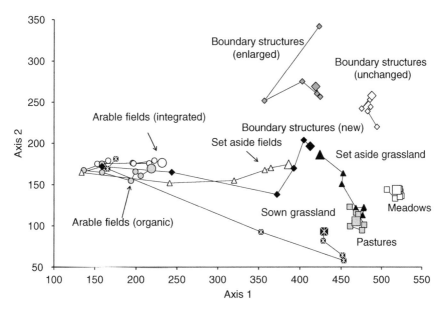

Figure 5 Development of species composition calculated using the mean values of the relevé coordinates from a DCA ordination. Every symbol represents one vegetation period in the development of a certain ecosystem type from 1 year before until 5 years after the change of land use. The last year is emphasised by particularly large symbols. An after the fact evaluation resulted in a coefficient of determination (r^2) of 0.156 for axis 1 and 0.059 for axis 2, respectively.

Table 2 Changes in cover abundance of single plant species 5 years after the change of land use.

					Changes in cover abundance						
Plant community	Arable fields, organic	Arable fields, integrated	Set-aside fields	Set-aside grassland	Sown grassland	Pastures, organic	Meadows, organic	Boundary structures, unchanged	Boundary structures, enlarged	Boundary structures, new	Total
P Anagallis arvensis	n.s.	−***	−**								−**
P Anthemis arvensis	+***	n.s.	−***		n.s.				n.s.		+*
P Apera spica - venti	+***	+***	−**		−***						+***
P Aphanes arvensis	+***	−***	n.s.		−**						n.s.
P Atriplex patula	+***	n.s.	+*		+**						+***
P Bromus hordeaceus			+*	n.s.	+**	+*	n.s.				+***
P Capsella bursa - pastoris	+***	−***	−***		n.s.	n.s.	n.s.		n.s.	n.s.	−
P Cerastium glomeratum		n.s.			+***	+**	n.s.				+***
P Chenopodium album	+***	n.s.	−**		−**						+*
P Chenopodium polyspermum	+**	n.s.	n.s.		−***						n.s.
P Fallopia convolvulus	+***	−*	−**		−**						n.s.
P Geranium dissectum	+***	n.s.	n.s.		n.s.	n.s.	n.s.				n.s.

	1	2	3	4	5	6	7
P *Lamium amplexicaule*	n.s.					n.s.	–***
P *Lamium purpureum*	+***	–***	–*	–**		n.s.	–***
P *Lollum multiflorum*	+***	+***	+*	+**	n.s.		+***
P *Matricaria recutita*	+***	–***	–***	–*		n.s.	n.s.
P *Myosotis arvensis*	+***	–***	–**	–*		n.s.	n.s.
P *Sonchus asper*	+***	+***		+***			+***
P *Stellaria media*	+***	–***	n.s.	n.s.		n.s.	n.s.
P *Thlaspi arvense*	+**	–***	–*	–*		n.s.	–***
P *Veronica arvensis*	+***	–***	–**	+*	n.s.		n.s.
P *Veronica hederifolia*	n.s.	–***		–***			–**
P *Veronica persica*	+***	–**	+**	+**	n.s.	n.s.	+***
P *Vicia hirsuta*	+***	+**	+*	n.s.		–*	+***
P *Viola arvensis*	n.s.	–***	–***	–***			–***
R *Cirsium arvense*	+***	+***	+***	n.s.		n.s.	+***
R *Convolvlus arvensis*	–*	n.s.				n.s.	n.s.
R *Daucus carota*	+*	n.s.					+**
R *Elymus repens*	n.s.	+***	+**	–**	+*	+*	+*
R *Epilobium parviflorum*		+**					+***
R *Equisetum arvense*	+*	n.s.	n.s.			n.s.	+***
R *Galium aparine*	+***	–*	–***	–***		n.s.	+*
R *Geum urbanum*	–*	+**	+*				+***
R *Glechoma hederacea*	+**			+*	n.s.		+*

Continued

Table 2 (*Continued*)

					Changes in cover abundance						
Plant community	Arable fields, organic	Arable fields, integrated	Set-aside fields	Set-aside grassland	Sown grassland	Pastures, organic	Meadows, organic	Boundary structures, unchanged	Boundary structures, enlarged	Boundary structures, new	Total
R *Lapsana communis*	+***	−***	−**						n.s.		+*
R *Medicago sativa*	+***		+**			+**	+*				+***
R *Poa angustifolia*			+*			+**	+*				+***
R *Torilis japonica*		n.s.	+**	+*		n.s.	n.s.	n.s.	n.s.		n.s.
R *Urtica dioica*			+***	+*		+**	n.s.		n.s.		+***
G *Achillea millefolium*	+**		+		+**	+	n.s.				+***
G *Agrostis stolonifera*		+***	+***	n.s.		−***	+**		n.s.		n.s.
G *Agrostis tenuis*			+***		+**	n.s.	+*		+*		+**
G *Alchemilla vulgaris*						n.s.	n.s.				n.s.
G *Alopecurus pratensis*		+*	+**	n.s.	+***	+*	n.s.	+*	n.s.		+***
G *Anthoxanthum odoratum*			+*		+**						+***
G *Anthriscus sylvestris*						n.s.	n.s.	n.s.			n.s.
G *Arrhenatherum elatius*	+***	+**	+***		+**	n.s.	+*		+*		+***
G *Bellis perennis*				−*	+***	+***	n.s.				+***
G *Cardamine pratensis*						n.s.	n.s.				n.s.

Species								
G Carum carvi	+***		+*				+***	+***
G Cerastium holosteoides	+*	n.s.	−*	n.s.		n.s.	n.s.	+***
G Crepis biennis	+***		+***	+**		+***		+***
G Crepis capillaris	+**		+**					+**
G Cynosurus cristatus				n.s.		n.s.	n.s.	n.s.
G Dactylis glomerata	+***	+***	+***	+***		+*	+*	+***
G Festuca arundinacea	+**		+**	+*		n.s.		+**
G Festuca pratensis	+*	n.s.	+***	n.s.		n.s.	n.s.	+***
G Festuca rubra			+*			n.s.	n.s.	+***
G Galium mollugo				+**	+*	n.s.		+***
G Heracleum sphondylium		+*	n.s.	n.s.		n.s.		+**
G Holcus lanatus	+*	n.s.	+*	+**		+**		+***
G Lathyrus pratensis				n.s.		n.s.		n.s.
G Leontodon autumnalis				n.s.		n.s.		n.s.
G Lolium perenne	+***	n.s.	+***	n.s.		n.s.		+***
G Lotus corniculatus	+***		+*					+***
G Phleum pratense	+***	n.s.	+***	+**		n.s.		+***
G Plantago lanceolata	n.s.	+*	+***	+*		n.s.		+***
G Poa pratensis	n.s.	n.s.	+***	n.s.		n.s.	+*	+***
G Poa trivialis	+***	n.s.	+***	+**		+***	n.s.	+***
G Ranunculus acris		n.s.	+**	+***		n.s.		+***

Continued

Table 2 (*Continued*)

	Changes in cover abundance										
Plant community	Arable fields, organic	Arable fields, integrated	Set-aside fields	Set-aside grassland	Sown grassland	Pastures, organic	Meadows, organic	Boundary structures, unchanged	Boundary structures, enlarged	Boundary structures, new	Total
G *Ranunculus repens*	+***	n.s.	+*	n.s.	+**	n.s.	+**				+***
G *Rumex acetosa*			+**	+**	+**	+***	n.s.				+***
G *Rumex obtusifolius*	+***	+***	+**	n.s.	+***	n.s.	n.s.				+***
G *Stellaria graminea*						n.s.		n.s.			+*
G *Taraxacum officinale*	+***	+***	+***	n.s.	+***	+***	+**	n.s.	n.s.	+*	+***
G *Trifolium dubium*							+*				+***
G *Trifolium pratense*	+***	+*	+**		+***	+*	+*	n.s.		n.s.	+***
G *Trifolium repens*	+***	+***	+***	−*	+***	+***	+**		n.s.	n.s.	+***
G *Trisetum flavescens*			+**		+***	n.s.	+*	n.s.			+***
G *Veronica serpyllifolia*		n.s.				n.s.	+**				+***
O *Betula pendula*	n.s.		+***								+***
O *Brassica napus*	n.s.	−***	−***		−*						−***
O *Epilobium adenocaulon*	+***	+***	+***	+**							+***
O *Epilobium angustifolium*											+***

O *Holcus mollis*		–*							n.s.	n.s.
O *Hypericum perforatum*	n.s.	+**		+**		n.s.		n.s.	n.s.	+***
O *Picea abies*		+**	n.s.	+**	n.s.				n.s.	+*
O *Plantago major*		+**			+**					+*
O *Poa annua*	+***	+***	n.s.	n.s.	+***	n.s.			n.s.	+***
O *Polygonum aviculare*	+***	n.s.	–**		+***	+***				+***
O *Quercus robur*		+*								+**
O *Ranunculus ficaria*		+*				+***		+***		+***
O *Veronica chameadrys*					n.s.	n.s.	+**	n.s.		+***
O *Vicia angustifolia*										+**
Species with increased cover abundance	42	25	44	4	32	17	24	3	4	69
Species remaining constant	9	17	7	15	6	33	25	14	23	18
Species with decreased cover abundance	1	17	16	4	14	2	0	0	1	8

Only species which occurred in at least 6 relevés and which reached a constancy of 40% in one vegetation type are listed. Significance according to the sign test by Dixon and Mood. For the arable fields records made in winter wheat in 1991 were compared to relevés sampled in the same crop in 1995, 1996 or 1997.

P = Pioneer plant community with a high percentage of therophytic species (Class of Stellarietea mediae). R = Ruderal plant communities with a high percentage of geo- and hemicryptophytic species (Class of Artemisietea and Agropyretea). G = Grassland plant communities (Class of Molinio – Arrhenatheretea and Agrostietea stoloniferae). O = Other plant communities.

significant increases exceeded unchanged ones and decreases. In integrated farming system, 13 of the 17 decreasing species were dicotyledonous annuals. In contrast, 20 of the 25 species with increasing cover abundance were perennials. In set-aside fields and in seeded grassland where soil tillage was abandoned, therophytic weeds declined significantly while herbaceous and woody perennials became more frequent.

In meadows and pastures as well as in all types of boundary structures, species remaining constant prevail over those with a significant change. In meadows and pastures predominantly grassland species increased. As the number of recording sites was low in the different types of boundary structures, species had to change in a high percentage of observation plots to reach the significance level. Only few species met this demand. Species with an increasing tendency are predominantly perennial grasses; the only species with a significant decrease was the annual weed *Viola arvensis*. Fallow grassland was the only vegetation type in which a balance between increases and decreases was observed.

During the first 5 years after the change of land use altogether 50 species were newly found in the research area. In Figure 6 these species are classified according to habitat types and the duration of their occurrence. Twenty six of them were observed in only 1 year ('transient species') and 24 taxa were recorded in more than 1 year ('persistent species'). Among the vegetation types, most newcomers were found in arable fields. Here, 26 new species occurred. Most of them were 'transient' annuals and only

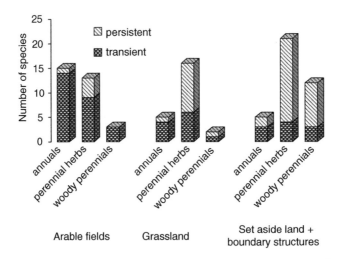

Figure 6 Species of different life forms newly occurring in the predominant habitat types 5 years after the change of land use. Species observed in only 1 year were classified as 'transient' and those which were recorded in more than 1 year as 'persistent'.

Vicia villosa, *Helianthus tuberosus* and *Campanula rapunculoides* were observed in at least 2 years. In grassland, 6 out of 12 new species were classified as 'persistent', but most of them were only observed with a few individuals at one grid point. Thus, it cannot be clarified whether they had previously been overlooked or whether they are genuine new findings. The vegetation type where newcomers established most successfully are undisturbed plots in fallow land and boundary structures. Here, 14 of the total 20 new species persisted for a longer term.

Rare species

Only in the arable fields rare and endangered species occurred frequently enough to evaluate their changes with statistical tests (Albrecht and Mattheis, 1998). Five years after the conversion to integrated farming system with reduced tillage rare species had decreased at 28 grid points, at 4 points their number remained unchanged and at 12 points it increased. According to the Wilcoxon matched pairs signed rank test this decrease is significant ($p < 0.01$). In the organic farming area, rare weed species were observed at 79 grid points. At 21% of these points their cover abundance decreased, at 13% it remained unaffected and at 66% it increased. Thus, increases significantly overweight the decreases in the organic farming system ($p < 0.001$, Wilcoxon test). Among the species occurring frequently enough for a statistical analysis, three did not change significantly while *Legousia speculum-veneris*, *Sherardia arvensis* and *Odontites vernus* were found more often in the fields under organic farming (Table 3). In integrated farming system, *Legousia speculum-veneris* decreased significantly and *Sherardia arvensis* increased.

3.1.5 Discussion

Development of the numbers of species on the grid point scale

After the change of land use and the re-designing of the landscape the total number of species per grid point clearly increased in most agricultural ecosystems investigated on the FAM research station. In fallow land and in the newly established boundary structures, abandoning the cultivation of competitive crops and discontinuing the use of herbicides obviously favoured the establishment of weeds already present in the diaspore pool of the soil. This species spectrum was expanded by taxa having flying devices which facilitate colonisation from outside like *Taraxacum officinale*, *Epilobium ciliatum* and *Cirsium vulgare* (Albrecht et al., 1998). In the arable fields under organic farming, reduced efficacy of mechanical versus herbicide control, the ban on synthetic fertilisers and higher crop diversity (Alfoeldi et al., 2002) may have favoured the development of species-rich weed communities. The increasing number of species in grassland may be

Table 3 Changes of constancy and cover abundance of rare arable weeds in winter cereals 5 years after the change from conventional to integrated and organic farming systems.

	Constancy (%)		Significance	Constancy (%)		Significance
	Conventional	Integrated		Conventional	Organic	
Anagallis minima	3	0				
Anchusa arvensis				2	2	n.s
Centaurea cyanus	3	0		26	23	n.s
Legousia speculum-veneris	13	2	−**	25	32	+***
Myosurus minimus	3	0		19	16	n.s
Odontites vernus				0	8	+***
Papaver argemone				1	0	
Ranunculus arvensis				3	2	
Scleranthus annuus	2	1		2	3	
Sherardia arvensis	2	10	+**	9	14	+**
Silene noctiflora				0	1	
Valerianella rimosa				0	2	
Veronica triphyllos				5	5	n.s

Significance of changes in cover abundance was calculated using the Dixon and Mood sign test. Species were classified to be 'rare' when at least two of the following criteria were fulfilled: (i) listed in the Red Data Book of Germany (Korneck et al., 1996), (ii) listed in the Red Data Book of Bavaria (Scheuerer and Ahlmer, 2003), (iii) listed in the Red Data Book of the investigated landscape (Scheuerer and Ahlmer, 2003), clearly decreasing in Germany during the last decades (Albrecht, 1995).

caused by a more regular grazing after the conversion to organic farming alongside with a very moderate fertilisation. Before the FAM project started grassland was extensively stocked or leased to other farmers and plant communities had a more ruderal character.

During the years following this initial phase, however, the number of species in some of the ecosystems declined again. Thus, in the organic farm fields species numbers fell from 34.5 in the 5th to 25 in the 10th year after the conversion. In arable fallow this value decreased from 39 in the 5th to 25 in the 8th year and in the newly established boundary structures it even dropped from 42 in the 1st to 25 in the 4th year after the change of management. In the different ecosystems this development may have different reasons.

In fallow land and newly established boundary structures, the high amount of light which is essential for both therophytic weeds and small perennials was increasingly absorbed by tall and competitive perennials. In fallows, complete vegetation cover ($\geq 95\%$) was attained within 4 years after the abandonment. Species which particularly profited from this development were *Elymus repens*, *Cirsium arvense* and *Urtica dioica*. They produce vegetative runners able to penetrate even closed litter layers and a dense canopy. In addition, also the percentages of grassland species such as *Arrhenaterum elatius* and of late successional trees and shrubs increased. As the decreases of light requiring species prevail over the increases of competitive perennials, the total number of species per grid point declined. All these results well correspond to the studies of Schmidt (1981), Osbornova et al. (1990), Fisher and Davies (1991) and Schmiedeknecht (1995) who found a corresponding development in species diversity and composition on fallow land.

The decline in the number of species in arable fields which were converted from conventional to organic farming, however, does not meet the expectations from literature. Becker and Hurle (1998), for example, found an average of 34 species in 49 fields which were exploited organically for 1–2 years. In both fields which were under organic farming for 3–8 years ($n = 69$) and for more than 8 years ($n = 46$) the corresponding value was 35 species. The authors concluded that species diversity was not significantly affected by the duration of organic farming practice. In the course of the present study, weed infestation in fields under organic farming was significantly reduced when the percentage of competitive cover crops in the rotation was increased (Albrecht and Sprenger, 2008). This practice has proved to be even more effective than the application of herbicides in the integrated system. And it is presumed to be the major cause for the decline in species numbers.

In terms of species conservation these results mean that the measures which were initiated to increase plant species diversity were quite successful during the initial phase of the project but thereafter they failed in some of the investigated ecosystems. Therefore, if high species diversity is an objective for the agricultural landscape of tomorrow, measures will become necessary to counteract these tendencies.

In fields under organic farming, maintaining the species diversity of the arable weed flora may be reached by the reduction of cover crops. This change would, however, also diminish the supply with forage. As a lack of forage also limits the stocking rate, this can finally affect the whole management concept of a farm. Therefore, diminishing the percentage of cover crops in the rotation would stand for reduced production intensity on the FAM research station which actually is – compared to other organic farms in Germany – on a very high level (Hülsbergen and Küstermann, 2005).

Another opportunity to favour species-rich annual weed communities could be to integrate 1 or 2 years of fallow into the crop rotation. In the present study, the number of species rapidly increased during 2 years of undisturbed set-aside. However, also the populations of noxious weeds can dramatically increase under these conditions (Davies et al., 1997; Squire et al., 2000; Salonen et al., 2001). Therefore, many farmers considering to return fallow land into arable fields try to prevent future weed infestation problems by cultivating cover crops. Discussing these negative effects of cover crops on the species diversity, Turley et al. (1998) proposed to restrict unsown fallow to those sites where rare species occur. The actual farming intensity, however, makes it very difficult to find those sites. Therefore, such actions should not be focused too closely on the known habitats of rare species. Moreover, temporary abandonment could also be used at locations where a high number of species is already present or where their development is intended.

A third opportunity to conserve the species-rich plant communities of the early successional phase could be to plough fallow fields every couple of years. This practice may suppress the dominance of competitive perennials and re-activate the seed banks of arable weeds. However, this practice would increase soil erosion, nutrient leaching and water runoff. As these effects contradict the objectives of soil protection, maintaining species-rich fallow by ploughing should stay restricted to a few areas which are inhabited by a weed flora with a high nature conservation value.

When fallow land remains undisturbed for a longer time, vegetation development strongly depends on the fertility of the substrate (Tilman, 1988; Osbornova et al., 1990; Hansson and Fogelfors, 1998). In regions having fertile and productive soils, secondary succession usually leads to species poor plant communities which are dominated by few common and highly competitive taxa. This type of succession also occurred in most of the investigated area. In terms of plant species conservation, the value of the resulting vegetation is low. Only on very unfertile soils a low productivity and a patchy vegetation structure may facilitate the establishment of plant communities with high conservation value. Such a development was documented by Wurbs and Glemnitz (1997) who investigated secondary succession on pleistocenic sands in the German Federal state of Brandenburg near Berlin. On fields where arable use was abandoned 20–60 years ago, near natural plant communities of open sand habitats were established which are so rare that they are listed in the German Red Data Book of endangered plant communities (Rennwald, 2000).

Presumably the most important function of boundary structures in arable landscapes is the protection of non-biotic resources (e.g. Pollard et al., 1974; Wischmeier and Smith, 1978). Therefore, the aim of conserving high plant species diversity must be carefully balanced with this objective. To attain quick development of the ground cover in newly established boundary structures, sawing is frequently preferred to spontaneous development. There, sawing autochthonous material collected from the region can lead to site-specific plant communities with a high nature conservation value (Anderlik-Wesinger et al., 1999). Another management measure which is highly recommended for the maintenance of herbaceous boundary structures is regular mowing. Although mulching was done only one time after a 4 years period of undisturbed succession, it caused a clear increase in the numbers of species (see Figure 3). Additional advantages of regular mowing are a facilitated access and use of the cultivated land and a reduced infestation of adjacent fields by problematic weeds like *Cirsium arvense* or *Elymus repens* (Mayer and Albrecht, 2008). Furthermore, regular management of boundary structures also favours various animal groups which prefer open agricultural landscapes (Kretschmar et al., 1995).

Development of the number of species on the farm scale

In general, data show that the numbers of species on the scale of the grid points and on the whole farm level react differently to changes in management and landscape design. The increase in the number of species on the grid point scale may be caused by both; diaspores stocked in soil for which emergence conditions were unsuitable under the former management and species which newly colonised the 10×10 m recording areas by dispersal. Thus, small-scale habitat capacities were not fully exploited before management and landscape design were changed. In a similar way, the nearly constant number of species in the whole research area could be interpreted by assuming that all species able to live in the research area are present. This assumption is supported by the fact that a considerable number of species was newly observed but failed to establish permanently. What is contradicting this assumption is the literature. Publications on plants and plant communities of the Tertiary Hills landscape suggest that there are quite a lot of species which are able to grow in the research area but are not found there (e.g. Popp, 1887–1891; Rodi, 1966; Otte, 1984; Ruthsatz, 1985; Ruthsatz and Otte, 1987). Thus, it seems likely that species with a potential to establish in the investigated area occur in the landscape but have no chance to reach the research farm.

The most common natural media which spread seeds in terrestrial ecosystems are wind and animals (Luftensteiner, 1982). Nevertheless, their practical importance in agroecosystems seems to be restricted. The limiting factor for wind dispersal is that it applies only to a small number of species which have specific flying devices. This group includes species from the Asteraceae genus which produce pappi like *Sonchus* spp. and *Cirsium* spp. and light seeded grasses like *Arrhenatherum elatius*. The second way, the transport by animals (zoochory), mostly occurs in the immediate surrounding of the parent plant and plays no important role for a wide-ranging spread (Bonn and Poschlod, 1998). Consequently, most species from agroecosystems possess no natural strategy for far distance dispersal. Thus, their capability to reach new habitats depends above all on human activities (hemerochory).

All strategies of dispersal have in common that the number of dispersed seeds exponentially declines with the distance to the source plant (Willson, 1983). Consequently, the chance to reach a new habitat in this way was severely reduced by increasing field and farm sizes during the last decades. In addition, the chances for a hemerochorous dispersal by harvesting material, machinery, and organic fertilisers decreased with an improved seed purification and an increasing self-sufficiency as regards the plant of the farms.

This lack of dispersal facilities does not explain why a considerable number of species newly appeared but were not found again during the investigation period. Of course, one reason is that the wrong species grew in the wrong vegetation type. For example, perennial shrubs cannot be expected to develop successfully in arable fields. However, there are some species recorded in only 1 year which must be presumed to be adapted to the living conditions in the target habitats. This may – for instance – apply to the annual weeds which were newly found. One reason why these species did not occur repeatedly may be a close adaptation to certain crops. Thus, *Abutilon theophrasti*, *Amaranthus retroflexus*, *Digitaria sanguinalis* and *Panicum capillare* are typical maize weeds (Hofmeister and Garve, 1998) because they germinate – in common with maize – at warm temperatures. Consequently, they only occurred in integrated farming when maize was grown. As maize occupies only 25% of the crop rotation there, favourable conditions for their development are met only every 4th year. As many arable weeds including *A. retroflexus* and *D. sanguinalis* can develop a persistent seed bank (Thompson et al., 1997), they may belong to the 'species pool' of the corresponding sites (Zobel et al., 1998) without being recovered in the actual vegetation.

Development of rare arable weeds

In the study area, rare species were predominantly found in arable fields. In the part of the research area where the integrated farming system was established, the number of rare weed species significantly declined. The reason for this change may be a modification in soil cultivation: while arable soils were regularly ploughed during the first phase of the project, minimum tillage has been practised since changing to integrated farming (Reents et al., 2008). Such tillage causes an accumulation of seeds in the upper soil and a decline at depths of >10 cm (see Albrecht and Sprenger, 2008). In consequence, a high proportion of seeds find good germination and establishment conditions on the soil surface (Cousens and Mortimer, 1995; Grundy et al., 1996). Thus, the median number of plants in the integrated farming area before herbicide application in cereals increased from 48 to 104 per m^2 (Albrecht and Sprenger, 2008). To suppress this increasing crop competition, weed control has had to become more efficient. This objective was achieved by splitting herbicide applications into several treatments with highly efficient compounds. The suppression of total weed cover illustrates the success of this technique; neither before nor after changing from conventional to integrated farming did weed cover exceed 5% (Albrecht and Sprenger, 2008). Consequently, the reduced soil tillage is presumably the main reason for the decrease of rare weed species in the integrated farming system.

At the beginning of the FAM project the investigated fields were largely and regularly ploughed. As the research site is characterised by marked differences in relief, enormous quantities of soil were lost by erosion. Concerning rare weed species, the study area is situated in a landscape which is poor in such plants. Consequently, the conservation of species was subordinated to the protection of the non-biotic resources and soil erosion was given highest priority for acting (Anderlik-Wesinger et al., 1995). A few years after introduction of reduced tillage and other soil protection measures erosion problems have diminished to a minimum (Auerswald et al., 1996). The fact that frequencies of rare weed species decreased at the same time is a common problem in nature conservation: Different environmental goals often make contradicting demands to nature management.

In contrast to the integrated farming system, the organic farming system showed a significant increase in the number of rare species per grid point. This result well agrees with Callauch (1981), Plakolm (1989), van Elsen (1989), Wolff-Straub (1989) and Frieben (1990) who compared numbers of endangered species in organic and conventional fields and found considerably higher numbers under organic farming practice. The authors concluded that organic farming has a beneficial influence on endangered weeds and they recommend this type of management to prevent further decline in rare weeds. The results of the present study suggest that such positive effects occurred also in the organic farming system on the FAM research station.

These tendencies found for rare species seem to apply to most of the individual species, too. In organic farming, three among six species occurring frequently enough for a statistical analysis increased and three did not show a significant change. In integrated farming with reduced tillage, one species increased and one decreased. Why *Sherardia arvensis* increased against the general decrease of rare species in fields with reduced tillage may depend on specific requirements of this taxon. Under Central European climate conditions, *S. arvensis* can germinate over the whole vegetation period but the seeds predominantly ripen in late summer and early autumn (Schneider et al., 1994). Thus, reduced ploughing during this time period may have favoured this significant increase. Remarkable was also the observation that some of the species in the organic farming area such as *Veronica triphyllos* did not increase despite their improved living conditions. It is suggested that this development was significantly influenced by the life strategies of the individual species. How features like population cycles or dispersal capacities complement each other to such species-specific life strategies and how they finally determine the long-term success of species in the cultural landscape is discussed by Mayer and Albrecht (2008).

Changes in the species composition

Different types and intensities of cultivation 'filter' species not adapted to such environmental influences (Keddy, 1992). The most important types of cultivation in agricultural ecosystems are regular tillage in arable fields, mowing and grazing in meadows and pastures, and abandoning management in fallow land and boundary structures. In the DCA ordination in Figure 5 these different living conditions form a triangle which reflects the composition of the vegetation. As axis 1 explains most of the interspecies variation and the arable field points are far away from the other two vegetation types, it may be supposed that boundary structures and grassland vegetation are more similar to each other than to the arable weed communities.

The greatest modifications in species composition were observed at those grid points where land use was changed. On arable fields which were turned into fallow land or in boundary structures, vegetation composition rapidly developed in the direction of the established boundary structures. The rate of changes decreased year by year, which corresponds to common succession models (e.g. Bornkamm, 1981; Osbornova et al., 1990). In newly established boundary structures this tendency ceased in the 4th year. In this year the stands were mulched and resemblance to fallow grassland increased. An extremely rapid change occurred from arable land to pasture like community in the new seeded grassland area during the first 2 years after sowing. Thereafter, the rate of change suddenly slowed down and a slight tendency occurred to develop back in the direction of the arable weed communities. The reason for this development might be a continuous re-establishment of arable weeds from the seed bank in the patches where the sward was disturbed by cattle trampling or grass harvesting machinery. Unpublished soil seed bank analyses show that even 5 years after sowing grassland, former arable fields contain between 2000 and 5000 arable weed seeds per m^2. In general, it can be said that species composition at sites where land use was changed shifts between the edges of the triangle built by the main ecosystem types (arable fields, grassland and boundary structures) and contains varying percentages of corresponding species. Thus, the ordination plot clearly shows the transition character of those biotopes.

Less pronounced changes in species composition were observed where the type of land use was only modified. After changes to both the integrated and the organic management, the arable field vegetation developed in the direction of fallow and grassland. Different reasons may be responsible for this process. In the organic farming system in addition to therophytic weeds many hemicryptophytic species increased which also grow in meadows and pastures. This may have been caused by volunteer crops emerging from precedent grass-clover mixtures and catch crops as well as from impure seeding material or from organic fertilisers which were only applied after the change of the management. In the integrated managed fields, minimum tillage favoured perennial plants which normally predominate in the perennial management types. Among the 25 species which showed a significant increase there, 20 are perennials such as *Cirsium* spp., *Elymus repens*, *Epilobium ciliatum*, *Equisetum arvense*, *Agrostis stolonifera*, *Poa trivialis* and *Lolium perenne*. Such species obviously profit from the lack of mechanical disturbance when direct drilling techniques are applied (Froud-Williams et al., 1981; Cousens and Mortimer, 1995). In contrast, the annual species many of which are closely restricted to arable field habitats (the 'characteristic species' in Table 1) were severely reduced by the intensification of weed control.

Species composition in unchanged boundary structures and in those which were enlarged differed from the beginning. A reason for this difference may be that the latter are situated on soils which have – on average – better nutrient supply than those unchanged (Anderlik-Wesinger et al., 1998). The extension of boundary structures by integrating neighbouring arable land initially led to a more weedy vegetation character. After that, species composition became increasingly similar to unchanged structures. The ordination plot in Figure 5 shows that vegetation development in the newly established boundary structures is rather similar to the succession observed in fallow fields. An important difference is, however, that the vegetation in boundary structures changed much faster from an annual to a perennial type plant community than that in abandoned arable fields. A reason for this accelerated development in the boundary structures may be the 'edge effect'. Edge effects are caused by narrow structures with a high boundary length to area ratio being invaded much faster by non-local plants than is the case in areas with lower boundary lengths (Forman, 1995). In a study on the spatial distribution of species in fallow fields Rew et al. (1992) correspondingly observed that perennials rapidly dominated the marginal zones while annuals were more frequently found in the areas which were at least 6 m away from field fringes.

The ordination diagram in Figure 5 also suggests that the species composition in meadows and pastures changed only marginally. This constancy corresponds to the observation that the greater percentage of species in both management types did not change their cover abundance significantly. Most of the few species which did show a change are characteristic grassland plants which significantly increased in cover abundance. This development may be caused by a more regular use of the grassland which was extensively stocked or leased to other farmers before the FAM project started. The position of pastures between meadows and arable field communities in Figure 5 may be caused by trampling that injures the sward and favours therophytic weeds.

In grassland and unchanged boundary structures where regular mowing was abandoned, a layer of litter was formed which shades the soil and leads to an accumulation of nutrients. Thus, low and light requiring species such as *Bellis perennis*, *Veronica arvensis*, *Trifolium repens* and *Carum carvi* (Ellenberg et al., 1991) gave way to tall perennials such as *Urtica dioica*, *Elymus repens*, *Dactylis glomerata* and *Heracleum sphondylium* (Table 2). These observations correspond to the results of Kornas and Dubiel (1991) who found the same development in abandoned *Arrhenatherum*-hay meadows and Schiefer (1981) who investigated succession on different types of *Arrhenatherum* grassland in south-western Germany.

3.1.6 Conclusions

The results of the present study show that changes in land use and re-designing the landscape led to significant increases in the number of species at the grid points in most of the ecosystem and management types. In terms of nature conservation issues this means that a lot can be done for species diversity in arable landscapes by changing to organic farming and increasing the number and size of boundary structures and set-aside fields. Only integrated farming which includes soil protection by reduced tillage as well as setting aside former grassland and abandoning maintenance of boundary structures showed no such positive effects. Although conditions for the development of plants in most of the investigated area are favourable, the lack of dispersal facilities prevents the establishment of new species from outside. Therefore, if promotion of species diversity in agricultural landscapes is intended, sophisticated strategies to intensify the interexchange of species between farms are necessary.

Acknowledgements

The scientific activities of the FAM Munich Research Network on Agroecosystems were financially supported by the German Federal Ministry of Education and Research (BMBF 0339370). Overhead costs of the Research Station Scheyern were funded by the Bavarian State Ministry for Science, Research and the Arts.

References

Agricola U, Barthel J, Laussmann H, Plachter H, 1996. Struktur und Dynamik der Fauna einer süddeutschen Agrarlandschaft nach Nutzungsumstellung auf ökologischen und integrierten Landbau. Verhandlungen der Gesellschaft für Ökologie 26, 681–692.

Albrecht H, 1995. Changes in the arable weed flora of Germany during the last five decades. Proceedings of the 9th European Weed Research Society Symposium, 'Challenges for Weed Science in a Changing Europe', Budapest, pp. 41–48.

Albrecht H, Mattheis A, 1998. The effect of organic and integrated farming on rare arable weeds on the Forschungsverbund Agrarökosysteme München (FAM) research station in southern Bavaria. Biological Conservation 86, 347–356.

Albrecht H, Sprenger B, 2008. Long term effects of reduced tillage on the populations of arable weeds. In: Schröder P, Pfadenhauer J, Munch JC (Eds.), Perspectives for agroecosystem management – balancing environmental and socioeconomic demands, pp. 237–256.

Albrecht H, Toetz P, Mattheis A, 1998. Untersuchungen zur Vegetationsentwicklung auf fünfjährigen Ackerbrachen. Zeitschrift für Pflanzenkrankheiten und Pflanzenschutz, Sonderheft XVI, 37–46.

Alfoeldi T, Fliessbach A, Geier U, Kilcher L, Niggli U, Pfiffner L, Stolze M, Willer H, 2002. Organic agriculture and the environment. In: El-Hage Scialabba N, Hattam C (Eds.), Organic Agriculture, Environment and Food Security, chapter 2. Environment and Natural Resources Series 4. Food and Agriculture Organisation of the United Nations (FAO), Rome, Italy.

Anderlik-Wesinger G, Albrecht H, Pfadenhauer J, 1998. Vegetationsentwicklung bestehender und neuangelegter Raine auf der FAM-Versuchsstation Klostergut Scheyern. Verhandlungen der Gesellschaft für Ökologie 28, 507–515.

Anderlik-Wesinger G, Albrecht H, Pfadenhauer J, 1999. Spontaneous and directed vegetation development on newly established boundary structures. Aspects of Applied Biology 54, 283–290.

Anderlik-Wesinger G, Kainz M, Pfadenhauer J, 1995. Integrierende Naturschutzplanung auf dem FAM-Versuchsgut Scheyern. Verhandlungen der Gesellschaft für Ökologie 24, 507–515.

Andreasen C, Stryhn H, Streibig JC, 1996. Decline in the flora of Danish arable fields. Journal of Applied Ecology 33, 619–626.

Anonymous, 2002. SPSS for WINDOWS 11.5.1 (16 Nov 2002), SPSS Inc., Chicago, USA.

Auerswald K, Kainz M, Schwertmann U, Beese F, Pfadenhauer J, 1996. Standards im Bodenschutz bei landwirtschaftlicher Nutzung – Das Fallbeispiel Scheyern. Verhandlungen der Gesellschaft für Ökologie 26, 663–669.

Baudry J, Burel F, 1984. Landscape Project: "Remembrement": Landscape consolidation in France. Landscape Planning 11, 235–241.

Becker B, Hurle K, 1998. Unkrautflora auf Feldern mit unterschiedlich langer ökologischer Bewirtschaftung. Zeitschrift für Pflanzenkrankheiten und Pflanzenschutz, Sonderheft XVI, 155–161.

Belde M, Sprenger B, Albrecht H, Pfadenhauer J, 2003. Bewertung, Prognose und Steuerung der Entwicklung von Ackerwildpflanzen. In: Schröder P, Huber B, Munch JC (Eds.), Jahresbericht 2002. FAM-Bericht 56, pp. 81–90.

Bonn S, Poschlod P, 1998. Ausbreitungsbiologie der Pflanzen Mitteleuropas. UTB für Wissenschaft, Wiesbaden, 404 pp.

Bornkamm R, 1981. Rates of changes in vegetation during secondary succession. Vegetatio 47, 213–220.

Boutin C, Jobin B, 1998. Intensity of agricultural practices and effects on adjacent habitats. Ecological Applications 8, 544–557.

Braun-Blanquet J, 1964. Pflanzensoziologie, 3rd ed. Springer Verlag, Wien/New York, 885 pp.

Callauch R, 1981. Ackerunkrautgesellschaften auf biologisch und konventionell bewirtschafteten Äckern in der weiteren Umgebung von Göttingen. Tuexenia 1, 25–38.

Chapman J, Sheail J, 1994. Field margins – an historical perspective. British Crop Protection Council CPC Monograph 58, 3–12.

Cousens R, Mortimer M, 1995. Dynamics of Weed Populations. Cambridge University Press, Cambridge, 332 pp.

Davies DJK, Christal A, Talbot M, Lawson HM, Wright GM, 1997. Changes in weed populations in the conversion of two arable farms to organic farming. In: Proceedings 1997 Brighton Crop Protection Conference-Weeds, Brighton, UK, pp. 973–978.

Dierschke H, Briemle G, 2002. Kulturgrassland. Wiesen, Weiden und verwandte Staudenfluren. Ulmer, Stuttgart, 230 pp.

Dierschke H, Wittig B, 1991. die Vegetation des Holtumer Moores (Nordwest-Deutschland). Veränderungen in 25 Jahren (1963–1988). Tuexenia 11, 171–190.

Dowdeswell WH, 1987. Hedgerows and Verges. Allen & Unwin Ltd., London, 190 pp.

Ellenberg H, Weber HE, Düll R, Wirth V, Werner W, Paulißen D, 1991. Zeigerwerte von Pflanzen in Mitteleuropa. Scripta Geobotanica 18. Goltze-Verlag, Göttingen, 248 pp.

European Communities, 1992. Council Regulation (EEC) 2078/92 of June 30, 1992 on agricultural production methods compatible with the requirements of the protection of the environment and the maintenance of the countryside. Official Journal of the European Communities L215, 85–90.

Fisher NM, Davies DHK, 1991. Effectiveness of sown covers for the management of weeds in set-aside fallows: the Bush trials. Proceedings of the 1991 British Crop Protection Conference in Brighton, Weeds 1, pp. 387–394.

Forman RTT, 1995. Land mosaics. The ecology of landscapes and regions. Cambridge University Press, Cambridge, 632 pp.

Frieben B, 1990. Bedeutung des Organischen Landbaues für den Erhalt von Ackerwildkräutern. Natur und Landschaft 65, 379–382.

Frieben B, 1995. Effizienz des Schutzprogrammes für Ackerwildkräuter. Mitteilungen der Landesanstalt für Ökologie, Landschaftsentwicklung und Forstplanung NRW 4/95, 14–19.

Froud-Williams RJ, Chancellor RJ, Drennan DSH, 1981. Potential changes in weed floras associated with reduced-cultivation systems for cereal production in temperate regions. Weed Research 21, 99–109.

Grundy AC, Mead A, Bond W, 1996. Modelling the effect of weed-seed distribution in the soil profile on seedling emergence. Weed Research 36, 375–384.

Hansson M, Fogelfors H, 1998. Management of permanent set-aside on arable land in Sweden. Journal of Applied Ecology 35, 758–771.

Hentschel A, 2001. Integration of agriculture and nature conservation in grassland regions of the West Eifel (North Rhine-Westphalia). PhD thesis, University of Bonn, Agricultural Faculty, 293 pp.

Hill MO, Gauch HG, 1980. Detrended correspondence analysis: an improved ordination technique. Vegetatio 42, 47–58.

Hofmeister H, Garve E, 1998. Lebensraum Acker, 2nd ed. Parey Verlag, Hamburg, 322 pp.

Hülsbergen KJ, Küstermann B, 2005. Development of an environmental management system for organic farms and its introduction into practice. In: Köpke U, Niggli U, Neuhoff D, Cornish P, Lockeretz W, Willer H (Eds.), Proceedings of the First Scientific Conference of the International Society of Organic Agriculture Research (ISOFAR), 21–23 September 2005, Adelaide, South Australia, pp. 460–463.

Keddy PA, 1992. Assembly and response rules: two goals for predictive community ecology. Journal of Vegetation Science 3, 157–164.

Kent M, Coker P, 1992. Vegetation Description and Analysis. Wiley, Chichester, UK, 363 pp.

Knauer N, 1995. Ökologie der Agrarräume und des Dorfes. In: Steubing L, Buchwald K, Braun E (Eds.), Natur und Umweltschutz. Fischer Verlag, Stuttgart, pp. 179–229.

Kornas J, Dubiel E, 1991. Land use and vegetation changes in the hay meadows of the Ojcow National Park during the last thirty years. Veröffentlichungen des Geobotanischen Institutes ETH Zürich 106, 208–231.

Korneck D, Schnittler DM, Vollmer I, 1996. Rote Liste der Farn- und Blütenpflanzen (Pteridophyta et Spermatophyta) Deutschlands. Schriftenreihe für Vegetationskunde 28, 21–187.

Korneck D, Schnittler M, Klingenstein F, Ludwig G, Takla M, Bohn U, May R, 1998. Warum verarmt unsere Flora? Auswertung der Roten Liste der Farn- und Blütenpflanzen Deutschlands. Schriftenreihe für Vegetationskunde 29, 299–444.

Kretschmar H, Pfeffer H, Hoffmann J, Schrödl G, Fux I, 1995. Strukturelemente in Agrarlandschaften Ostdeutschlands– Bedeutung für den Biotop- und Artenschutz. ZALF-Bericht 19. ZALF, Müncheberg, 164 pp.

Lang A, Barthel J, 2008. Spiders (Araneae) in arable land: species community, influence of land use on diversity, and biocontrol significance. In: Schröder P, Pfadenhauer J, Munch JC (Eds.), Perspectives for agroecosystem management – balancing environmental and socioeconomic demands, pp. 307–326.

Luftensteiner HW, 1982. Untersuchungen zur Verbreitungsbiographie von Pflanzengemeinschaften an vier Standorten in Niederösterreich. Bibliographia Botanica 135, 68 pp.

Marggraf R, 2003. Comparative assessment of agri-environment programmes in federal states of Germany. Agriculture, Ecosystems and Environment 98, 507–516.

Mattheis A, Otte A, 1994. Ergebnisse der Erfolgskontrollen zum Ackerrandstreifenprogramm im Regierungsbezirk Oberbayern, 1985–1991. Schriftenreihe der Stiftung zum Schutz gefährdeter Pflanzen 5, 56–71.

Mayer F, Albrecht H, 2008. Dispersal strategies: are they responsible for species success in arable ecosystems? In: Schröder P, Pfadenhauer J, Munch JC (Eds.), Perspectives for agroecosystem management – balancing environmental and socioeconomic demands, pp. 257–278.

McCune B, Mefford MJ, 1999. PC-ORD. Multivariate Analysis of Ecologic Data, Version 4. MjM Software Design, Glendan Beach, USA.

Meisel K, von Hübschmann A, 1976. Veränderungen der Acker- und Grünlandvegetation im nordwestdeutschen Flachland in jüngerer Zeit. Schriftenreihe für Vegetationskunde 10, 109–124.

Müller T, 1977. Trifolio-Geranietea sanguinei Th. Müller 61. In: Oberdorfer E. (Ed, 1978), Süddeutsche Pflanzengesellschaften Teil II, Fischer Verlag, Stuttgart, pp. 249–298.

Osbornova J, Kovarova M, Leps J, Prach K,1990. Succession in abandoned fields. Studies in Central Bohemia, Czechoslovakia. Geobotany 15, 168.

Otte A, 1984. Änderungen in Ackerwildkraut-Gesellschaften als Folge sich wandelnder Feldbaumethoden in den letzten 3 Jahrzehnten. Dissertationes Botanicae 78. Cramer, Vaduz, Liechtenstein, 165 pp.

Pfadenhauer J, Ganzert C, 1992. Konzept einer integrierten Naturschutzstrategie im Agrarraum. In: Bayerisches Staatsministerium für Landesentwicklung und Umweltfragen (Ed.), Untersuchung zur Definition von landespflegerischen Leistungen in der Landwirtschaft nach ökologischen und ökonomischen Kriterien. Materialen Umwelt und Entwicklung Bayern 84, 5–50.

Plakholm G, 1989. Unkrautuntersuchungen in biologisch und konventionell bewirtschafteten Getreideäckern Oberösterreichs. PhD thesis. Universität für Bodenkultur, Wien, 269 pp.

Pollard EH, Hooper MD, Moore NW, 1974. Hedges. Collins, London, 256 pp.

Popp B, 1887–1891. *Flora von Scheyern*. Programm der vollständigen Lateinschule im erzbischöf-lichen Knabenseminare zu Scheyern. Pfaffenhofen a. d. Ilm.

Reents HJ, Küstermann B, Kainz M, 2008. Sustainable land use by organic and integrated farming systems. In: Schröder P, Pfadenhauer J, Munch JC. (Eds), Perspectives for agroecosystem management – balancing environmental and socioeconomic demands, pp. 17–39.

Rennwald E, 2000. Verzeichnis und Rote Liste der Pflanzengesellschaften Deutschlands. Schriften-reihe für Vegetationskunde 35 (Bundesamt für Naturschutz, Landwirtschaftsverlag), 800 pp.

Rew LJ, Wilson PJ, Froud-Williams RJ, Boatman ND, 1992. Changes in vegetation composition and distribution within set-aside land. British Crop Protection Council Monograph 50 'Set-aside', pp. 79–84.

Robinson RA, Southerland WJ, 2002. Post-war changes in arable farming and biodiversity in Great Britain. Journal of Applied Ecology 39, 157–176.

Rodi D, 1966. Ackerunkrautgesellschaften und Böden des westlichen Tertiär-Hügellandes. Hoppea Denkschrift der Regensburgischen Botanischen Gesellschaft 26, 161–198.

Ruthsatz B, 1985. Die Pflanzengesellschaften des Grünlandes im Raum Ingolstadt und ihre Verarmung durch die sich wandelnde landwirtschaftliche Nutzung. Tuexenia 5, 273–301.

Ruthsatz B, Otte A, 1987. Kleinstrukturen im Raum Ingolstadt: Schutz und Zeigerwert. Teil III. Feldwegränder und Ackerraine. Tuexenia 7, 139–163.

Salonen J, Hyvönen T, Jalli H, 2001. Weed flora in organically grown spring cereal in Finland. Agricultural and Food Science in Finland 10, 231–242.

Scheuerer M, Ahlmer W, 2003. Rote Liste gefährdeter Gefäßpflanzen Bayerns mit regionalisierter Florenliste. Schriftenreihe des Bayerischen Landesamtes für Umweltschutz 165, 372.

Schiefer J, 1981. Bracheversuche in Baden-Württemberg. Beihefte der Veröffentlichungen zu Naturschutz und Landschaftspflege Baden-Württemberg 22, 325.

Schmidt W, 1981. Ungestörte und gelenkte Sukzession auf Ackerbrachen. Scripta Geobotanica XV. Goltze-Verlag, Göttingen, Germany, 120 pp.

Schmiedeknecht A, 1995. Untersuchungen zur Auswirkung von Flächenstillegungen auf die Vegetationsentwicklung von Acker- und Grünlandbrachen im mitteldeutschen Trockengebiet. Dissertationes Botanicae 245, 176.

Schneider C, Sukopp U, Sukopp H, 1994. Biologisch-ökologische Grundlagen des Schutzes gefährdeter Segetalpflanzen. Schriftenreihe für Vegetationskunde 26, 356.

Schrautzer J, Wiebe C, 1993. Geobotanische Charakterisierung und Entwicklung des Grünlandes in Schleswig-Holstein. Phytocoenologia 22, 105–144.

Schröder P, 2008. Mesofauna. In: Schröder P, Pfadenhauer J, Munch JC. (Eds), Perspectives for agroecosystem management – balancing environmental and socioeconomic demands, pp. 293–306.

Schröder P, Huber B, Reents HC, Munch JC, Pfadenhauer J, 2008. Outline of the Scheyern project. In: Schröder P, Pfadenhauer J, Munch JC (Eds.), Perspectives for agroecosystem management – balancing environmental and socioeconomic demands, pp. 3–16.

Sinowski W, 1995. Die dreidimensionale Variabilität von Bodeneigenschaften – Ausmaß, Ursachen und Interpolation. PhD thesis, TU München, Shaker Verlag, Aachen, ISBN 3-8265-0994-3, FAM-Bericht 7, 158 pp.

Sokal RR, Rohlf FJ, 1998. Biometry: The Principles and Practice of Statistics in Biological Research, 3rd Ed. Freeman and Company, New York, 887 pp.

Squire GR, Rodger S, Wright G, 2000. Community-scale seedbank response to less intense rotation and reduced herbicide input at three sites. Annals of Applied Biology 136, 47–57.

Steidl I, Ringler A, 1997. Agrotope (1. Teil) – Landschaftspflegekonzept Bayern, Band II.11. Eds. Bayerisches Staatsministerium für Landesentwicklung und Umweltfragen and Bayerische Akademie für Naturschutz und Landschaftspflege, München, 253 pp.

Sutcliffe OL, Kay QON, 2000. Changes in the arable flora of central southern England since the 1960s. Biological Conservation 93, 1–8.

Thompson K, Bakker JP, Bekker R, 1997. The Soil Seed Banks of North West Europe. Cambridge, Cambridge University Press, 276 pp.

Tilman D, 1988. Plant Strategies and the Dynamics and Structure of Plant Communities. Monographs in Population Biology. Princeton University Press, Princeton, 360 pp.

Turley DB, Wright G, Hebden P, 1998. Changes in weed seedbanks during long-term set-aside. Aspects of Applied Biology 51, 265–272.

van Elsen T, 1989. Ackerwildkraut-Bestände biologisch-dynamisch und konventionell bewirtschafte-ter Hackfruchtäcker in der Niederrheinischen Bucht. Lebendige Erde 4, 277–282.

Wicke G, 1998. Stand der Ackerrandstreifenprogramme in Deutschland. Schriftenreihe – Landesanstalt für Pflanzenbau und Pflanzenschutz (Mainz) 6, 55–84.

Willems JH, 1990. Calcareous grasslands in Continental Europe. In: Hillier SH, Walton DHW, Wells DA (Eds.), Calcareous Grasslands. Ecology Management. Bluntisham Books, Bluntisham, pp. 3–30.

Willson FM, 1983. Dispersal mode, seed shadows, and colonization patterns. Vegetatio 107/108, 261–280.

Wischmeier WH, Smith DD, 1978. Predicting Rainfall Erosion Losses – A Guide to Conservation Planning. US Department of Agriculture Agricultural Handbook 537, 58 pp.

Wisskirchen R, Haeupler H, 1998. Standardliste der Farn- und Blütenpflanzen Deutschlands. Ulmer Verlag, Stuttgart, 765 pp.

Wolff-Straub R, 1989. Vergleich der Ackerwildkraut-Vegetation alternativ und konventionell bewirtschafteter Äcker. Schriftenreihe der Landesanstalt für Ökologie, Landschaftsentwicklung und Forstplanung NRW 11, 70–111.

Wurbs A, Glemnitz M, 1997. Nährstoffgehalte alter Ackerbrachen auf Sandböden und ihre Bedeutung für die Vegetationsentwicklung. Zeitschrift für Ökologie und Naturschutz 6, 233–245.

Zobel M, van der Maarel E, Dupré C, 1998. Species pool: the concept, its determination and significance for community relation. Applied Vegetation Science 1, 55–66.

Chapter 3.2

Long-Term Effects of Reduced Tillage on the Populations of Arable Weeds

H. Albrecht and B. Sprenger

3.2.1 Introduction

Weeds can significantly reduce the yield of arable crops. Hence, the worldwide losses by weed infestation are estimated to amount to 14% of the potential yields (Oerke et al., 1994). Analysing the losses in single crops, Oerke and Steiner (1996) found weed competition to limit the yields in the majority of the most frequently cultivated species. On the German scale, experts estimate the decline without any weed control in winter wheat to attain 50% of the potential yield (Zwerger et al., 2004). To minimise these losses, combining inversion tillage with herbicide applications became the common practice. This type of management was quite successful and clearly decreased the weed populations during the last decades (Kutzelnigg, 1984; Erviö and Salonen, 1987; Andreasen et al., 1996). However, it also affected the abiotic resources and the organism communities in arable ecosystems by increasing soil erosion (Agassi, 1995; Bork, 1988) and pollution (Cooper, 1993) and by decreasing the biodiversity

Perspectives for Agroecosystem Management
Edited by P. Schröder, J. Pfadenhauer and J.C. Munch

(Wilson et al., 1999; Robinson and Southerland, 2002). Currently, reduced tillage gains importance because it diminishes these negative effects by protecting the abiotic resources and the soil organisms (Friebe, 1993; Tebrügge, 2001).

The objective of the present study was to analyse the long-term effects that a change from conventional to reduced tillage has on the populations of arable weeds. To describe how the weed vegetation reacts to this modification, the number of weed plants in both the aboveground vegetation and the soil seed bank were recorded. The weed density is frequently studied because it can provide reliable estimates of the potential yield losses by weed competition (Cousens, 1985). On the other hand, the seed banks reflect the overall weed potential much better than the aboveground plants because they represent a much higher percentage of the whole weed populations and they are less affected by actual environmental conditions. The results of these studies should lead to recommendations for the practical weed management and could help to estimate the chances to predict future weed infestation from seed bank data.

3.2.2 Research area and management

The investigations were carried out on the experimental farm of the FAM Munich Research Network on Agroecosystems at Scheyern where the consequences of changes in the agricultural management were analysed to develop sustainable land use strategies (Schröder et al., 2008; Reents et al., 2008). Before FAM took over, most of this area was cultivated with cash crops such as wheat (*Triricum aestivum L.*), barley (*Hordeum vulgare L.*) and oil seed rape (*Brassica napus L.*). Mineral fertilisation, chemical weed control and inversion tillage were the characteristic features of the corresponding exploitation. During the 2-year long inventory phase, all fields were cultivated equally with winter wheat in 1991 and spring barley in 1992 to level out different starting conditions. After harvest in autumn 1992, the second and main phase of the project was started by redesigning the area and establishing the integrated management farm on 30.0 ha of arable land. There, ploughing was replaced by non-inversion tillage with a rotary tiller. The crop rotation contains potatoes, winter wheat, maize and winter wheat. To prevent erosion, water runoff and nutrient leaching, soil was covered as long as possible with growing plants or plant residues. This was accomplished, for instance, by mulch seeding or by sowing cover crops under the preceding main crops (Auerswald et al., 2000). Mineral fertilisers and slurry were applied according to the requirements of the cultivated crops. Weed control in winter wheat was done with isoproturone (against grasses), sulfonyle urea (against dicotyledonous species), fluroxypyr (against *Galium aparine*), diflufenican (against *Veronica* spp. and *Viola arvensis*) and urea (against *Cirsium arvense*) herbicides.

To analyse the effects of different soil cultivation strategies in detail, an additional plot experiment was established in the northern part of the FAM Research Station in 1992 (Sprenger, 2004). There, the following treatments of primary tillage were tested: (i) no primary tillage, (ii) reduced (=non-inversion) tillage with a rotary tiller to a depth of approximately 15 cm, (iii) inversion tillage with a mouldboard plough to a depth of approximately 20 cm. Fertilisation, weed control and the crop rotation were practiced just as in the management system on the farm scale.

3.2.3 Materials and methods

The investigations in the arable weed vegetation were carried out on two scales. The long-term changes in the weed vegetation were recorded using the 50 × 50 m grid of reference points, which was fixed throughout the experimental area (Schröder et al., 2008). At each of the 132 grid points, a 10 × 10 m area was defined to monitor the development. In a time series of 9 years from 1992 to 2000, two relevés were carried out per year in accordance with the Zurich–Montpellier School method (Braun-Blanquet, 1964). Thus, the cover abundance of each single species and the total cover of weeds and crops were estimated. In addition, the density of individuals was recorded by counting the number of plants and shoots emerging in four rectangles of 625 cm^2 for each quadrat. The first set of records was made in spring before weed control; the second one was carried out in summer before harvest. The seed bank analysis was started by sampling 20–30 soil cores from the whole plough layer of each grid point and mixing them to attain one sample of 1 kg fresh weight. The samples were collected by using soil borers with core diameters of 17 mm.

In the experimental plots where the three different tillage regimes were established in 1992, effects of these treatments were studied from 1999 to 2002 (Sprenger, 2004). The density of individuals was estimated by counting the plants in 18 rectangles per treatment, each of 0.05 m^2 in size. The number of seedlings was recorded – depending on the crop – four to eight times a year. Weed phytomass was recorded each year before harvest. Soil samples in the experimental plots were taken to the depths of 0–5, 5–10 and 10–20 cm. There, the corresponding core diameter was 7.7 cm. Fifteen samples per treatment and depth were analysed separately. Sampling was generally done in winter at the end of the vegetation rest. The analysis of the seed numbers was carried out using the seedling emergence method in which seeds were given 2 years to germinate (Albrecht and Forster, 1996). To record the total 'diaspore pool', i.e. both true seeds and clonal fragments capable of regeneration, samples were not agitated through sieves.

As the background populations from which the samples were drawn cannot be assumed to be normally distributed, Box and Whisker plots, the 95% interval of confidence, the Wilcoxon matched pairs signed rank test, the sign test by Dixon and Mood, the Friedman test and the Kruskal–Wallis ANOVA, which all are non-parametric methods, were used to describe the vegetation characteristics (Sokal and Rohlf, 1998). To make the results comparable to other references, the arithmetic means are additionally mentioned. In a few cases, normal distributed data allowed the application of the LSD test. The analyses were computed using SPSS 11.5 statistical package for Windows (Anonymous, 2002), nomenclature of species follows Wisskirchen and Haeupler (1998).

3.2.4 Results and discussion

Population development after the introduction of reduced tillage

Eight years after the introduction of reduced tillage, the median number of seedlings in cereal crops before weed control in the spring had significantly increased from 48 to 104 plants m^{-2}. Within this time span, seedling numbers showed a remarkable variation. In the initial 2 years, the values decreased to only 36 plants m^{-2}, but during the third and fourth year they increased again to 88 plants m^{-2} (Figure 1). In the fifth and sixth year, plant numbers re-decreased to 56 m^{-2} and in the final 2 years they reached the level of 108 m^{-2}. The numbers of seeds in soil increased from 3500 to 6090 m^{-2} during the first 3 years of reduced tillage. Until the sixth year, however, this value declined again to 3190 seeds m^{-2}. Therefore, the increase in the number of weed seedlings in the spring was highly significant while the number of seeds in soil showed no corresponding change.

This change in the relationship between the seed bank and the above-ground vegetation may be a characteristic feature of weed population development after the introduction of reduced tillage. Hence, Sprenger (2004), who investigated the vertical distribution of weed seeds in the experimental plots of the FAM research station after 8 years of reduced tillage, found that 80% of the viable seeds were situated in the upper 10 cm, while only 20% occurred in depths between 10 and 20 cm. These results agree with Froud-Williams et al. (1983), Cavers and Benoit (1989), Cousens and Moss (1990), Feldman et al. (1997) and O'Donovan and McAndrew (2000), who also observed that non-inversion tillage leads to an accumulation of seeds at the soil surface. Thus, the proportion of seeds with a chance to germinate and to establish increased, despite the fact that there was no significant change in the total number of seeds in soil. In their review of North American publications on 'conservation tillage', Swanton et al. (1993) stated that the development of the weed vegetation after abandoning inversion tillage strongly depends on the weed management. Hence, weed infestation could be minimised if population development

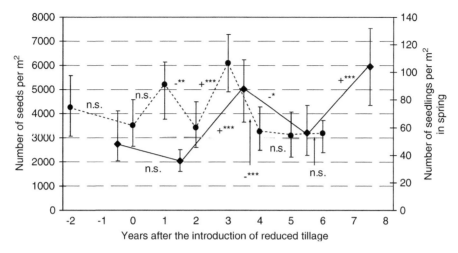

Figure 1 Development of the soil seed bank (●) and of the seedling density (◆) in spring before weed control in course of 8 years after the change to reduced tillage (median values; bars represent 95% confidence intervals). Before the change, seedlings were recorded in spring-sown barley, thereafter in winter wheat. As winter wheat had a percentage of about 50% in the crop rotation, data from 2 years were combined to calculate the changes for the whole area. The Friedman test showed that the median values generally differed ($P < 0.001$), and the Wilcoxon test was used to test changes from year to year. Seed bank data from Albrecht (2004).

is consequently prevented by weed control. On the FAM research station, infestation problems with increasing weed numbers were counteracted by an improved weed control strategy. With winter wheat this was realised by splitting the application of herbicides into two treatments. The first one was carried out in early spring by using a herbicide combination which contained isoproturone to suppress both dicots and grasses. Late emerging species like *G. aparine* and *Cirsium arvense* frequently necessitated a second application of fluroxypyr and urea herbicides. Before fields were cultivated with potatoes and maize, they were treated with glyphosate. To minimise the input of this herbicide, the application was done by spraying Roundup® in a rather low dose of $2.2 \, \mathrm{l \, ha^{-1}}$ ($360 \, \mathrm{g \, glyphosate \, l^{-1}}$) during the initial phase of the project. As many monocotyledonous plants survived this treatment, the dose was increased thereafter to $3.5 \, \mathrm{l \, ha^{-1}}$. The suppression of weed infestation problems illustrates the success of this practice; neither before nor after changing to integrated farming did weed cover exceed 5%.

The introduction of reduced tillage resulted in severe changes in the species composition. Thus, the vegetation relevés and the seed bank analyses both revealed that particularly the perennial species *Epilobium ciliatum*,

Table 1 Survey of references (seed bank analyses and field investigations, includes all species, the development of which is mentioned in at least three publications) on the effect of 'conservation tillage' versus plough tillage on the populations of different weed species.

Species/groups	Number of references			Trend
	+	±	−	
Monocotyledonous annuals	9	0	1	+ +
Poa annua	7	0	0	+ +
Apera spica-venti	4	0	1	+
Alopecusus myosuroides	3	0	0	+
Dicotyledonous annuals	1	1	6	− −
Galium aparine	7	0	0	+ +
Matricaria spp.	6	0	0	+ +
Sonchus spp.	4	0	0	+
Stellaria media	4	1	1	+
Thlaspi arvense	3	0	1	+
Amaranthus spp.	3	0	2	±
Fallopia convolvulus	2	0	1	±
Veronica persica	2	0	1	±
Chenopodium album	6	2	6	±
Perennials	9	1	1	+ +
Elymus repens	8	0	0	+ +
Cirsium arvense	4	0	0	+
Taraxacum officinale	4	0	0	+
Epilobium spp.	3	0	0	+

The term 'conservation tillage' comprises different types and intensities of 'reduced' and 'minimum' tillage. The numbers in columns represent the sum of references which report increasing (+), constant (±) or decreasing (−) frequencies. 'Trend' gives an estimate of the general development. References: Albrecht (2004), Amann (1991), Belde et al. (2000), Bilalis et al. (2001), Boström and Fogelfors (1999), Cardina et al. (1991), Derksen et al. (1993), Fogelfors and Boström (1998), Frick and Thomas (1992), Froud-Williams et al. (1983), Gruber et al. (2000), Knab (1988), Knab and Hurle (1986), Kobayashi et al. (2003), Legere and Samson (1999), Mayor and Maillard (1995), McClosky et al. (1996), Mulugeta and Stoltenberg (1997), O'Donovan and McAndrew (2000), Pallutt (1999), Pollard and Cussans (1981), Pollard et al. (1982), Schwerdle (1977), Skuterud et al. (1996), Sprenger (2004), Streit et al. (2002), Streit et al. (2003), Swanton et al. (1999), Teasdale et al. (1991), Thomas and Frick (1993), Tørresen and Skuterud (2002), Tuesca et al. (2001), Vanhala and Pitkänen (1998) and Zanin et al. (1997).

Poa trivialis, Cirsium arvense, Taraxacum officinale, Equisetum arvense and *Elymus repens* gained from this change (Belde et al., 2000; Albrecht, 2004; Albrecht et al., 2008; Sprenger, 2004). This result well agrees with the observations of other authors (Table 1). As a great part of these species produce vegetative runners or vigorous roots, reduced soil disturbance may have favoured their spread (Pekrun and Claupein, 1998). On the other hand, dicotyledonous annuals such as *Aphanes arvensis*, *Capsella bursa-pastoris*, *Matricaria recutita*, *Myosotis arvensis*, *Veronica arvensis* and *Viola arvensis* were particularly diminished. One reason for their decline may be that these plants are adapted to low germination temperatures and therefore preferably occur in autumn-sown crops (Otte, 1996). Consequently, the intensified herbicide treatments and the increased cultivation of spring-sown crops may have enforced their decline. This observation confirms the general trend that broadleaved annuals tend to decline under reduced tillage (Table 1). These results suggest that these species cope with regular ploughing better than with disturbance during the germination and early establishment period. However, not all annuals declined. Hence, *Stellaria media*, *Poa annua* and *Veronica persica* are annual species which were found significantly more often. Carrying out seed germination tests in climate chambers, Otte (1996) observed that all these species can germinate over a wide span of temperatures. In the fields, these species were frequently found between the stubbles of winter wheat after harvest or before the cultivation of the subsequent crop. These species are obviously able to use autumn and early spring for their development and therefore they profit from omitting primary tillage. Reports on an increase of monocotyledonous species (Table 1) could not be confirmed for all species. Here, *Apera spica-venti*, which is a typical weed for autumn-sown cereals, may have suffered from the introduction of maize and potatoes into the crop rotation.

Effects of different tillage treatments

The investigations in the experimental plots showed that the type of tillage clearly differentiated the weed populations. Hence, the numbers of seeds in soils which were treated with a rotary tiller for 8–10 years were five times higher than in plots which were cultivated with a mouldboard plough (Figure 2). The number of seedlings emerging from this diaspore pool in winter wheat before weed control in spring was 40 m^{-2} compared to 20 m^{-2} under the plough tillage. Summing up all seedlings which emerged over the whole vegetation period, 280 plants m^{-2} were recorded when tillage was done with the rotary tiller and only 80 m^{-2} were counted when the soil was tilled with a plough. Correspondingly, also the weed phytomass significantly increased in plots with reduced tillage (Figure 3). However, the highest seed numbers and phytomasses were recorded in the plots without primary tillage (Figures 2 and 3).

The management regime which was practiced in the experimental plots under plough tillage cannot be compared to the one which was applied on the farm scale before reduced tillage was introduced. Hence, the weed control on the farm scale was rather simple compared to the two treatments with highly efficient herbicide compounds which were sprayed in the experimental plots. In addition, varying environmental conditions like field edges or soil properties may have interfered with the

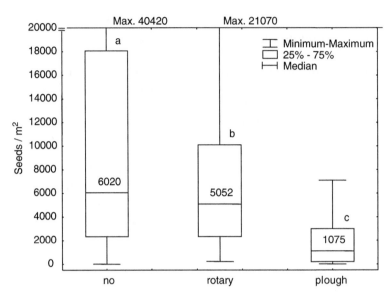

Figure 2 Box plots for the numbers of seeds in soils after 8–10 years without primary tillage, with rotary tillage and with mouldboard ploughing ($n = 450$). Differences were analysed with the LSD test (adapted from Sprenger, 2004).

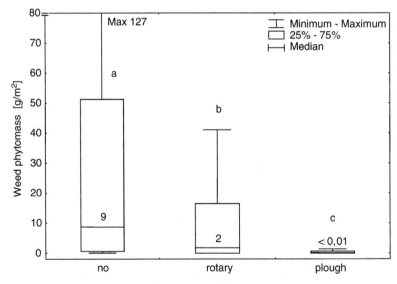

Figure 3 Box plots for the weed phytomass after 8–10 years without primary tillage, with rotary tillage and with mouldboard ploughing ($n = 396$). Differences were analysed with the Mann–Whitney U-test (adapted from Sprenger, 2004).

Table 2 Relationship between the soil seed bank and the number of seedlings emerging in spring.

Crop species	Scale	Spearman rank correlation		Seedling emergence (%)		
		Coefficient	P	Median	Mean	n
Winter wheat	Farm	0.31	<0.001	1.75	6.61	404
	Experimental plots	0.26	<0.050	1.23	4.88	72
Spring wheat	Farm	0.14	ns	1.30	4.09	33
	Experimental plots	0.12	ns	1.31	5.25	36
Maize	Farm	0.44	<0.001	1.26	3.35	178
	Experimental plots	0.75	<0.001	1.54	2.45	36
Potatoes	Farm	0.29	<0.001	0.66	2.29	190
	Experimental plots	0.58	<0.001	0.31	0.60	36
Total	Farm	0.34	<0.001	1.32	3.60	805
	Experimental plots	0.35	<0.001	1.14	3.61	180

On the farm scale, the correlation was calculated for the first 7 years after the introduction of reduced tillage, and on the scale of the experimental plots this was done for the time span between the 8th and the 10th year.

weed populations on the different scales. However, the results from the experimental plots clearly indicate that ploughing in combination with an intensive weed control can be highly efficient in reducing weed population densities and in controlling weed infestation problems.

The 'relative abundance index' is a quantity to find out the species which particularly profit from a certain type of management (Streit et al., 2003). In the experimental plots, omitting primary tillage particularly favoured *Elymus repens* and *P. annua*, while using the rotary tiller instead of the plough significantly increased *G. aparine*. The other species frequently occurring in the research area, i.e. *Chenopodium album*, *M. recutita*, *S. media* and *Viola arvensis*, were not affected by the tillage treatments. The rotary tiller favouring *G. aparine* may be explained by the fact that seeds are buried deeper into the soil than without primary tillage, but nearer to the surface than in a ploughed substrate (Sprenger, 2004). Investigations by Röttele and Koch (1981) and Benvenuti et al. (2001) showed that seedlings of *G. aparine* possess – in contrast to smaller seeded species – enough vigour to emerge successfully even if they are buried to a depth of several centimetres. The references cited in Table 2 also reveal that 'conservation tillage' favours monocotyledonous weeds.

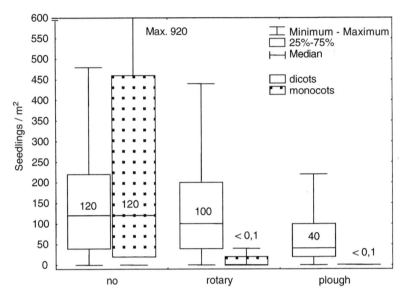

Figure 4 Number of dicotyledonous and monocotyledonous weed plants
recorded over the vegetation periods after 8–10 years without
primary tillage, with rotary tillage and with mouldboard ploughing
($n = 432$). From Sprenger (2004).

In the present study this was confirmed for the plots without primary
tillage but not in the area where 'reduced tillage' was practiced with a
rotary tiller (Figure 4).

Relationship of the soil seed bank to the aboveground vegetation

The number of weed seedlings which emerge in spring before weed
control is an important indicator to predict the weed competition the cultivated crops are exposed to during the vegetation period. Therefore,
Gerowitt and Heitefuss (1990) have taken the relationship between the
weed seedling density and the crop yield to develop economic thresholds
which provide a key to estimate the necessity of weed control measures.
As weed seedlings directly emerge from the soil seed bank, it was an aim
of the present study to examine the chances to predict weed seedling
emergence from soil seed bank data. An early availability of corresponding
information could provide good time to integrate effective weed control
and requirements of environmental conservation into sophisticated weed
management strategies.

A value which is highly suitable to describe this relationship is the correlation coefficient between the number of seeds in soil and the density of
seedlings emerging at the soil surface before weed control (Cardina et al.,
1996). In the present study, this correlation was highly significant (Table 2).

The percentage of seeds which successfully established seedlings was 3.6% on both the farm and the experimental plot scale (mean value for different crops; median values were 1.3 and 1.1%, respectively). This result corresponded well to the range of 0.3–1.7% found by Froud-Williams et al. (1983) in untilled fields in England, the 2.7–6.7% recorded by Tørresen (2003) for reduced tillage in spring cereals in Norway and the 11.8% Dessaint et al. (1997) observed in plots without any chemical weed control in France.

Using the data from the experimental plots to calculate the H values in a Kruskal–Wallis ANOVA revealed that the importance of environmental variables for seedling emergence in spring decreased according to the following order: actual crop > preceding crop > year > type of tillage (Sprenger, 2004). This means that the cultivated crop type had the most important influence on the emergence of seedlings. Particularly high emergence rates were recorded in winter cereals, whereas particularly low values occurred in potatoes (Table 2). The low numbers for the potato fields may have been caused by both the use of herbicides and the cultivation of ridges. In contrast, seedlings emerging in winter cereals were only confronted with chemical weed control. In the investigations on the farm scale, the cultivation of potatoes and maize significantly decreased the number of seeds in soils, the corresponding decline being 22% in potatoes and 16% in maize, respectively. In contrast, the cultivation of winter wheat moderately increased the seed bank. This change, however, was not significant ($P > 0.05$).

Despite this correlation between the soil seed bank and the number of seedlings on the soil surface was significant, the correlation coefficient was low and the variation among the corresponding data was great (Figure 5). Hence, the economic threshold of 50 plants m^{-2} in winter wheat (Gerowitt, 1992) was surpassed at many grid points even when the corresponding seed numbers were distinctly low. Even in those areas where the seed numbers in soil fell below a level of 1500 m^{-2} – this applied for only 20% of the investigated fields – seedling density in spring exceeded the threshold in 44% of this area. If herbicides would have been applied according to this database, weed infestation could have caused significant yield losses in a large proportion of the unsprayed area.

Several factors contribute to this variation in the relationship between the seed bank and the soil surface vegetation. One of them is the choice and the sequence of crops in the crop rotation with their specific cultivation practice. This may have affected seedling emergence in several ways, e.g. by the time of cultivation (Roberts, 1984), the depth of burial (Grundy et al., 1996; Grundy et al., 2003) and the selectivity of herbicides (Berger, 2002). Furthermore, varying weather conditions can influence this relationship (Vleeshouwers, 1997; Grundy, 2001; Tørresen, 2003; Sprenger, 2004). This factor can modify weed populations directly, i.e. by reducing

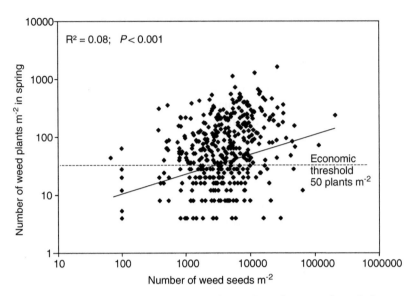

Figure 5 Relationship between the number of seeds in soil and the number
         of plants in winter wheat before weed management in early spring.
         The axes are given in logarithmic scales ($n = 406$).

the germination at low temperatures. On the other hand, it may work indirectly, e.g. by delaying weed control measures. Furthermore, also the time
span in which fields are cultivated under a certain management regime
may contribute to this variation. Thus, a weak correlation between the seed
bank and the number of seedlings was predominantly found when reduced
tillage systems were only just introduced (e.g. Ball and Miller, 1989; Derksen
et al., 1998). This relationship may increase with the duration of this management because plant communities gradually adapt to this new type of
cultivation.

Another reason for these vague predictions is the great variation
among the seedling emergence rates of different species. It ranged from
below 10:1 to more than 100:1, and it was closely related to the population
ecology of the individual species (Albrecht and Pilgram, 1997). Many
species, the seed numbers of which were more than hundred times higher
than the plant density, belong to communities which are typical for
ephemerally moist soils. These species persist in dry seedbed conditions as
dormant seeds and do not emerge until the moisture offers good chances
for successful development (Hofmeister and Garve, 1998). Taxa belonging
to this group are *Juncus bufonius*, *Plantago intermedia*, *Sagina procum-
bens*, *Anagallis minima*, *Cerastium glomeratum*, *P. trivialis*, *Gnaphalium
uliginosum* and *Veronica serpyllifolia*. Other species which were scarcely
found in the aboveground vegetation were *Trifolium repens*, *Rumex
obtusifolius*, *Urtica dioica* and *Cerastium holosteoides*. In Central Europe,

these species predominantly grow in pastures. Their seeds may have been transferred to the fields by manure where they obviously lack suitable conditions to establish. A second group of species is characterised by particularly low seed numbers and a high density of plants and shoots. Some of them have vegetative propagation and generative reproduction seems to play a minor role in their life cycle (*Convolvulus arvensis*, *Equisetum arvense*, *Elymus repens*). Furthermore, this group contains species with particularly large seeds such as *Centaurea cyanus*, *Sherardia arvensis*, *G. aparine*, *B. napus* and *Raphanus raphanistrum*. As their seeds have a high capacity for nutrient storage, they can germinate from a depth of several centimetres and produce vigorous seedlings which are able to establish even in a dense vegetation layer (Lambelet-Haueter, 1986; Grundy et al., 2003). Most of the remaining species, the relation between the potential and the apparent density of which ranges between 10:1 and 100:1, are 'normal', therophytic weeds with small-to-medium-sized seeds.

3.2.5 Conclusions

The results of the present study clearly indicate that the change from plough to rotary tillage causes an accumulation of weed seeds at the soil surface. If weed control practice is not adapted to this new situation, these seeds can germinate and may develop plants which cause yield losses and increase the soil seed bank furthermore. To prevent this development, an adaptation of weed control is needed. In the research area, this was achieved by splitting the herbicide applications into several treatments with highly efficient compounds. This practice led to a moderate decrease of the total number of seeds in soil, the number of seedlings emerging in the field, however, still increased. Another option to counteract corresponding weed infestation problems is to re-introduce occasional ploughing. By practising this option, seeds accumulated at the soil surface would be buried to deeper soil layers where many of them would die by fungal and microbial infection, predation, fatal germination, etc. (Pekrun et al., 2003). In contrast, the subsoil which is turned to the surface contains very low number of seeds. Therefore, seedling emergence from soils which were under reduced tillage for a long time can be expected to be low after ploughing. However, ploughing may also increase soil erosion and nitrogen leaching, and it can deplete the activity and diversity of soil organisms. For this purpose, corresponding measures must be carefully balanced with the requirements of non-biotic resource protection.

Another point of concern for practical weed control is that the change to reduced tillage diminished most of the annual weeds including the problematic species *M. recutita* and *Apera spica-venti*. On the other hand, perennials such as *Cirsium arvense* and *Elymus repens*, both of which produce vegetative runners, were favoured by this type of management. An additional species which is difficult to control and which increased after the introduction of rotary tillage is *G. aparine*. Our results suggest that the development of these species demands particular attention and may occasionally require separate control measures.

For the farmers, prediction of weed emergence from seed bank data could be particularly helpful to estimate the need for future weed control measures in good time. The results of the present study have shown that the corresponding relationship between the soil seed bank and the number of weed seedlings in the aboveground vegetation distinctly varies from site to site. Reasons for this variation are differences in the population ecology of the individual species, and complex interactions of the seeds with site and management conditions. Therefore, an early and accurate prediction of weed seedling emergence from seed bank data would necessitate detailed information on the population ecology of the single species and on the influence of environmental factors on their germination and establishment. Costs and time expense for the development of a corresponding model may be too high to be realised.

Acknowledgements

The scientific activities of the FAM Munich Research Network on Agroecosystems were financially supported by the German Federal Ministry of Education and Research (BMBF 0339370). Overhead costs of the Research Station Scheyern were funded by the Bavarian State Ministry for Science, Research and the Arts.

References

Agassi M, 1995. Soil Erosion, Conservation and Rehabilitation. Marcel Dekker, New York, 424 pp.

Albrecht H, 2004. Langfristige Veränderung des Bodensamenvorrates bei pflugloser Bodenbearbeitung. Zeitschrift für Pflanzenkrankheiten und Pflanzenschutz, Special Issue XIX, 97–104.

Albrecht H, Anderlik-Wesinger G, Kühn N, Mattheis A, Pfadenhauer J, 2008. Effects of land use changes on the plant species diversity in agricultural ecosystems. In: Schröder P, Pfadenhauer J, Munch JC (Eds.), Perspectives for agroecosystem management – balancing environmental and socioeconomic demands, pp. 203–235.

Albrecht H, Forster EM, 1996. The weed seed bank of soils in a landscape segment in southern Bavaria. Part I: Experimental site, seed content, species composition and spatial variability. Vegetatio 125, 1–10.

Albrecht H, Pilgram M, 1997. The weed seed bank in a landscape segment in southern Bavaria – II. Relation to environmental factors and to the soil surface vegetation. Plant Ecology 131, 31–43.

Amann A, 1991. Einfluss von Saattermin und Grundbodenbearbeitung auf die Verunkrautung in verschiedenen Kulturen. PhD Thesis, Univ. Hohenheim, 148 pp.

Andreasen C, Stryhn H, Streibig JC, 1996. Decline in the flora of Danish arable fields. Journal of Applied Ecology 33, 619–626.

Anonymous, 2002. SPSS for Windows 11.5. SPSS Inc., Chicago.

Auerswald K, Albrecht H, Kainz M, Pfadenhauer J, 2000. Principles of sustainable land-use systems developed and evaluated by the Munich Research Alliance on Agro-Ecosystems (FAM). Petermanns Geographische Mitteilungen 144(2), 16–25.

Ball DA, Miller SD, 1989. A comparison of techniques for estimation of arable soil seedbanks and their relationship to weed flora. Weed Research 29, 365–373.

Belde M, Mattheis A, Sprenger B, Albrecht H, 2000. Langfristige Entwicklung ertragsrelevanter Ackerwildpflanzen nach Umstellung von konventionellem auf integrierten und ökologischen Landbau. Zeitschrift für Pflanzenkrankheiten und Pflanzenschutz, Special Issue XVII, 291–301.

Benvenuti S, Macchia M, Miele S, 2001. Quantitative analysis of emergence of seedlings from buried weed seeds with increasing soil depth. Weed Science 49, 528–535.

Berger B, 2002. Chenische Verfahren. In: Zwerger P, Ammon HU (Eds.), Unkraut – Ökologie und Bekämpfung. Ulmer Verlag, Stuttgart, pp. 140–204.

Bilalis D, Efthimiadis P, Sidiras N, 2001. Effect of three tillage systems on weed flora in a 3-year rotation with four crops. Journal of Agronomy & Crop Science 186, 135–141.

Bork HR, 1988. Bodenerosion und Umwelt – Verlauf, Ursachen und Folgen der mittelalterlichen und neuzeitlichen Bodenerosion, Bodenerosion- sprozesse, Modelle und Simulationen. Landschaftsgenese Landschaft- sökologie, 13, 249.

Boström U, Fogelfors H, 1999. Type and time of autumn tillage with and without herbicides at reduced rates in southern Sweden. 2. Weed flora and diversity. Soil & Tillage Research 50, 283–293.

Braun-Blanquet J, 1964. Pflanzensoziologie, 3rd ed. Springer Verlag, Wien, 865 pp.

Cardina J, Regnier E, Harrison K, 1991. Long-term tillage effects on seed banks in three Ohio soils. Weed Science 39, 186–194.

Cardina J, Sparrow DH, McCoy EL, 1996. Spatial relationship between seedbank and seedling populations of common lambsquarters (*Chenopodium album*) and annual grasses. Weed Science 44, 298–308.

Cavers PB, Benoit DL, 1989. Seed banks in arable land. In: Leck MA, Parker VT, Simpson RL (Eds.), Ecology of Soil Seed Banks. Academic Press, San Diego, pp. 309–328.

Cooper CM, 1993. Biological effects of agriculturally derived surface water pollutant on aquatic systems – A review. Journal of Environmental Quality 22, 402–408.

Cousens R, 1985. An empirical model relating crop yield to weed and crop density and a statistical comparison with other models. Journal of Agricultural Science 105, 513–521.

Cousens R, Moss SR, 1990. A model of the effects of cultivation on the vertical distribution of weeds within the soil. Weed Research 30, 61–70.

Derksen DA, Lafond GP, Thomas AG, Loepky HA, Swanton CJ, 1993. Impact of agronomic practices on weed communities: Tillage systems. Weed Science 41, 409–417.

Derksen DA, Watson PR, Loeppky HA, 1998. Weed community composition in seedbanks, seedling, and mature plant communities in a multi-year trial in western Canada. Aspects of Applied Biology 51, 43–50.

Dessaint F, Chadoeuf R, Barralis G, 1997. Nine years' soil seed bank and weed vegetation relationships in an arable field without weed control. Journal of Applied Ecology 34, 123–130.

Erviö LR, Salonen J, 1987. Changes in the weed population of spring cereals in Finland. Annales Agriculturae Fenniae 26, 201–226.

Feldman SR, Alzugary C, Torres P, Lewis P, 1997. The effect of different tillage systems on the composition of the seedbank. Weed Research 37, 71–76.

Fogelfors H, Boström U, 1998. Effects of autumn tillage and reduced herbicide doses on the part of the weed seedbank that produce seedlings. Aspects of Applied Biology 51, 229–236.

Frick BL, Thomas AG, 1992. Weed surveys in different tillage systems in southwestern Ontario field crops. Canadian Journal of Plant Sciences 72, 1337–1347.

Friebe B, 1993. Auswirkungen verschiedener Bodenbearbeitungsverfahren auf die Bodentiere und deren Abbauleistung. Informationen zu Naturschutz und Landschaftspflege in Nordwestdeutschland 6, 171–187.

Froud-Williams RJ, Drennan DSH, Chancellor RJ, 1983. Influence of cultivation regime on weed floras of arable cropping systems. Journal of Applied Ecology 20, 187–197.

Gerowitt B, 1992. Dreijährige Versuche zur Anwendung eines Entscheidungsmodells für die Unkrautbekämpfung nach Schadensschwellen in Winterweizen. Zeitschrift für Pflanzenkrankheiten und Pflanzenschutz, Special Issue XIII, 301–310.

Gerowitt B, Heitefuss R, 1990. Weed economic thresholds in the Federal Republic of Germany. Crop Protection 9, 323–331.

Gruber H, Händel K, Broschewitz B, 2000. Einfluss der Wirtschaftsweise auf die Unkrautflora in Mähdruschfrüchten einer sechsfeldrigen Fruchtfolge. Zeitschrift für Pflanzenkrankheiten und Pflanzenschutz, Special Issue XVII, 33–40.

Grundy AC, 2001. Weed emergence and the weather. British Crop Protection Conference – Weeds, Brighton, UK, pp. 75–81.

Grundy AC, Mead A, Bond W, 1996. Modelling the effect of weed-seed distribution in the soil profile on seedling emergence. Weed Research 36, 375–384.

Grundy AC, Mead A, Burston S, 2003. Modelling the emergence response of weed seeds to burial depth: Interactions with seed density, weight and shape. Journal of Applied Ecology 40, 757–770.

Hofmeister H, Garve E, 1998. Lebensraum Acker, 2nd ed. Parey Buchverlag, Berlin, 322 pp.

Knab W, 1988. Auswirkung wendender und nichtwendender Grundbodenbearbeitung auf die Verunkrautung in Abhängigkeit von Fruchtfolge und Unkrautbekämpfung. PhD Thesis, Univ. Hohenheim, 148 pp.

Knab W, Hurle K, 1986. Influence of soil cultivation on weed populations. In: EWRS Symposium on Economic Weed Control, Hohenheim, pp. 309–316.

Kobayashi H, Nakamura Y, Watanabe Y, 2003. Analysis of weed vegetation of no-tillage upland fields based on the multiplied dominance ratio. Weed Biology and Management 3, 77–92.

Kutzelnigg H, 1984. Veränderungen der Ackerwildkrautflora im Gebiet um Moers/Niederrhein und ihre Ursachen. Tuexenia N.S. 4, 81–102.

Lambelet-Haueter C, 1986. Analyse de la flore potentielle, en relation avec la flore réelle, en grandes cultures de la région genevoise. Candollea 41, 299–323.

Legere A, Samson DN, 1999. Relative influence of crop rotation, tillage, and weed management on weed associations in spring barley cropping systems. Weed Science 47, 112–122.

Mayor JP, Maillard A, 1995. Results from an over 20-years-old ploughless tillage experiment at Changins. IV. Seed bank and weed control. Revue Suisse d'Agriculture 27, 229–236.

McClosky M, Firebank LG, Watkinson AR, Webb DJ, 1996. The dynamics of experimental arable weed communities under different management practices. Journal of Vegetation Science 7, 799–808.

Mulugeta D, Stoltenberg DE, 1997. Weed and seedbank management with integrated methods as influenced by tillage. Weed Science 45, 706–715.

O'Donovan JT, McAndrew DW, 2000. Effect of tillage on weed populations in continuous barley (*Hordeum vulgare*). Weed Technology 14, 726–733.

Oerke EC, Dehne HW, Schönbeck F, Weber A, 1994. Crop Production and Crop Protection – Estimated Crop Losses in Major Food and Cash Crops. Elsevier, Dordrecht, 830 pp.

Oerke EC, Steiner U, 1996. Ertragsverluste und Pflanzenschutz. Die Anbausituation für die wirtschaftlich wichtigsten Kulturpflanzen. Schriftenreihe der Deutschen Phytomedizinischen Gesellschaft 6. Ulmer Verlag, Stuttgart, 165 pp.

Otte A, 1996. Populationsbiologische Parameter zur Keimung von Ackerwildkräutern. Zeitschrift für Pflanzenkrankheiten und Pflanzenschutz, Special Issue XV, 45–60.

Pallutt B, 1999. Einfluß von Fruchtfolge, Bodenbearbeitung und Herbizidanwendung auf Populationsdynamik und Konkurrenz von Unkräutern in Wintergetreide. Gesunde Pflanzen 51, 109–120.

Pekrun C, Claupein W, 1998. Forschung zur reduzierten Bodenbearbeitung in Mitteleuropa: eine Literaturübersicht. Pflanzenbauwissenschaften 2, 160–175.

Pekrun C, El Titi A, Claupein W, 2003. Implications of soil tillage for crop and weed seeds. In: El Titi A (Ed.), Soil Tillage in Agroecosystems. CRC Press, Boca Raton, FL, pp. 116–146.

Pollard F, Cussans GW, 1981. The influence of tillage on the weed flora in a succession of winter cereal crops on a sandy loam soil. Weed Research 21, 185–190.

Pollard F, Moss SR, Cussans GW, Froud-Williams RJ, 1982. The influence of tillage on the weed flora in a succession of winter wheat crops on a clay loam soil and a silt loam soil. Weed Research 22, 129–136.

Reents HJ, Küstermann B, Kainz M, 2008. Sustainable land use by organic and integrated farming systems. In: Schröder P, Pfadenhauer J, Munch JC (Eds.), Perspectives for agroecosystem management – balancing environmental and socioeconomic demands, pp. 17–39.

Roberts HA, 1984. Crop and weed emergence patterns in relation to time of cultivation and rainfall. Annuals of Applied Biology 105, 263–275.

Robinson RA, Southerland WJ, 2002. Post-war changes in arable farming and biodiversity in Great Britain. Journal of Applied Ecology 39, 157–176.

Röttele M, Koch W, 1981. Verteilung von Unkrautsamen im Boden und Konsequenzen für die Bestimmung der Samendichte. Zeitschrift für Pflanzenkrankheiten und Pflanzenschutz, Special Issue IX, 383–391.

Schröder P, Huber B, Reents HC, Munch JC, Pfadenhauer J, 2008. Outline of the Scheyern project. In: Schröder P, Pfadenhauer J, Munch JC (Eds.), Perspectives for agroecosystem management – balancing environmental and socioeconomic demands, pp. 3–16.

Schwerdtle F, 1977. Der Einfluss des Direktsäverfahrens auf die Verunkrautung. Zeitschrift für Pflanzenkrankheiten und Pflanzenschutz, Special Issue 8, 155–163.

Skuterud R, Semb K, Saur J, Mygland S, 1996. Impact of reduced tillage on the weed flora in spring cereals. NJAS – Wageningen Journal of Life Sciences 10, 519–532.

Sokal RR, Rohlf FJ, 1998. Biometry: The Principles and Practice of Statistics in Biological Research, 3rd ed. Freeman and Company, New York, 887 pp.

Sprenger B, 2004. Populationsdynamik von Ackerwildpflanzen im integrierten und organischen Anbausystem. PhD Thesis, TU München-Weihenstephan, http://mediatum.ub.tum.de, 151 pp.

Streit B, Rieger SB, Stamp P, Richner W, 2002. The effect of tillage intensity and time of herbicide application on weed communities and populations in maize in central Europe. Agriculture, Ecosystems and Environment 92, 211–224.

Streit B, Rieger SB, Stamp P, Richner W, 2003. Weed populations in winter wheat as affected by crop sequence, intensity of tillage and time of herbicide application in a cool and humid climate. Weed Research 43, 20–32.

Swanton CJ, Clements DR, Derksen DA, 1993. Weed succession under conservation tillage: A hierarchical framework for research and management. Weed Technology 7, 286–297.

Swanton CJ, Shresta A, Roy R, Ball-Coelho BR, Knezevic SZ, 1999. Effect of tillage systems, N, and cover crop on the composition of weed flora. Weed Science 47, 454–461.

Teasdale JR, Beste CE, Potts WE, 1991. Response of weeds to tillage and cover crops residue. Weed Science 39, 195–199.

Tebrügge F, 2001. Conservation tillage-protection of soil, water, and climate and influence on management and farm income. In: Garcia-Torres L, Benites J, Martinez-Vilela A (Eds.), Conservation Agriculture – A Worldwide Challenge, ECAF and FAO, Madrid, pp. 303–316.

Thomas AG, Frick BL, 1993. Influence of tillage systems in weed abundance in southwestern Ontario. Weed Technology 7, 699–705.

Tørresen KS, 2003. Relationship between seedbanks and emerged seeds in long term tillage experiments. Aspects of Applied Biology 51, 55–62.

Tørresen KS, Skuterud R, 2002. Plant protection in spring cereal production with reduced tillage. IV. Changes in the weed flora and weed seedbank. Crop Protection 21, 179–193.

Tuesca D, Puricelli E, Papa JC, 2001. A long-term study of weed flora shifts in different tillage systems. Weed Research 41, 369–382.

Vanhala P, Pitkänen J, 1998. Long-term effects of primary tillage on aboveground weed flora and on the weed seedbank. Aspects of Applied Biology 51, 99–104.

Vleeshouwers LM, 1997. Modelling Weed Emergence Patterns. PhD Thesis, Wageningen Agricultural University.

Wilson JD, Morris AJ, Arroyo BE, Clark SC, Bradbury RB, 1999. A review of the abundance and diversity of invertebrate and plant foods of granivorous birds in northern Europe in relation to agricultural change. Agriculture, Ecosystems and Environment 75, 13–30.

Wisskirchen R, Haeupler H, 1998. Standardliste der Farn- und Blütenpflanzen Deutschlands. Ulmer Verlag, Stuttgart, 765 pp.

Zanin G, Otto S, Riello L, Borin M, 1997. Ecological interpretation of weed flora dynamics under different tillage systems. Agriculture, Ecosystems and Environment 66, 177–188.

Zwerger P, Malkomes HP, Nordmeyer H, Söchting HP, Verschwele A, 2004. Unkrautbekämpfung: Gegenwart und Zukunft – aus deutscher Sicht. Zeitschrift für Pflanzenkrankheiten und Pflanzenschutz, Special Issue XIX, 27–38.

Chapter 3.3

Dispersal Strategies: Are They Responsible for Species Success in Arable Ecosystems?

F. Mayer and H. Albrecht

3.3.1 Introduction

A greater knowledge of diaspore dispersal may lead to a better understanding of species' success in agricultural landscapes. To be successful means to establish stable populations on each potential habitat. Perhaps the most successful species are those whose dispersal is promoted or at least not disadvantaged by human activity. Hodgson and Grime (1990) suggested the existence of a strong selection pressure for high dispersal ability due to disruptive effects of today's farming. They found an increasing number of species with efficient dispersal strategies either in space (long distance dispersal) or in time (persistent seed bank). Are species which are able to use efficient dispersal mechanisms the more successful ones in arable ecosystems?

Perspectives for Agroecosystem Management
Edited by P. Schröder, J. Pfadenhauer and J.C. Munch

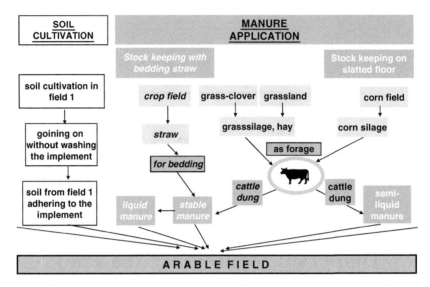

Figure 1 Dispersal opportunities for weed diaspores on a cattle farm. The arrows represent the direction of dispersal (Mayer, 2000).

There are two crucial factors which determine dispersal success: the species' dispersal ability and its opportunities to 'catch' a dispersal agent. The first depends on the species' morphology and physiology, the second on the individual growing situation of a plant. In agricultural landscapes this growing situation is heavily influenced by human activity. The diversity of man-made ways of dispersal is shown in Figure 1. Analysing the dispersal agents which cause the dispersal processes inevitably leads to the entire spectrum of dispersed species and, therefore, to the knowledge which species benefit from this agent and which do not. Such an analysis results in an estimation of the efficiency of a dispersal agent within a system. An estimation of the dispersal ability of a certain species requires an analysis of all its dispersal strategies and potential dispersal opportunities.

3.3.2 Specific ways of dispersal in arable ecosystems

Dispersal by manure application

One agricultural process which was suggested to be important for plant species dispersal is the application of organic fertiliser. The number and species of viable seeds in manure, forage and bedding straw from farms of different farming intensities were determined (Hodgson and Grime, 1990; Mayer and Albrecht, 1998; Mayer, 2000). The results allowed (i) to estimate the general dispersal efficiency of this process by calculating a seed-output-to-input-ratio representing the percentage of seeds which viably passed through the system cattle farm (Table 1), (ii) to compare the dispersal efficiency of different farming systems (Table 1), (iii) to estimate the seed input into the arable field (Table 2) and (iv) to assess the chance for species to profit from this process. Depending on the farm, 3% to 40% of the seed input by forage and bedding remained viable in manure (Table 1). The efficiency of seed transport within the material-cycle of a farm was higher when cattle were kept on litter than when keeping them on a slatted floor, at a low intensity of weed control, when the main forage was hay instead of silage, when the seed numbers in the original material were high especially in straw and of course it was dependent on the manuring regime. Species richness in manure was better correlated with species richness in straw than in forage. A maximum potential input into the field of 2080 seeds m^{-2} was calculated (Mayer, 2000).

Three groups of species were found in manure (Table 4): (i) Grassland species which originated mainly from forage are less important as arable weeds (except *Rumex obtusifolius*); studies of changes in abundance of species on the research farm showed an increase of grassland species in arable fields (Albrecht et al., 2008); in former times the special adaptations of grassland species to digestion were used by farmers [e.g. clover species such as *Trifolium repens, Medicago lupulina, Lotus corniculatus* (Boeker, 1959; Gardener et al., 1993) or *Plantago-* and *Juncus* species (Boesewinkel and Bouman, 1995; Grime et al., 1981; Müller-Schneider, 1986)]; they rendered grazing animals into seed distributors (Barrow and Havstad, 1992; Gardener et al., 1993; Lennartz, 1957; Özer, 1979; Russi et al., 1992). (ii) Arable weeds with mature seeds at harvest time at a harvestable height (*Apera spica-venti, Tripleurospermum inodorum, Elymus repens*) made most of the seeds in manure. (iii) For those species whose seeds reached the manure heap and the semi-liquid manure pit supplementary [e.g. *Tussilago farfara, Clematis vitalba, Solidago-* and *Erigeron* species (Müller-Schneider, 1986)] this way of dispersal means an elongation of their anyway long dispersal distances.

The analysis of organic manure failed to demonstrate the presence of seeds of rare weed species, although some of them were present as weeds in the fields of the research farm (Albrecht and Mattheis, 1998; Albrecht et al., 2008). This observation rose the question whether germinable seeds of rare species such as *Papaver argemone, P. dubium, Legousia speculum-veneris, Centaurea cyanus, Spergula arvensis* and *Trifolium arvense* were not found because of being too few to be sampled or because of their loss of germinability during storage in silage, stable or semi-liquid manure or during digestion (Mayer et al., 2000). If the latter proves to be true for such species the chance of long distance dispersal is not increased by manure application. The germination rates of the six species varied between only 0.4% and 76% and each species showed a decrease after

Table 1 Material consumption and manure production on four farms of increasing farming intensity (1, 2, 3, 4) and the calculated seed numbers, transported on the farms in one season.

Farm	Consumed material and produced manure/farm, year (t dw)				Calculated seeds/farm, year (in thousand)			
	1	2	3	4	1	2	3	4
Maize silage (1)	–	–	21.3	89.3	–	–	243	0
Grass silage (2)	48	95.3	12.5	43.2	313	3 358	0	0
Hay (3)	16.5	12.6	12.5	53.2	30 360	1 259	20 563	444 742
Straw (4)	43	50.3	18.5	–	64 787	6 608	12 991	–
Total input [Σ(1–4)]	107.5	158.2	64.8	185.7	95 460	11 225	33 797	444 742
Stable manure (5)	42.3	64.2	23.4	–	9 400	3 162	1 401	–
Stored stable manure (6)	14.1	20.2	7.8	–	1 876	1 079	134	–
Liquid manure (7)	1.8	–	3.3	–	37	–	22	–
Semi-liquid manure (8)	–	7.5	–	76.3	–	267	–	11 316
Total output [Σ(5–8)]	58.2	91.9	34.5	76.3	11 313	4 508	1 557	11 316
Output/input	0.54	0.58	0.53	0.41	0.12	0.40	0.05	0.03

Rows (1) to (4) represent the input into the system, rows (5) to (8), the output; the left 4 columns show the materials' input–output-system, the right ones the corresponding numbers of seeds. The last row shows how much of the consumed material and how many of the introduced seeds leave the system again. No sampling (from Mayer, 2000).

Table 2   Numbers of seeds that potentially reached 1 m^2 in one season by manure application.

Farm	Crop / field	Organic fertilisation 1998			Seeds m^{-2}
		Stable manure t ha^{-1}	Stored stable manure t ha^{-1}	Semi-liquid manure m^3 ha^{-1}	Mean
1	Winter-wheat	–	40	55	136
1	Winter-rye	40	–	–	223
1	Grass-clover/1	20	–	–	111
1	Grass-clover/2	40	–	–	223
2	Winter-wheat/1	–	16.9	22	26
2	Winter-wheat/2	–	18.1	39	30
2	Winter-rye	–	23.9	–	32
2	Grass-clover/1	–	–	26	4
2	Grass-clover/2	9.7	–	38	17
2	Potatoes/1	28.8	–	–	35
2	Potatoes/2	28.1	–	–	34
4	Winter-wheat/1	–	–	32	19
4	Winter-wheat/2	–	–	15	9
4	Winter-wheat/3	–	–	30	18
4	Maize	–	–	38	23

Data: applied manure on farms 1, 2 and 4 in the year 1998 (on farm 3 the exact manure applications per field were not recorded) (with permission from Mayer, 2000).

storage (see also Elema et al., 1990; Özer, 1979) or digestion (Figure 2; see also Eisele, 1997; Elema et al., 1990; Kempski, 1906; Lennartz, 1957; Sarapatka et al., 1993). After 6 months storage in silage all species except *Papaver dubium* kept their germinability. Three months later *Centaurea cyanus* did not germinate any more and after 12 months only *Trifolium arvense* seedlings emerged. Some *Trifolium arvense* seeds also remained germinable after 1-year storage in semi-liquid and stable manure and even after digestion. Russi et al. (1992) even found a higher germination rate of seeds of three *Trifolium* species after digestion due to a higher percentage of soft seeds. Besides *Trifolium arvense, Spergula arvensis* and *Papaver dubium* succeeded in germinating after being stored in semi-liquid manure for 3 months (Mayer et al., 2000). Species relatively resistant against digestion are *Chenopodium album, Atriplex hortense* (Atkeson et al., 1934; Kempski, 1906), *Plantago lanceolata, P. media* (Kempski, 1906; Lennartz, 1957), *Polygonum lapathifolium, Rumex acetosella, R. obtusifolius* (Chytil, 1986; Kempski, 1906; Lennartz, 1957; Sarapatka et al., 1993), *Medicago sativa* (Atkeson et al., 1934), *Agrostis alba, A. tenuis* and *Phleum pratense* (Lennartz, 1957; Özer, 1979). In semi-liquid manure

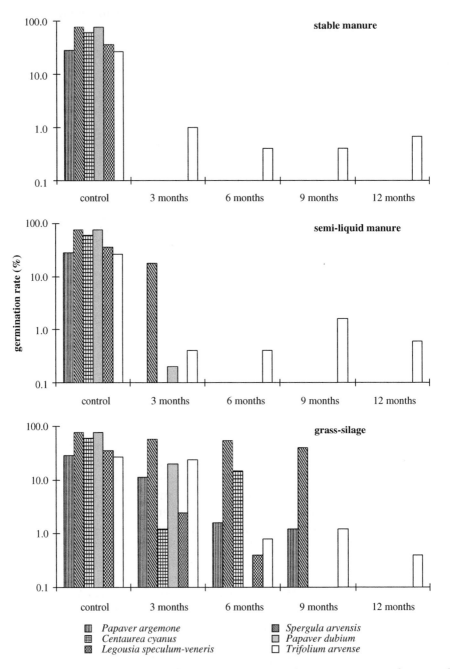

Figure 2 Germination rates of the test species after storage in silage and stable and semi-liquid manure (Mayer, 2000).

seeds of *Chenopodium album, Rumex obtusifolius* and *Echinochloa crus-galli* stayed viable for the longest time referring to Besson et al. (1986), Elema et al. (1990) and Sarapatka et al. (1993). Chytil (1986) found seeds of *Polygonum lapathifolium* and *Amaranthus retroflexus* to be resistant in this medium. The effect of storage in silage was only studied by Chytil (1986) and Elema et al. (1990).

Only seeds of *Abutilon theophrasti*, which was scarcely affected by the procedure, and a few of *Chenopodium album* survived the treatment. Obviously, the species tested by Mayer et al. (2000) are less sensitive to ensiling.

Dispersal by soil cultivation

Tillage is considered to be important in secondary long distance dispersal of weed seeds in two ways: of seeds which appear on the soil surface due to seed rain and of seeds contained in the soil seed bank. A sowing experiment showed that each of five implements (plough, heavy cultivator, rotary tiller, rotary harrow, curry comb) and even the tractor alone carried seeds 23 m along and further out of a plot (Mayer et al., 1998; Mayer, 2000). Howard (in Cousens and Mortimer, 1995) documented a seed transport by tillage of up to 2 m, Moss (1988) of 15 m and Fogelfors (1985) of 35 m. Rew and Cussans' (1997) experiments resulted after 2 years of customary farming in a maximum dispersal distance of 4.5 m. Seed export out of the plot occurred mainly under moist soil conditions and by the plough and the heavy cultivator, called the 'soil-adhering-group' (SAG) of implements, as seed transport mainly resulted from adhering soil. Within the plot the 'pure-seeds-group' (PSG: seeds do not have to be enclosed in soil to be transported: curry comb, rotary harrow) was more successful in seed dispersal, in particular when a mulching layer was present on the soil surface. This was also shown by Fogelfors (1985). In general, the PSG was most efficient in the transport of seeds within the field and led to a high evenness of seed distribution; whereas the SAG showed high seed export rates and caused an uneven, hence, an unpredictable seed distribution. Similar results were found by Schippers et al. (1993).

Seed attributes also play a role in dispersal by tillage. Big seeds benefit from high amounts of adhering soil and also from mulching material. Elongated seeds get even easier entangled in mulching material and, thus, are supported by litter to the highest degree. Small seeds require only little soil clods to adhere to the surface of implements, but fall easily through dragged litter (Mayer, 2000).

The soil samples collected from the implements contained both seeds which were sown and seeds which were incorporated in the soil seed bank. The seed bank species which emerged from the collected soil had a lower average seed weight than the species of the recorded aboveground

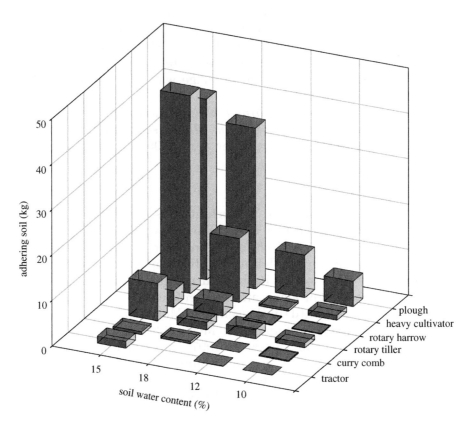

Figure 3 Amounts of soil adhering to the implements after cultivation under different soil moisture conditions (Mayer, 2000).

vegetation (Mayer et al., 2002). However, referring to Bakker et al. (1996) and Thompson et al. (1998), this only may reflect the seed size distribution typical for a soil seed bank as small and compact seeds are most easily buried. The total number of carried seeds was highly correlated with the amount of adhering soil as already shown by Schippers et al. (1993) (Figure 3, Table 3).

3.3.3 Dispersal by human activity – a dispersal strategy or a random process?

Plant species benefit most from a dispersal agent when their seed morphology is adapted to the specific agent. They are more likely to be taken up by this agent than those seeds which are devoid of adaptation. Some dispersal processes occur completely independently of seed attributes – both

Table 3   Collected soil, seed numbers and species by five implements and a tractor at four different soil moistures.

		Plough	Heavy cultivator	Rotary tiller	Rotary harrow	Curry comb	Tractor
18% swc	Collected soil (g)	35 390	14 024	1602	2988	1601	–
	Number of seeds	1 945	202	320	658	460	–
	Seeds/kg soil	55	14	200	220	866	–
	Number of species	35	19	30	25	11	–
15% swc	Collected soil (g)	39 752	43 560	8206	3650	602	1504
	Number of seeds	1 765	844	161	119	20	6
	Seeds/kg soil	44	19	20	33	33	4
	Number of species	37	28	26	15	3	4
12% swc	Collected soil (kg)	9 312	500	1962	129	7	13
	Number of seeds	29	10	69	3	5	2
	Seeds/kg soil	3	20	35	23	760	152
	Number of species	9	3	21	2	3	2
10% swc	Collected soil (kg)	5 592	1 060	990	161	235	–
	Number of seeds	29	13	11	13	2	–
	Seeds/kg soil	5	12	11	81	9	–
	Number of species	11	11	7	8	2	–

swc: soil water content (with permission from Mayer, 2002).

natural and human-caused processes such as dispersal of seeds in the soil seed bank. As such dispersal events occur accidentally a species is more likely to be disseminated, the more seeds it offers the agent. Therefore, random dispersal mechanisms promote species of high frequency and/or with a high seed production.

Higgins et al. (2003) concluded that in general there is only a poor relationship between potential species' dispersal strategies derived from its seed morphology and the actual way of dispersal. Whereas Eriksson (2000) suggested that species with a high dispersal and colonisation ability are characterised by an intermediate seed size.

The analysis of various farming activities according to their efficiency as dispersal opportunities for plant diaspores yielded the species listed in Table 4. That some species are specifically adapted e.g. to dispersal by wind (e.g. plumed seeds) is already known; but are there adaptations to such human-caused ways of dispersal? If not, dispersal is equally likely for each present diaspore. Thus, the more often a species occurs and the more diaspores it produces the bigger the chance to be involved in such a random process. This would mean that rare species are disadvantaged as against abundant species.

Tables 5 and 6 show plant species attributes which turned out to be conducive or obstructive to dispersal by manure application or tillage. Of course, of those species which are suitable for such dispersal agents the ones which are more abundant will be involved more often than the rare ones. This may be a reason why *Trifolium arvense* despite its resistance shown in the storage experiment was not found when the dispersal media were analysed. Seeds of the six rare test species are up to 1.5 mm (except *Centaurea cyanus*: 4 mm) in size and more or less spherical. Only *Centaurea cyanus* achenes are elongated with a short haired, bristly pappus. This morphology enables *Centaurea cyanus* seeds which are lying on the soil surface to be dispersed by soil cultivation and mechanical weed control particularly when mulching material is present. The tiny seeds of the other five species only need little soil to adhere to implements. Thus, even with relatively dry soil they have the chance to be dragged away during cultivation. Certainly, under moist conditions there is a higher probability for all species to adhere to the implements. The process is then similar to dispersal of seeds present in the soil seed bank which is a rather random event, independent of seed attributes. Such random dispersal events certainly occur more often to a species, the more seeds it makes available to the agent. Consequently, due to their seed attributes the test species (except *Trifolium arvense*) are not suitable for dispersal in manure, although they may be present in feed and straw. Their attributes foster them to be dispersed by soil cultivation when lying on the soil surface.

Table 4   The species which were found in manure and in soil samples adhering to tillage implements, to a tractor or a curry comb and their dispersal strategies (except the hemerochorous ones) according to Müller-Schneider (1986).

	Manure			Tillage				Long distance dispersal strategies								Other dispersal strategies							
	a	b	c	d	e	f	g	1	2	3	4	5	6	7	8	9	1C	11	12	13	14	15	16
Agropyron repens	X	X					X	X							X								
Agrostis stolonifera		X													X		X						
Agrostis tenuis			X												X		X						
Alopecurus pratensis			X		X										X								
Amaranthus retroflexus					X			X															
Anagallis arvensis				X				X														X	
Anthemis arvensis							X	X															
Apera spica-venti		X					X	X							X								
Aphanes arvensis		X					X								X								
Arabidopsis thaliana							X											X					
Arenaria serpyllifolia			X														X	X					
Arrhenatherum elatius		X													X	X	X						
Atriplex patula						X									X	X	X						
Avena pubescens					X										X	X							
Bromus hordeaceus							X																
Capsella bursa-pastoris			X				X					X				X	X						
Cardamine pratensis	X																			X			
Cerastium glomeratum							X											X					
Cerastium holosteoides			X				X											X					
Chenopodium album			X				X	X	X							X	X						
Chenopodium polyspermum								X	X						X								

(Continued)

Table 4 (Continued)

	Manure			Tillage				Long distance dispersal strategies								Other dispersal strategies							
	a	b	c	d	e	f	g	1	2	3	4	5	6	7	8	9	10	11	12	13	14	15	16
Clematis vitalba	X							X			X				X	X							
Dactylis glomerata			X				X								X	X							
Echinochloa crus-galli	X		X		X					X													
Erigeron annuus											X												X
Euphorbia exigua				X			X	X								X	X			X			
Fallopia convolvulus		X													X	X	X						
Festuca pratensis		X													X	X							
Festuca rubra		X						X								X	X						
Galeopsis tetrahit					X											X							
Galinsoga ciliata							X																
Galium aparine		X				X	X			X													
Geranium dissectum				X							X									X			
Gnaphalium uliginosum						X									X								
Holcus lanatus	X						X									X							
Juncus bufonius							X								X			X					
Juncus effusus				X																			
Lactuca serriola				X			X				X												
Lamium purpureum							X								X								
Lolium multiflorum			X		X		X								X	X							
Lolium perenne			X				X	X								X							
Lotus corniculatus			X																X	X			
Matricaria chamomilla		X					X										X		X				
Matricaria discoidea				X													X	X					
Medicago lupulina		X						X															

Species	1	2	3	4	5	6	7	8	9	10	11	12
Mentha arvensis			×									
Myosotis arvensis	×			×	×	×			×			
Phleum pratense	×			×		×		×		×	×	×
Plantago intermedia		×		×		×					×	
Plantago lanceolata	×				×						×	×
Plantago major		×		×	×						×	×
Poa annua		×		×	×	×	×	×			×	×
Poa pratensis	×			×			×				×	
Poa trivialis		×		×			×				×	
Polygonum aviculare	×			×	×	×					×	
Polygonum lapathifolium	×			×	×						×	×
Polygonum persicaria			×		×						×	
Raphanus raphanistrum				×	×							
Rumex crispus	×				×					×		
Rumex obtusifolius		×		×						×		×
Solidago virgaurea	×						×					
Sonchus arvensis			×		×		×					
Sonchus asper				×		×	×				×	
Sonchus oleraceus			×		×	×	×				×	
Stellaria media		×		×	×	×					×	×
Taraxacum officinalis				×			×					
Thlaspi arvense				×							×	
Trifolium campestre	×									×	×	
Trifolium pratense		×				×				×	×	×
Trifolium repens		×		×	×	×					×	×
Tripleurospermum inodorum		×	×									
Tussilago farfara	×		×		×		×					
Urtica dioica		×		×	×					×	×	×
Veronica arvensis				×				×			×	
Veronica chamaedris	×										×	

(Continued)

Table 4 (Continued)

	Manure			Tillage				Long distance dispersal strategies								Other dispersal strategies							
	a	b	c	d	e	f	g	1	2	3	4	5	6	7	8	9	10	11	12	13	14	15	16
Veronica hederifolia					×																	×	×
Veronica persica							×															×	×
Veronica serpyllifolia			×									×					×						
Vicia hirsuta					×			×								×	×			×			
Viola arvensis						×																	×

a: only semi-liquid manure; b: only stable manure; c: all kinds of manure; d: only at soil dry matter (dm) of 85%; e: only at soil dm of 82%; f: only at soil dm of 90%; g: at various soil moistures; 1–7: natural LDDS; 1: dysochorous (birds); 2: endochorous (deer); 3: epichorous; 4: trichometeorochorous; 5: hydrochorous; 6: dysochorous (mice); 7: endochorous (not specified); 8: pterometeorochorous; 9: endochorous (domestic animals); 10: endochorous (cattle); 11: boleochorous; 12: dysochorous (ants);13: ballochorous; 14: endochorous (earth worm); 15: blastochorous; 16: stomatochorous (with permission from Mayer, 2000).

Table 5   Species attributes which are conducive or obstructive to dispersal by manure application (Mayer, 2000).

Species attributes conducive to dispersal by manure application	Way of dispersal and conditions	Species attributes obstructive to dispersal by manure application
Mature at harvest time height > 15 cm	Diaspores in straw or hay	Seeds not yet mature or already shed at harvest time Very small growing
Resistant to ensiling, due to hardseededness, small spherical seeds with a smooth surface	Diaspores in silage	Not resistant to ensiling due to unsuitable attributes
Diaspores get into straw	Diaspores in stable manure	No diaspores get into straw
Resistant to digestion, due to hardseededness, small spherical seeds with a smooth surface or mucigenous seeds	Diaspores in stable manure and slurry	Not resistant to digestion due to unsuitable attributes
Natural long distance dispersal strategy (LDDS) (e.g. anemochory)	External factors to get into manure	No natural LDDS
Resistant to storage in manure, due to hardseededness, small spherical seeds with a smooth surface	Being spread with manure	Resistant to storage in manure due to unsuitable attributes

3.3.4 Dispersal opportunity and species success

Table 4 shows all potential dispersal strategies of species found in manure or in the soil adhering to tillage implements. A correlation analysis between the development of species abundance in arable fields of the research farm (Albrecht et al., 2008) and species' opportunities for long distance dispersal yielded the Pearson correlation coefficients shown in Table 7. In general, the organic farming system showed better correlations between dispersal and the development of species abundance than the integrated system. In the latter one only potential dispersal by manure seems to have an effect on species success. Whereas in the organic farming system the possibility of being dispersed by manure application and by tillage obviously influenced species success. As there is no correlation between natural long distance dispersal opportunities and species success the high correlation coefficient between the sum of human-caused and

Table 6   Species attributes which are conducive or obstructive to dispersal by tillage.

Species attributes conducive to dispersal by tillage	Way of dispersal and conditions	Species attributes obstructive to dispersal by tillage
Seeds present in the soil seed bank (SSB), seeds just shed	Dispersal by tillage (in general)	Not yet mature seeds at time of tillage, no seeds in the SSB
	Seed rain:	
High seed production	Presence on the soil	Low seed production
High abundance	Surface	Rareness
Independent of seed shape	Dispersal by the SAG	Independent of seed shape
Big, elongated seeds	Dispersal by the PSG	Small, spherical seeds
Independent of seed shape	Dispersal at moist soil	Independent of seed shape
Small, elongated seeds	Dispersal at dry soil	Big, spherical seeds
Big, elongated seeds	With mulching material	Small, spherical seeds
	Seed bank:	
High seed production	Presence of seeds in adhering soil	Low seed production
High abundance		Rareness
Persistent seed bank		Transient seed bank
Gnaphalium uliginosum, Juncus bufonius, Veronica hederifolia able to regenerate out of root fragments (*Cirsium arvense*) or rhizomes (*Agropyron repens*)	Transport of vegetative diaspores	No vegetative reproduction

SAG: soil-adhering-group; PSG: pure-seeds-group (explanation see text) (Mayer, 2000).

Table 7   Pearson correlation coefficients between changes in species abundance and dispersal opportunities in the organic and the integrated farming system at the research station.

Pearson correlation coefficient	Changes in abundance	
	Organic	Integrated
Dispersal by manure (DM)	0.32	0.34
Dispersal by tillage (DT)	0.41	−0.09
Natural LDDS	0.03	0.19
Other dispersal strategies	0.20	0.06
Σ (DM, DT, natural LDDS)	0.43	0.13

natural ways of dispersal and species success must be caused by human activity. As one principle of integrated farming is minimum tillage it is not surprising that there is no correlation (even a slightly negative one) between dispersal by tillage and species abundance. The very low correlation between natural long distance dispersal strategy (LDDS) and species success, particularly in the organic system, may be an indication that natural dispersal opportunities cannot be used efficiently. Endochorous, epichorous or dysochorous dispersal only can occur when the specific animals pass and contact the plants. Dispersal by wind only can be efficient when the plants expose their diaspores to the wind and when the diaspores can cover a distance without hindrance. Within an arable field conditions often are not suitable for such dispersal agents especially when species are small growing. Whereas Lake and Leishman (2004) showed that in other habitats wind dispersal leads to a successful invasion of exotic species.

Obviously, on the research station a species' success is influenced by the chance to be dispersed by farming operations.

3.3.5 *Cirsium arvense* and *Veronica triphyllos* – a successful and a stagnant species

In order to show the situation of a very successful species (in terms of habitat maintenance and colonisation) the long distance dispersal ability and conditions of *Cirsium arvense* on the research farm were estimated. Abundance, generative reproduction, generative dispersal potential and long distance dispersal ability of diaspores of *Cirsium arvense* were determined (Mayer, 2000; Mayer and Albrecht, 2003). On the fallow lands the ratio of female to male stands was 2.5:1 and the distance between them in most cases was shorter than 100 m. Generative reproduction varied between 7500 and 21 000 well developed achenes m^{-2} within a thistle stand. Approximately 85% of this reproduction potential represented the dispersal potential (achenes that leave the head with pappus). Long distance dispersal was documented to 139 m for generative dispersal and to 24 m for root transport by tillage. An example of application – dispersal from a fallow land into adjacent arable fields on the research farm – showed, that under defined conditions, up to 10 well developed achenes m^{-2} can be found at a distance of 139 m (Mayer, 2000; Mayer and Albrecht, 2003). These results may be seen as specific for the FAM-research farm as other authors found different data. Oesau (1992) e.g. recorded a female to male ratio of 300:1 and concluded that pollination cannot be sufficient for building germinable seeds. Bakker (1960) doubted the generative dispersal ability of *Cirsium arvense* as he found a very weak connection between pappus and achene. A dispersal distance of achenes of 50 m is documented by Oesau (1998). These differences in results may be caused by the great variety of ecotypes of *Cirsium arvense* (Hodgson, 1964).

Within the last 13 years this species succeeded in colonising most of its potential habitats on the research farm. Firstly, as *Cirsium arvense* grows taller than most crop species it is able to use its wind dispersal strategy efficiently even when growing in a crop field; all the more as the efficiency of wind dispersal primarily depends on the frequency of updrafts more than on actual horizontal wind speed (Mayer, 2000; Nathan et al., 2002; Tackenberg et al., 2003). Secondly, this natural dispersal strategy rather is promoted by human activity than disadvantaged and, last, human activity offers additional opportunities for long distance dispersal. Wind is relatively little affected by man and rather less in modern land-use systems where hedgerows are the victims of maximising arable field sizes. A factor indirectly contributing to the expansion of this species' distribution is the increased area of set aside fields where for example *Cirsium arvense* found optimal conditions for colonisation. These set aside fields are seed sources for surrounding arable fields, in particular, in case no hedgerows exist which would prevent seeds from flying across the boundaries.

The roots of *Cirsium arvense* are responsible for vegetative growth and, therefore, for the maintenance of the thistle stand. Human activity, in particular soil cultivation, resulted in a further opportunity of long distance dispersal of *Cirsium arvense*: the transport of root fragments.

Manure application obviously is a farming operation which does not contribute to the thistle's dispersal. No thistle seeds were found in either material of the 'cattle-farm-cycle'. Certainly, the thistle as competitive species (Grime et al., 1988) benefits from the nutrient rich sites resulting from manuring where it drives out less productive species.

The rare species *Veronica triphyllos* was found neither in manure nor in soil adhering to tillage implements. As it is growing only about 20 cm high and seeds are already shed at crop harvest there is almost no chance for seeds to get into straw and later on into manure. A transport by soil working implements which under certain conditions occurs accidentally is also unlikely because of the species' rareness. For the stagnation of the distribution of *Veronica triphyllos* on the research farm despite the existence of appropriate unoccupied habitats (Albrecht et al., 2000) lacking dispersal opportunities may be the reason. Even if Frank and Klotz (1990) document dispersal by wind, water and ants (Müller-Schneider, 1986, documents only boleochorie) for *Veronica triphyllos* it is growing too small to really use the wind efficiently. Dispersal by ants is only of short range and water may only be an efficient dispersal agent on an arable field situated on a slope which anyway is not appropriate for arable agriculture.

Cirsium arvense and *Veronica triphyllos* are two contrasting species both in success and in dispersal opportunities. *Cirsium arvense* is a species with efficient natural LDDSs and its distribution is – both indirectly and directly – supported by modern farming systems. In contrast, *Veronica triphyllos* neither is adapted to efficient long distance dispersal mechanisms (neither natural nor human-caused) nor is it abundant enough to be involved in accidental dispersal processes.

3.3.6 Conclusions

In conclusion the relation between species success and dispersal can be seen as a feedback process: efficient dispersal increases the species' success which again increases the species' chance to be dispersed. Thus, to be really successful a species first needs specific adaptations to dispersal media to get abundant enough to get involved in accidental dispersal processes.

Acknowledgements

The scientific activities of the *FAM Munich Research Network on Agroecosystems* were financially supported by the German Federal Ministry of Education and Research (BMBF 0339370). Overhead costs of the Research Station Scheyern are funded by the Bavarian State Ministry for Science, Research and the Arts.

References

Albrecht H, Anderlik-Wesinger G, Kühn N, Mattheis A, Pfadenhauer J, 2007. Effects of land use changes on the plant species diversity in agricultural ecosystems. In: Schröder P, Pfadenhauer J, Munch JC (Eds.), Perspectives for agroecosystem management – balancing environmental and socioeconomic demands, pp. 203–235.

Albrecht H, Mattheis A, 1998.The effects of organic and integrated farming on rare arable weeds on the Forschungsverbund Agrarökosysteme München (FAM) research station in southern Bavaria. Biological Conservation 86, 347–356.

Albrecht H, Mayer F, Mattheis A, 2000. *Veronica triphyllos L.* in the Tertiärhügelland landscape in southern Bavaria – An example for habitat isolation of a stenoeceous plant species in agroecosystems. Zeitschrift für Ökologie und Naturschutz 8, 219–226.

Atkeson FW, Hulbert HW, Warren TR, 1934. Effect of bovine digestion and of manure storage on the viability of weed seeds. Agronomy 26, 390–397.

Bakker D, 1960. A comparative life-history study of *Cirsium arvense* L. scop. and *Tussilago farfara* L., the most troublesome weeds in the newly reclaimed polders of the former zuiderzee. In: Harper JL (Ed.), The Biology of Weeds. Blackwell Scientific Publications, Oxford, pp. 205–222.

Bakker JP, Poschlod P, Strykstra RJ, Bekker RM, Thompson K, 1996. Seed banks and seed dispersal – Important topics in restoration ecology. Acta Botanica Neerlandica 45, 461–490.

Barrow JR, Havstad KM, 1992. Recovery and germination of gelatin-encapsulated seeds fed to cattle. Journal of Arid Environments 22, 395–399.

Besson JM, Schmitt R, Lehmann V, Soder M, 1986. Unterschiede im Keimungsverhalten von Unkrautsamen nach Behandlung mit gelagerter, belüfteter und methanvergorener Gülle. Mitt. für Schweizer Landwirte 35, 73–80.

Boeker P, 1959. Samenauflauf aus Mist und Erde von Triebwegen und Ruheplätzen. Zeitschrift für Acker- und Pflanzenbau 108, 77–92.

Boesewinkel FD, Bouman F, 1995. The seed: Structure and function. In: Kigel J, Galili G (Eds.), Seed Development and Germination. Marcel Dekker, Inc., New York, Basel, Hong Kong, pp. 853.

Chytil K, 1986. Untersuchungen zur Verschleppung von Ackerunkräutern, insbesondere durch Müllkomposte und Klärschlämme. Veröffentlichungen der Bundesanstalt für alpenländische Landwirtschaft Gumpenstein 6, pp. 136.

Cousens R, Mortimer M, 1995. Dynamics of Weed Populations. Cambridge University Press, Cambridge.

Eisele JA, 1997. The influence of composting on seeds of *Vicia hirsuta (L.)* S.F. Gray. Seed Science and Technology 25, 325–328.

Elema AG, Bloemhard CMJ, Scheepens PC, 1990. Risk-analysis for the dissemination of weeds by liquid cattle manure. Mededelingen van de Faculteit Landbouwwetenschappen, Rijksuniversiteit Gent (Belgium), 55, 1203–1208.

Eriksson O, 2000. Seed dispersal and colonization ability of plants – Assessment and implications for conservation. Folia Geobotanica 35, 115–123.

Frank D, Klotz S, 1990. Biologisch-ökologische Daten der Flora der DDR. Wissenschaftliche Beiträge der Martin-Luther-Universität Halle-Wittenberg 32, pp. 167.

Fogelfors H, 1985. The importance of the field edge as a spreader of seed-propagated weeds. Swedish Weed Conference 26(1), 178–189.

Gardener CJ, McIvor JG, Jansen A, 1993. Passage of legume and grass seeds through the digestive tract of cattle and their survival in faeces. Journal of Applied Ecology 30, 63–74.

Grime JP, Hodgson JG, Hunt R, 1988. Comparative Plant Ecology: A Functional Approach to Common British Species. Unwin Hyman, London, p. 742.

Grime JP, Mason G, Curtis AV, Rodman J, Band SR, Mowforth MAG, Neal AM, Shaw S, 1981. A comparative study of germination characteristics in a local flora. Journal of Ecology 69, 1017–1059.

Higgins SI, Nathan R, Cain ML, 2003. Are long-distance dispersal events in plants usually caused by nonstandard means of dispersal? Ecology 84, 1945–1956.

Hodgson JG, Grime JP, 1990. The role of dispersal mechanisms, regenerative strategies and seed banks in the vegetation dynamics of the British landscape. In: Bunce RGH, Howard DC (Eds.), Species Dispersal in Agricultural Habitats. Belhaven Press, London, pp. 65–81.

Hodgson JM, 1964. Variations in ecotypes of Canada thistle. Weeds 12, 167–171.

Kempski E, 1906. Über endozoische Samenverbreitung und speziell die Verbreitung von Unkräutern durch Tiere auf dem Weg des Darmkanals. PhD thesis, Universität Rostock. Carl Georgi, Bonn, p. 172.

Lake JC, Leishman MR, 2004. Invasion success of exotic species in natural ecosystems: The role of dispersal, plant attributes and freedom from herbivores. Biological Conservation 117, 215–226.

Lennartz H, 1957. Über die Beeinflussung der Keimfähigkeit der Samen von Grünlandpflanzen beim Durchgang durch den Verdauungstrakt des Rindes. Zeitschrift für Acker- und Pflanzenbau 103(4), 427–453.

Mayer F, 2000. Long distance dispersal of weed diaspores in agricultural landscapes – The Scheyern approach. PhD thesis, TU München, Shaker Verlag, ISBN 3-8265-8185-7, FAM-Bericht 47, p. 206.

Mayer F, Albrecht H, 1998. Die Bedeutung der organischen Düngung für die Ausbreitung der Wildkrautsamen. Zeitschrift für Pflanzenkrankheiten und Pflanzenschutz Sonderheft XVI, pp. 175–182.

Mayer F, Albrecht H, 2003. Ausbreitungsbiologie der Ackerkratzdistel (*Cirsium arvense* (*L.*) *Scop.*). Landbauforschung Völkenrode Sonderheft 255, 9–18.

Mayer F, Albrecht H, Pfadenhauer J, 1998. The transport of seeds by soil-working implements. In: Champion GT, Grundy AC, Jones NE, Mashall EJP, Froud-Williams RJ (Eds.), Weed seedbanks: Determination, dynamics & manipulation. Association of Applied Biologists, c/o Horticulture Research International, Wellesbourne, Warwick CV35 9EF, UK, Oxford, pp. 83–90.

Mayer F, Albrecht H, Pfadenhauer J, 2000. The influence of digestion and storage in silage and organic manure on the germinative ability of six weed species (*Papaver argemone, P. dubium, Legousia speculum-veneris, Centaurea cyanus, Spergula arvensis, Trifolium arvense*). Zeitschrift für Pflanzenkrankheiten und Pflanzenschutz, Sonderhaft XVII, 47–54.

Mayer F, Albrecht H, Pfadenhauer J, 2002. Secondary dispersal of seeds in the soil seed bank by cultivation. Zeitschrift für Pflanzenkrankheiten und Pflanzenschutz, Sonderheft XVIII, 551–560.

Moss SR, 1988. Influence of cultivations on the vertical distribution of weed seeds in the soil. VIII Colloque International sur la Biologie, l'Ecologie et la Systematique des Mauvais Herbes, pp. 71–80.

Müller-Schneider P, 1986. Verbreitungsbiologie der Blütenpflanzen Graubündens. Veröffentlichungen des Geobotanischen Instituts der ETH, Stiftung Rübel, Zürich 85, p. 263.

Nathan R, Katul GG, Horn HS, Thomas SM, Oren R, Avissar R, Pacala SW, Levin SA, 2002. Mechanisms of long-distance dispersal of seeds by wind. Nature 418, 409–413.

Oesau A, 1992. Erhebung zur Verunkrautungsgefährdung bewirtschafteter Äcker durch stillgelegte Nachbarflächen. Zeitschrift für Pflanzenkrankheiten und Pflanzenschutz, Sonderheft XIII, 61–68.

Oesau A, 1998. Untersuchungen zur generativen Propagation der Ackerkratzdistel (*Cirsium arvense (L.) Scop.*). Zeitschrift für Pflanzenkrankheiten und Pflanzenschutz, Sonderheft XVI, 75–82.

Özer Z, 1979. Über die Beeinflussung der Keimfähigkeit der Samen mancher Grünlandpflanzen beim Durchgang durch den Verdauungstrakt des Schafes und nach Mistgärung. Weed Research 19, 247–254.

Rew LJ, Cussans GW, 1997. Horizontal movement of seeds following tine and plough cultivation: Implicants for spatial dynamics of weed infestations. Weed Research 37, 247–256.

Russi L, Cocks PS, Roberts EH, 1992. The fate of legume seeds eaten by sheep from a Mediterranean grassland. Journal of Applied Ecology 29, 772–778.

Sarapatka B, Holub M, Lhotska M, 1993. The effect of farmyard manure anaerobic treatment on weed seed viability. Biological Agriculture and Horticulture 10, 1–8.

Schippers P, Ter Borg SJ, van Groenendal JM, Habekotte B, 1993. What makes *Cyperus esculentus* (yellow nutsedge) an invasive species? Brighton Crop Protection Conference – Weeds, Brighton, 5A-2, pp. 495–504.

Tackenberg O, Poschlod P, Kahmen S, 2003. Dandelion seed dispersal: The horizontal wind speed does not matter for long-distance dispersal – it is updraft! Plant Biology 5, 451–454.

Thompson K, Bekker RM, Bakker JP, 1998. Weed seed banks; evidence from the north-west European seed bank database. In: Champion GT, Grundy AC, Jones NE, Marshall EJP, Froud-Williams RJ (Eds.), Weed Seedbanks: Determination, Dynamics & Manipulation. Association of Applied Biologists, c/o Horticulture Research International Wellesbourne, Warwick CV35 9EF, UK, Oxford, pp. 105–112.

Chapter 3.4

Soil Microbial Communities and Related Functions

A. Gattinger, A. Palojärvi and M. Schloter

3.4.1 Introduction

Soil microbial communities play an important role in agroecosystem functioning and are on the field scale essential for plant nutrition and health. On a larger scale, they contribute to global element cycling. Furthermore, they are involved in turnover processes of organic matter, breakdown of xenobiotics and formation of soil aggregates. In contrast to plant diversity, and the macro and meso fauna, the aspect of soil microbial diversity is a rather new approach. A major problem in soil microbial analysis has been that most soil microorganisms cannot be characterised by classical microbiological cultivation techniques. It is estimated that approximately 80–90% of soil microorganisms are not yet cultured by means of classical methods (Amann et al., 1995).

At the beginning of the last decades, cultivation-independent techniques became more frequent, leading to a broader view of microbial life in soil. It is now widely accepted that a given soil consists of members belonging to all three domains of the biosphere, the *Bacteria*, the *Eucarya* and the *Archaea* (e.g. Liesack et al., 1997; Gattinger et al., 2002a). Cultivation-independent molecular approaches are based on the direct extraction of DNA from soil and a subsequent analysis molecular marker genes (e.g. genes coding for ribosomal RNA). These genes are functionally conserved in all organisms and contain conserved, variable and highly variable regions (Amann et al., 1995) and can therefore be used as marker genes to describe structural microbial diversity. At present, more than 15 000 sequences of rRNA genes are available in databases. Apart from structural genetic analyses, soil microbial communities can also be studied in terms of their in situ functions by means of molecular techniques. There are several PCR-based assays available for the detection of functional genes with respect to key processes during C and N cycling (see Emmerling et al., 2002).

Another cultivation-independent technique for the analysis of the structural diversity of soil microorganisms is the phospholipid approach. Phospholipids are essential components of membranes of all living cells, and their fatty acid (PLFA: phospholipid fatty acids) or ether-linked isoprenoid side chains (PLEL: phospholipid etherlipid) allow a taxonomic differentiation within complex microbial communities (Zelles, 1999; Gattinger et al., 2002a, b, 2003). This approach is now well established in soil ecology and serves as a phenotypic and thus complementary tool to genotypic (molecular genetic) approaches.

Soil microbial communities and related functions were studied under the constraints of agricultural land use in general and the specific aspects of Research Station Scheyern in particular. Hence, the impact of copper from previous hop cultivation in Scheyern was also investigated as well as the more general aspects of herbicide degradation and soil tillage practices.

3.4.2 Materials and methods

Soil material from different fields from Research Station Scheyern was sampled for the different soil-microbiological studies, which were sometimes accompanied by laboratory experiments for more process-orientated research.

The determination of soil microbial communities and related functions was based on a variety of different methods from classical microbial isolation and cultivation to molecular tools.

The following designations and classifications for functional groups of PLFA were used: ester-linked fatty acids (EL-PLFA) composed of saturated (SATFA), monounsaturated (MUFA), polyunsaturated (PUFA) and hydroxy substituted (PLOH). The subunits of non–ester-linked phospholipids fatty acids (NEL-PLFA) are the unsubstituted fatty acids (UNSFA) and hydroxy substituted fatty acids (UNOH).

3.4.3 Singular environmental and anthropogenic effects

Soil cultivation

Soil cultivation in Latin 'agri cultura' is the basic treatment of soil with the objective of improving crop-growing conditions. It can be drainage of water-logged sites, ploughing, digging, tilling or the formation of ridges during potato farming.

Ploughing, an effective tillage technique for weed control in low-input agriculture is likely to result in an increased decomposition of soil organic matter compared to zero tillage (Doran, 1980). However, since plant nutrition in low-input systems is maintained primarily through the soil organic matter, soils in such systems should be rich in organic substances. In the hilly landscape of Scheyern, zero tillage has the additional advantage of reducing soil erosion through the release of plant residues at the soil surface.

To investigate the effects of conventional and minimal tillage on microbial properties (von Lützow and Palojärvi, 1995), maize straw was added to topsoil samples from plot A18. Conventional tillage (CT) was simulated by incorporating straw into the soil, and in the case of minimum tillage (MT), straw was applied on to the soil surface. The treated samples were then incubated under aerobic conditions (80% N_2, 20% O_2) for 360 days in a batch experiment. The results revealed an increased C mineralisation in samples under CT conditions as compared to samples simulating the MT practice. Specifically, higher C amounts are respired as CO_2 in CT than in MT. After 240 days, 45 and 60% of the straw-derived C was respired in MT and CT, respectively. In the same period of time, there was also a higher increase in dissolved organic carbon (DOC) in MT than in CT samples, whereas CT showed a higher C_{mic} (microbial carbon) content than MT, whose level remained unchanged until the end of the experiment. In MT, a lower metabolic quotient (qCO_2) was measured, which indicates a more efficient substrate use in the MT samples than in CT. PLFA analyses according to Frostegård et al. (1993) of the two treatment conditions revealed different microbial communities. Further, MT samples contained a higher percentage of the fungal lipid marker (18:2Δ9,12) (Palojärvi et al., 1997). This may have been due to the release of maize straw on the soil surface favouring a higher proportion of fungi in the upper soil layer in MT than in CT.

As a result of ridge-till practice, potato fields are a spatially very heterogeneous soil-ecological system. The different bulk density and pore size distribution in the ridge soil (RS), uncompacted interrow soil (IS) and IS compacted by tractor traffic (CS) significantly affect chemical and biological properties (Ruser et al., 1998; Flessa et al., 2002; Gattinger et al., 2002a). Analyses of PLFA and (PLEL) have revealed differences in size and structure of microbial communities in the three different soil zones of a potato field (Gattinger et al., 2002a). Marked differences in the quantity of phospholipid biomarker concentrations (corresponding to microbial biomass) among these zones were noted, whereby lipid contents were related to fresh soil volume instead of soil dry matter. Compaction of IS caused an increase in bacterial and eukaryotic biomass, expressed as total PLFA concentration, as well as an increase in total archaeal biomass, expressed as total PLEL concentration, causing a decrease in the fungi:bacteria ratio. Because of the higher water-filled pore space in CS (an indirect measure for reduced O_2 availability), a more pronounced anaerobic microbial community was estimated than in IS, which may explain the elevated N_2O fluxes in this soil zone. Flessa et al. (2002) documented a total N_2O–N emission of 2.50 kg ha^{-1} in the same soil zone (CS) and emissions of 2.00 and 1.64 kg ha^{-1} in the compartments RS and IS, respectively, during the cropping period 1998. The higher N_2O emission rate in RS than in IS cannot be solely explained by soil porosity. According to Gattinger et al. (2002a), rhizodeposits and plant debris in RS favour an active heterotrophic microflora, which delivers adequate denitrifying conditions.

Effects of pesticides on soil microbial communities

Long-term application of copper-based pesticides has led to accumulations of 1500 ppm in French vineyard soils (Flores-Vélez et al., 1996), 1000 ppm in Japanese horticultural soils (Aoyama and Nagumo, 1997) and 450 ppm in German hop plantations (Mölter, 2000). During establishment of the FAM research network in 1990, fields under hop cultivation were converted to arable farming.

Comparisons between soils formerly used as arable crop rotation and as hop plantation revealed higher microbial biomass in the crop rotation system than in hop plantation (Zelles et al., 1994). Eco-physiological parameters in hop soil were significantly different from soil with crop rotation: A low microbial biomass, a low qCO_2 and a high polyhydroxybutyrate-to-PLFA ratio give indication for environmental stressor to the hop soil. It is most likely that these effects are due to the application of Cu, which has been used as part of a fungicide formulation during hop cultivation. PLFA profiles in former hop yard soils indicated an increase in the proportion of Gram-negative bacteria and a decrease in the proportion of Gram-positive bacteria in comparison to the control (Zelles et al., 1994). These findings led to subsequent studies on the toxic effects of copper. During a microcosm experiment, Mölter et al. (1995a) determined lower microbial biomasses in Cu-amended soils (300, 600 and 3000 ppm Cu) than in control soils (18 ppm Cu and 200 ppm Cu). Differences in community structure were most pronounced between the control and the 3000 ppm treatment and between the control and the 600 ppm treatment respectively. The most striking effect, however, was found in the ratio of different ratio of the SATFAs, which was lowest in the 3000 ppm Cu treatments. This indicates a physiological response of the bacterial community towards higher membrane stability.

Since the German ban on the maize herbicide atrazine in 1990, other triazine compounds (e.g. terbuthylazine) are currently being used. Mölter et al. (1995b) observed only marginal effects of terbuthylazine application on microbial biomass and community structure on two different arable soils from Scheyern during a microcosm experiment. Although in both soils significant shifts were determined after 2 days of incubation for ω-PLOHs and 18:1 MUFAs (both constituents of Gram-negative bacteria; Zelles, 1997), terbuthylazine application did not influence the overall community structure. Using the ratio of the SATFAs as an indicator for membrane physiology, different results were obtained: Increasing pesticide addition resulted in a decrease in some plots and in an increase in other plots. This effect might be explained by the contrasting cropping histories of these two soils and by Cu contents of 18 and 200 ppm, respectively (Mölter et al. 1995a, b).

Degradation of pesticides by soil microorganisms

Most agricultural pesticides are degraded by microbial processes (Torstensson, 1980). In this manner, information on the biomineralisation of pesticides becomes important for estimating their fate in the environment. While the impact of abiotic factors on pesticide degradation in soils has been considered in many studies (as reviewed by Hurle, 1982), the relationship between microbial soil properties and biodegradation processes remains insufficiently studied. To develop concepts for biological decontamination for soils and water, knowledge of the influence of biotic impact factors on the degradation of organic pesticides is essential (Fournier, 1993). Soils from four different crop systems were sampled for their mineralisation capacity of two different free and plant-incorporated pesticides (von Wirén-Lehr et al., 1997). Two of these soils originated from the FAM research station, one from a former hop plantation (Hop) and the other one from an arable field managed organically for the past two years (Bio-2y). Additional soil samples were taken from an arable field managed organically for 15 years (Bio-15y), as well as a conventionally managed arable field (Conv.) originating from Ottmaring near Augsburg (von Wirén-Lehr et al., 1997). Free isoproturon and glyphosate were mineralised to a high extent in all four soils, as shown by degradation experiments of ^{14}C-labelled compounds (Lehr et al., 1996b; von Wirén-Lehr et al., 1997). Whereas 45–55% of the plant cell wall–bound glyphosphate was mineralised in the four soils within 26 days (von Wirén-Lehr et al., 1997), only 2–3% of the plant cell wall–bound isoproturon was degraded to CO_2 within the same time period (Lehr et al., 1996b). These findings indicate the importance of bioavailability to metabolising soil microorganisms in the context of pesticide degradation. Plant-associated residues derived from glyphosate were associated non-specifically to the plant matrix, whereas isoproturon residues were bound to plant cell wall fractions mainly covalently (von Wirén-Lehr et al., 1997). Interestingly, the two organically managed soils (Bio-2y and Bio-15y) showed the highest mineralisation activity of free, ^{14}C-labeled isoproturon. Thus, even after long term of organic farming in the case of soil Bio-15y, the microbial communities possess the ability for rapid isoproturon degradation. Moreover, degradation studies of bacterial isolates from the four soils demonstrate that mixed bacterial cultures metabolise isoproturon much more effectively than pure cultures alone (Lehr et al., 1996a). These findings underscore the importance of complex microbial consortia for effective pesticide degradation in the environment.

3.4.4 Effect of land use and farming systems

Land use and farming systems influence soil microbial communities and related functions through several factors. Within the FAM network, the land use systems arable land, grassland and fallow were investigated, and among others, soil microbiological properties as affected by the farming system, integrated or organic, were compared.

Soil microbial communities as affected by type of land use
Comparison between different land use systems in Scheyern revealed differences in microbial biomass and community structure. Zelles et al. (1994, 1995) found in grassland soils a higher microbial biomass and a community structure different from that in arable soils with comparable soil texture. Following the distributions of signature PLFA, a higher proportion of fungi and Gram-positive bacteria can be assumed for grassland. The $q\mathrm{CO}_2$ is lower in grassland than in arable soils as determined by Zelles et al. (1994) and indicates a more energy-efficient microflora in the permanent grassland (Anderson and Domsch, 1990). Dilly et al. (2001) investigated functional aspects of the soil microflora in 11 arable, 5 grassland and 3 fallow soils and did not find any difference in the relative microbial colonisation expressed as C_{mic}/C_{org}, whereas in four of the five grassland soils and in the three fallow sites, a significantly lower $q\mathrm{CO}_2$ was determined than in the 11 arable soils.

Schloter et al. (1998) investigated the intragenus diversity of a widely distributed soil bacterium (*Ochrobactrum*) in arable and fallow soil and found the same genotypes in both systems but differing in the abundances. This leads to the assumption that a change in land use does not necessarily manipulate the genetic pool of the microflora.

Geue (2002) investigated arbuscular mycorrhiza fungi (AMF) in the rhizosphere of grassland and arable soils by applying classical and molecular techniques. He found similar community structures for both land use systems whereby *Acaulospora longula* and species belonging to the genus *Glomus* (*G. mosseae*, *G. caledonium* and *G. geosporum*) were the most dominating. Rhizosphere phosphate availability correlated with fungal colonisation. However, no significant relationship was detected between mycorrhization, land use type, season, plant species or phosphate availability. The results further showed that annual hyphal colonisation increased during the vegetation period, whereas the number of vesicles decreased. Geue (2002) attributed this phenomenon to the allocation of resources from the vesicles to extraradical spores, which were formed at the end of the year.

Integrated versus organic farming

Soil microbial communities play a key role in the functioning of organic farming systems. Because of the constraints of the closed-cycle principle (Lampkin, 1992), their contribution is essential, particularly for plant nutrition through the mineralisation of organic matter. Mäder et al. (2002) found higher microbial biomasses (C_{mic}), enzymatic activities and lower qCO_2 values in organic than in plots under integrated management in the Suisse long-term DOK field trial. However, this was not the case when soils under organic and integrated farming from the research station in Scheyern were compared: No difference was found in C_{mic}, C_{mic}/C_{org} ratio and qCO_2 in soil samples from organic and integrated farming systems (von Lützow et al., 2002; Dilly et al., 2001). When the qCO_2 was related to organic carbon, the corresponding qCO_2/C_{org} ratio was significantly lower in soils under organic than under integrated management (Dilly et al., 2001) (Figure 1).

Sehy et al. (2003) were the first one who performed a comparative study on trace gas fluxes in soils under integrated and organic management. High N_2O flux rates were measured in the fields following N fertilisation, precipitation, soil tillage and freeze-thaw events. The most important prerequisite for the occurrence of high N_2O flux rates was sufficient soil mineral N content. High N_2O emissions occurred only when the soil nitrate content surpassed a critical value of 5 kg N ha^{-1}. This became obvious especially in the organic farming system, where very low soil nitrate contents limited N_2O emissions during the vegetation period. After harvest of the main crop, the soil was tilled and N limitations were thus removed. This resulted in substantial N_2O emissions from the organically managed field. When soil nitrate contents were high, soil water status was the most important factor controlling N_2O emissions from all sites. Soil water content varied depending on landscape position and the amount and temporal distribution of precipitation. Eighty-five per cent of all elevated flux rates (>50 µg N_2O–N m^{-2} h^{-1}) occurred at soil water contents between 55 and 90% WFPS, with maximum emissions at 65% WFPS. In

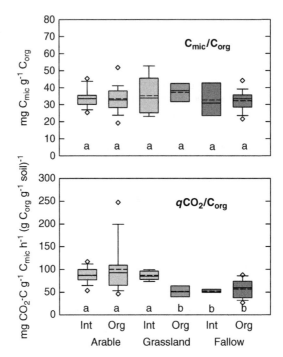

Figure 1   Mean C_{mic}/C_{org} and qCO_2/C_{org} ratios (for 0–20 cm soil depth) in integrated-conventional (Int) and organic (Org) managed arable, grassland and fallow soils at the experimental farm in Scheyern (different letters indicate significant differences, Student–Newman–Keuls Method, $p < 0.05$; boxes encompass 25 and 75% quartiles; central and dashed lines represent mean and median, respectively; bars and rhombus exceed 90 and 95% confidence interval, respectively; grey intensity denotes organic C content) (Dilly et al., 2001).

general, high N_2O flux rates were only measured when critical levels of nitrate content (>5 kg N ha^{-1}), WFPS ($>60\%$) and soil temperature ($>7°C$) were surpassed at the same time. This suggests that microbial denitrification is the most important process for high N_2O emissions from the investigated sites (Davidson, 1991). Elevated flux rates from all sites were measured following freeze-thaw events. Annual N_2O emissions varied between 2.8 and 8.9 kg N_2O–N ha^{-1} in the integrated farming system and between 1.0 and 3.3 kg N_2O–N ha^{-1} in the organic farming system. In both investigated years, area-related N_2O emissions were lower in the organic than in the integrated farming system. However, if the N_2O emissions were related to crop yield, no difference between the two farming systems was found.

3.4.5 Precision farming: site-specific fertilisation

Site-specific fertilisation, a measure of precision farming, takes into account that soil physical, chemical and biological heterogeneity within a field may result in spatial and temporal variability in nutrient cycling, which is likely to have an impact on crop yield as well as on N losses through leaching and gas emission. With precision farming, variations in soil or crop characteristics within a field are identified and mapped, and management actions are spatially and temporally adjusted to crop requirements in the respective areas (Dawson and Johnston, 1997; Mandal and Ghosh, 2000).

At the research station in Scheyern, microbial communities were investigated for their response to site-specific fertilisation (precision farming), spatial heterogeneity (high- and low-yield areas) and season in an arable soil under integrated management.

PLFA analyses showed that bacterial and fungal biomass and community structure in topsoil were not influenced by precision farming practices (Schloter et al., 2003). Instead, bacterial and fungal biomass and community structure followed a seasonal course. Microbial biomass was reduced during the summer month, due to the dryness and hot temperatures. The microbial community structure changed in late springtime probably due to the application of fertilisers and high amounts of root exudates in the rhizosphere.

Morphological characterisation of fungal communities revealed differences in the diversity of active fungal communities among sampling dates, high-yield areas and low-yield areas and among plots treated with and without site-specific fertilisation (Hagn et al., 2003). Mainly in spring, significantly higher values for the Shannon–Weaver index indicated a higher diversity at the low-yield sites. Interestingly, at the end of the vegetation period, in the plots under precision farming, higher values for Shannon diversity and evenness were observed. It turned out that the most frequently found isolates at all sampling dates and plots belonged to the genus *Trichoderma*. Although some of the isolated *Trichoderma* showed biocontrol activity towards a soil-borne plant pathogen (Hagn et al., 2003), it remains to be shown in a current study whether precision farming after 5 years of its introduction favours the growth of biocontrol-active *Trichoderma* or other species.

Schloter et al. (2003) also determined enzymatic activities of soil microorganisms and found a seasonal variation with the highest measured values in spring and early summer, as well as clear effects based on the investigated plot (high yield or low yield) and the used farming management system (conventional or precision) (Figure 2).

Proteolytic activity was significantly higher on the high-yield plots at all measured times. Mainly on the low-yield sites, precision farming caused a higher proteolytic activity compared to the conventional management.

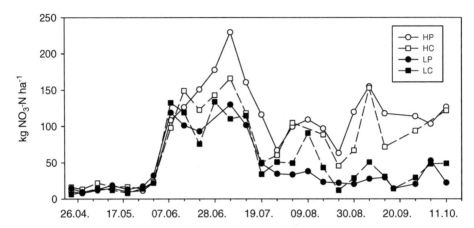

Figure 2 Potential proteolytic activities from April to October 1999. HC: high-yield area, fertilised according to conventional practices; HP: high-yield area, fertilised according to precision farming practices; LC: low-yield area, fertilised according to conventional practices; LP: low-yield area, fertilised according to precision farming practices (Schloter et al., 2003).

Nitrification and denitrification activities were mainly affected after the application of N fertiliser. Significant higher values were found on the low-yield plots, where conventional farming was applied. Whereas nitrification and denitrification activities responded to differentiated N fertilisation, proteolytic activity responded to spatial differences in terms of high-yield and low-yield sites within the arable field (Schloter et al., 2003).

It was further investigated if the observed influence of site-specific fertilisation on denitrification activity affected also the N_2O fluxes in situ (Sehy et al., 2003). Through the application of site-specific fertilisation, N input was increased by 17% in the high-yielding area and reduced by 17% in the low-yielding area as compared to the control. In the high-yielding area, neither crop yield nor N_2O emissions were affected by higher N input, presumably because soil nitrate contents were above levels limiting crop yield or N_2O production. However, in the low-yielding area, the lower precision-based N input in LP as compared to LC resulted in 34% less release of N_2O in roughly 10 months following differentiated fertilisation, whereas crop yield was not affected.

Acknowledgements

The scientific activities of the FAM Munich Research Network on Agroecosystems were financially supported by the German Federal Ministry of Education and Research (BMBF 0339370). Overhead costs of

Research Station Scheyern were funded by the Bavarian State Ministry for Science, Research and the Arts.

References

Amann RI, Ludwig W, Schleifer KH, 1995. Phylogenetic identification and in situ detection of individual microbial cells without cultivation. Microbiology Review 59, 143–169.

Anderson TH, Domsch KH, 1990. Application of eco-physiological quotients (qCO$_2$ and qD) on microbial biomass from soils of different cropping histories. Soil Biology and Biochemistry 22, 251–255.

Aoyama M, Nagumo T, 1997. Effects of heavy metal accumulation in apple orchard soils on microbial biomass and microbial activities. Soil Science Plant Nutrition 43, 601–612.

Davidson EA, 1991. Fluxes of nitrous oxide and nitric oxide from terrestrial ecosystems. In: Rogers JE, Whitman WB (Eds.), Microbial Production and Consumption of Greenhouse Gases: Methane, Nitrogen Oxides, and Halomethanes. American Society for Microbiology, Washington DC, pp. 219–235.

Dawson CJ, Johnston AE, 1997. Aspects of soil fertility in the interpretation of yield maps as an aid to precision farming. In: Stafford JV (Ed.), Precision Agriculture '97. Bios Scientific Publishers, Warwick, pp. 87–94.

Dilly O, Winter K, Lang A, Munch JC, 2001. Energetic eco-physiology of the soil microbiota in two landscapes of southern and northern Germany. Journal of Plant Nutrition and Soil Science 164, 407–413.

Doran JW, 1980. Soil microbial and biochemical changes associated with reduced tillage. Soil Science Society of American Journal 44, 765–771.

Emmerling C, Schloter M, Hartmann A, Kandeler E, 2002. Functional diversity of soil organisms – a review of recent research activities in Germany. Journal of Plant Nutrition and Soil Science 165, 408–420.

Flessa H, Ruser R, Schilling R, Loftfield N, Munch JC, Kaiser EA, Beese F, 2002. N$_2$O and CH$_4$ fluxes in potato fields: Automated measurement, management effects and temporal variation. Geoderma 105, 307–325.

Flores-Vélez LM, Ducaroir J, Jaunet AM, Robert M, 1996. Study of the distribution of copper in an acid sandy vineyard soil by three different methods. European Journal of Soil Science 47, 523–532.

Fournier C, 1993. Microbial degradation of pesticides in soil and subsoil: Mechanisms and problems of control. In: DelRe AAM, Capi E, Evens SP, Natali P, Trevisan M (Eds.), Mobility and Degradation of Xenobiotics. Proceedings of the IX Symposium of Pesticide Chemistry. Piacenza, Italy, pp. 343–355.

Frostegård A, Baath E, Tunlid A, 1993. Shifts in the structure of soil microbial communities in limes forests as revealed by phospholipid fatty acid analysis. Soil Biology & Biochemistry 25, 723–730.

Gattinger A, Günthner A, Schloter M, Munch JC, 2003. Characterization of Archaea in soils by polar lipid analysis. Acta Biotechnologica 23, 21–28.

Gattinger A, Ruser R, Schloter M, Munch JC, 2002a. Microbial community structure varies in different soil zones of a potato field. Journal of Plant Nutrition and Soil Science 165, 421–428.

Gattinger A, Schloter M, Munch JC, 2002b. Phospholipid etherlipid and phospholipid fatty acid fingerprints in selected euryarchaeotal monocultures for taxonomic profiling. FEMS Microbiology Letters 213, 133–139.

Geue H, 2002. Molecular Biological Analysis for the detection of arbuscular mycorrhizal fungi on wild plant populations of agricultural areas. PhD Thesis, TU München, Sharker Verlag, ISBN 3-8322-0940-9, FAM-Bericht 54, 122 pp. (http://tumb1.biblio.tu-muenchen.de/publ/diss/ww/2002/geue.html).

Hagn A, Pritsch K, Schloter M, Munch JC, 2003. Fungal diversity in agricultural soil under different farming management systems, with special reference to biocontrol strains of *Trichoderma* spp. Biology and Fertility of Soils 38, 236–244.

Hurle K, 1982. Untersuchungen zum Abbau von Herbiziden in Böden. In: Colhoun J, Heitefuss R, Kern H (Eds.), Acta Phytomedica Vol. 8, Paul Parey, Berlin, Hamburg, pp. 1–116.

Lampkin N, 1992. Organic Farming. Farming Press Books. Ipswich, 1–701.

Lehr S, Gläßgen WE, Sandermann H Jr, Scheunert I, 1996a. Metabolism of isoproturon in soils originating from different agricultural management systems and in cultures of isolated soil bacteria. International Journal of Environmental Analytical Chemistry 65, 231–245.

Lehr S, Scheunert I, Beese F, 1996b. Mineralization of free and cell-wall-bound isoproturon in soils in relation to soil microbial parameters. Soil Biology & Biochemistry 28, 1–8.

Liesack W, Janssen PH, Rainey FA, Ward-Rainey NL, Stackebrandt E, 1997. Microbial diversity in soil: The need for a combined approach using molecular and cultivation techniques. In: v. Elsas JD, Trevors JT, Wellington EMH (Eds.), Modern Soil Microbiology. Marcel Dekker, New York, pp. 375–439.

Mäder P, Fließbach A, Dubois D, Gunst L, Fried P, Niggli U, 2002. Soil Fertility and Biodiversity in Organic Farming. Science 296, 1694–1697.

Mandal D, Ghosh SK, 2000. Precision farming – The emerging concept of agriculture for today and tomorrow. Current Science 79, 1644–1647.

Mölter K, 2000. Strukturelle und ökopysiologische Charakterisierung von Bodenmikroorganis-menpopulationen in kupferbelasteten

Agrarflächen anhand von Phospholipidfettsäure (PLFA-) Profilen. PhD Thesis, TU München, 158 pp.

Mölter K, Zelles L, Bai QY, Hartmann A, 1995a. Verwendung von Phospholipid-Biomarkern zur Untersuchung der Wirkung von Kupferbelastungen auf die mikrobielle Populationsstruktur zweier landwirtschaftlich genutzter Böden. Mitteilgn Dtsch Bodenkundl Gesellsch 75, 91–94.

Mölter K, Zelles L, Bai QY, Hartmann A, 1995b. Lagerungseffekt und Schadstoffwirkung – ein Kurzzeit-Laborversuch zur Wirkung von Terbuthylazin auf die Phospholipid-Fettsäure (PLFA) – Zusammensetzung zweier Böden. Mitteilgn Dtsch Bodenkundl Gesellsch 76, 673–676.

Palojärvi A, Sharma S, Rangger A, von Lützow M, Insam H, 1997. Comparison of Biolog and phospholipid fatty acid patterns to detect changes in microbial community. In: Insam H, Rangger A (Eds.), Microbial Communities. Functional versus Structural Approaches. Springer Verlag, Berlin, Germany, pp. 37–48.

Ruser R, Flessa H, Schilling R, Steindl H, Beese F, 1998. Soil compaction and fertilization effects on nitrous oxide and methane fluxes in potato fields. Soil Science Society of American Journal 62, 1587–1595.

Schloter M, Bach HJ, Metz S, Sehy U, Munch JC, 2003. Influence of precision farming on the microbial community structure and functions in nitrogen turnover. Agriculture, Ecosystems & Environment 98, 295–304.

Schloter M, Zelles L, Hartmann A, Munch JC, 1998. New quality of assessment of microbial diversity and biochemical methods. Plant Nutrition and Soil Science 161, 425–431.

Sehy U, 2003. N_2O-Freisetzungen landwirtschaftlich genutzter Böden unter dem Einfluss von Bewirtschaftungs-, Witterungs- und Standortfaktoren. PhD Thesis, TU München, oekom verlag München, ISBN 3-936581-40-1, 130 pp.

Sehy U, Ruser R, Munch JC, 2003. Nitrous oxide fluxes from maize fields: Relationship to yield, site-specific fertilization, and soil conditions. Agriculture, Ecosystem & Environment 99, 97–111.

Torstensson L, 1980. Role of microorganisms in decomposition. In: Hance RJ (Ed.), Interaction between Herbicides and the Soil. Academic Press, London, pp. 159–177.

von Lützow M, Leifeld J, Kainz M, Kögel-Knabner I, Munch JC, 2002. Indications for soil organic matter quality in soils under different management. Geoderma 105, 243–258.

von Lützow M, Palojärvi A, 1995. Die Rolle der aktiven organischen Substanz als dynamischer Speicher im Stoffumsatz. FAM-Bericht 9, 119–125.

von Wirén-Lehr S, Komoßa D, Gläßgen WE, Sandermann H Jr, Scheunert I, 1997. Mineralization of [^{14}C]Glyphosate and its plant-associated residues in arable soils originating from different farming systems. Pesticide Science 51, 436–442.

Zelles L, Bai QY, Ma RX, Rackwitz R, Winter K, Beese F, 1994. Microbial biomass, metabolic activity and nutritional status determined from fatty acid patterns and poly-hydroxybutyrate in agriculturally-managed soils. Soil Biology & Biochemistry 26, 439–446.

Zelles L, Bai QY, Rackwitz R, Chadwick D, Beese F, 1995. Determination of phospholipid- and lipopolysaccharide-derived fatty acids as an estimate of microbial biomass and community structure in soils. Biology and Fertility of Soils 19, 115–123.

Zelles L, 1997. Phospholipid fatty acid profiles in selected members of soil microbial communities. Chemosphere 35, 275–294.

Zelles L, 1999. Fatty acid patterns of phospholipids and lipopolysaccharides in the characterisation of microbial communities in soil: A review. Biology And Fertility of Soils 29, 111–129.

Chapter 3.5

Mesofauna

P. Schröder

3.5.1 Introduction

Both, soil mega- and mesofauna are of great importance for the turnover of organic matter and decomposition processes in soils (Hole, 1981; Dunger, 1983). These processes have also direct influence on the health and growth of crops and other plant species (Edwards and Lofty, 1977). Especially, Collembola and earthworms contribute to the decomposition of plant residues and fungi at the soil surface and underground. Hence, the dynamics of N- as well as C-pools are directly influenced by such soil animals, especially as they are able to move in the substrate and visit larger areas during their quest for food. Due to this role, Collembola and earthworms became the most intensively studied organisms of the mesofauna in the FAM. Their role in nutrient turnover, their interaction with microbial activities and their impact in the food web have been studied with respect to the redesign of the landscape in Scheyern and to the agricultural practises used.

Collembola are small (1–5 mm), wingless hexapods with antennae always present (Bellinger et al., 1996–2006). Their mouthparts are located

Perspectives for Agroecosystem Management
Edited by P. Schröder, J. Pfadenhauer and J.C. Munch

within a gnathal pouch and such hardly visible (entognathous). However, Collembola have been classified according to the form and use of the mouthparts. Those possessing a well-developed mandibular molar plate chew their food and are probably vegetarian, while those without molar plate are suctorial and probably carnivorous. As namegiving feature (springtails) distinguishing them from other species, almost all *Collembola* possess a ventral forked abdominal appendage, the so-called furca (see also http://www.collembola.org). To date, there are approx. 7500 described species worldwide. Collembolan fossils date back to the Devonian time and they are are among the oldest known records of terrestrial animals. Collembola are distributed virtually ubiquitously in soils and can be regarded as a very successful animal line in evolution.

Earthworms are an important group of organisms termed "decomposers" in agroecology, besides other destruents including springtails, nematodes, bacteria, protozoa and fungi (Werner, 1990). Systematics separate them into several subgroups based on their biota: they are regarded as epigeic, endogeic and anecic. The epigeic earthworms (e.g. *Eisenia foetida*) live on the surface or in the upper soil layers and feed on undecomposed plant litter. These worms are usually small and reproduce rapidly. Endogeic species forage below the soil surface in horizontal, often branched burrows. During forward movement through soil rich in organic matter, these species ingest large amounts of soil. Hence they have a major impact on the decomposition of dead plant roots, but they are not important in the incorporation of surface litter. The third group, the anecic earthworms build vertical burrows that extend deep into the soil and may be used for years. These worms exit their burrows and collect feed on the soil surface, e.g. manure, leaf litter and other organic matter. Well known among the anecics are *Lumbricus terrestris* and *Aporrectodea longa*. They are most important for agroecosystems, as they exert profound effects on the decomposition of organic matter and on the formation of soil. In an agroecosystem, earthworms may increase the metabolic activity in soils, yet nematode abundance and microbial biomass may decrease (Yeates, 1981; Ruz Jerez et al., 1988). This occurs because earthworms reduce the amount of substrate available to other decomposers, and because earthworms ingest other decomposer organisms as they feed. However, this process will accelerate nutrient cycling rates.

In the FAM, microorganisms and earthworm populations have increased after landscape redesign in all land use systems, and especially Collembola abundances declined. Biomass and numbers of Collembola may vary with water retention potential of the soil, and soil pH plays a significant role for abundance and species distribution. Interestingly, correlations are inverse between earthworms and Collembola, with a stronger significance for the euedaphic species.

Microfoauna has a huge impact on the N dynamics of the soil, much more than estimated for protozoa. Here, the amount of immobilised nitrogen in the system is positively correlated with Collembola numbers. Set aside land rapidly develops into a system with high biodiversity and, after the climax stadium, returns to reduced numbers.

One of the driving questions of the FAM project was to determine effects of agrochemical use on the ecosystem. Especially, the abundance of mesofauna was investigated with view to pesticide use or history of the site. The hypothesis was that detrivorous food webs are closely regulated, and different land use techniques would significantly influence the abundances and species distribution of Collembola, especially those living in the soil. A putative indicator function of the mesofauna was postulated and should be evaluated.

3.5.2 Occurrence and distribution of mesofauna

As mentioned above, soil fauna and microorganisms are of paramount importance for the turnover of organic matter in agroecosystems. Especially, earthworms and Collembola are abundant in agricultural soils where they develop a wide range of species with quite different environmental preferences.

Collembola

Of the presently described 7500 Collembola species worldwide, approx. 1500 are found in Central Europe. According to their vertical distribution in the soil, five different life forms are distinguished. Epedaphic Collembola live on the soil surface, have large eyes and are strongly pigmented. Hemiedaphic Collembola are inhabitants of the upper soil and litter layer. Euedaphic species are eyeless and white. All of them require high moisture but intermediate to low temperature. Drought or high temperatures are lethal to all species.

Only few studies are available on the influence of Collembola and other microarthropods on the mineralisation of leaf litter (Table 1).

On the Scheyern research farm, an inventory of the mesofauna was performed in 1991, including sampling at each of the 405 gridpoints of the 143 ha farm. The abundant Collembola species were found to be *Folsomia manolachei, Folsomia quadrioculata, Isotoma notabilis, Isotomurus palustris, Onychiurus armatus, Lepidocyrtus cyaneus* and *Folsomides parvulus*. The abundances varied between 1500 animals m^{-2} to 100 000 animals m^{-2} depending on the season and the type of land use. These species were not found equally distributed throughout the farm, but seemed to be dependent on the land use. *F. quadrioculata* was more abundant in arable fields

Table 1 Studies on the influence of microarthropods on litter degradation (adapted from Mebes, 1999).

Reference	Duration (days)	Size used	Taxa investigated	Effect observed
Mebes, 1999	117	Minicontainer	Collembola	+
Tian et al., 1998	98	Litter bags	Microarthropods	+
Lenz and Eisenbeis, 1998	266	Minicontainer	Nematodes	No effect
Vreeken-Buijs and Brussaard, 1996	365	Litter bags	Microarthropods	–
Reddy et al., 1994	330	Litter bags	Microarthropods, selected	+
Beare et al., 1992	300	Litterbags	Microarthropods	+ (< 5%)
Setala and Huhta, 1990	730	Microcosms (straw)	Microbes ± fauna	+
Seastedt, 1984	270–900	Various	Microarthropods	+ (23%)

but appeared also in non-agricultural sites, whereas *F. manolachei* occurred more frequently in fallows, and *F. parvulus* was almost restricted to the field. *I. palustris, I. notabilis* and *O. armatus* could be found more abundant in the fallow, and *L. cyaneus* was exclusively found there (Mebes and Filser, 1998).

Seasonal differences of abundances are significant in the fallow land, whereas the distribution of the Collembolan species in the investigated field sites seemed to be more or less stochastical. When estimating the abundances and distribution of soil arthropods in an agricultural landscape with the aim to assess functional diversity, neuronal networks were thought to be very effective. However, the main assessment problem was the non-linear nature of the relationship between abundances and habitat characteristics. In order to qualitatively and quantitatively assess Collembola on the Scheyern research farm, thorough investigations of the populations were performed. Besides microbial respiration and biomass the abundances of *F. quadrioculata* were recorded in each of the plots characterised by the sampling points of the inventory mentioned above. Measurements included also determination of soil acidity, moisture, C_t, N_t and particle size, to characterise the microenvironment of the animals. Microorganisms and Collembola were sampled with stainless steel cylinders (diameter 7.5 cm) to a depth of 20 cm. For the analysis of the microbial biomass the soil was sieved in a 2 mm mesh and analysed at room temperature (22°C). The CO_2 production was measured by infrared gas analysis with an automated system. Microbial biomass was determined according to Anderson and Domsch (1989). A specific respiration rate

(qCO_2) was calculated. For the analysis of the Collembola from the same soil sample, each soil core was subdivided into 5 cm samples and the Collembola were extracted with a Mac-Fayden-High-Gradient-Extractor. Earthworms were sampled by formalin extraction with 10 replicates. For more detailed information see Filser et al. (1995). Defaunated soils were used as controls and reinoculated with single species of Collembola. While single species had an immobilising effect on the nitrogen contents, species mixtures usually seem to increase nitrogen mineralisation. However, it has to be critically discussed that artificially defaunated soil plots were used to determine the leaching of nitrogen for abiotic reasons.

Arable field sites were found to have relatively low and homogeneous numbers of Collembola (3200 ± 3000). In pastures and meadows of the farm the numbers were significantly higher, up to 15 600 individuals m^{-2}, and a maximum of 100 000 individuals were found in the northern grassland of the site that was managed in the frame of the organic farming system (Fromm et al., 1992).

A possible explanation for the observed differences is that the field management and tillage will make the environment and the distribution more uniform. More diversity is found in the grassland, and the abstention from the use of synthetic agrochemicals was thought to increase the individual numbers. Hence, these agricultural measures and the observable changes in plant biodiversity could also be regarded as triggers for Collembola diversity.

To prove this assumption, the development of nematode, lumbricide and Collembolan communities was investigated in farming systems that had been converted from conventional management into organic farming, into integrated farming or into fallow (Filser et al., 1996). The comparison between the integrated field and the fallow indicated succession processes even after a number of years. After 2 years, microbial biomass and protozoa were found to be equal in both fields at the end of the vegetation period, but basal respiration in the fallow was increased. This was connected to a higher abundance of the megafauna in the fallow, whereas predators were found to be higher in the fields.

It was found that the organic farming site and the integrated fields clearly differed in the species composition of Collembola, however the investigation suffered from high variability of animal abundances, indicating that exact food web investigations have to be performed at narrow intervals. Another critical point might be that aggregations of Collembola may have sizes of 5 to 35 cm (Ponge, 1993; Kampichler, 1998; Kampichler et al., 2000), which makes it difficult to determine the true population densities in a given field site.

Overall, the result is in accordance with reports of several other authors who had reported shifts in the dominance structure of Collembola after agricultural management conversion (Brussaard et al., 1988; Filser et al., 1996).

Direct migration of Collembola has been shown to be triggered mainly by temperature, humidity, pheromones, competition and food availability. It had been shown that edaphic Collembola cover only small distances, and it had been reported, that two populations, separated by no more than 14 m, were clearly genetically different (Petersen, 1978). Epigeic species are able to migrate; however, their abundance is also controlled by arthropod predators.

Hemiedaphic Collembola, however, might well be influenced by direct surface dispersal (Filser, 1995b; Filser et al., 1995). This was found to be the case in *I. palustris* and Sminthuridae collected in the arable land of the farm (Mebes and Filser, 1997).

Despite this, the uneven distribution of one species, *L. cyaneus*, in the investigated field sites, was attributed more to higher reproduction rates or lower mortalities of the animals in the fallows as compared to the adjacent field site, than to direct migration of populations.

In microcosm experiments on nitrogen turnover under the influence of Collembola, it was found that microcosms inoculated with single species generally had lower turnover than those with mixed populations, but that nitrogen turnover decreased with increasing abundances of animals. This discrepancy is explained with the fact that specialised feeding is only possible at low population densities. Increasing animal density would increase competition for food sources and force the populations to switch to alternative fodder, thereby decreasing the efficiency of the decomposition. It is taken into consideration that under field conditions a certain degree of migration would be possible for the animals, enabling the species to remain specialised, even if their environmental conditions would change. This fact would also explain the differences in abundances between field and fallow plots. However, pore size is the limiting factor for the distribution of non-burrowing soil animals, and loamy soils might therefore possess smaller biomasses of mesofauna than others (Filser, 1995a).

Earthworms

The important role of earthworms for the turnover in soils has been recognised early. They are able to import organic matter into the soil, they chop and distribute it and enlarge the surface accessibility of nutritious organic matter for microbes. Their detritus, excrements and dead bodies are further valuable sources of nutrients for the food web of the soil. A crucial role has been assigned to earthworms regarding the soil structure and the distribution of water and air in soils.

On the Scheyern farm, the abundance, biomass, species diversity and frequency of earthworms was extremely heterogeneous (Table 2). Clearly a factor influencing their distribution was the soil heterogeneity of the farmland. Biomass was found to be a good marker over the years. The best method for earthworm sampling was the formalin extraction (Raw, 1959);

Table 2 Earthworm population on the Scheyern farm given in average abundance (individuals ha^{-1}); average biomass (g DM ha^{-1}) and frequency (%) of adults during two sampling years (1994 and 1995, data adapted from Nagel, 1996).

Species	Abundance	Biomass	Frequency
Allolobophora antipai vs. *tuberculata*	400	41	13
Aporrectodea caliginosa	120 000	27 160	93
Allolobophora chlorotica	750	35	7
Aporrectodea rosea	10 085	638	73
Lumbricus castaneus	10 600	720	80
Lumbricus rubellus	10 000	2 822	47
Lumbricus terrestris	38 865	46 190	80
Octolasium lacteum	535	85	13

however, due to the ecotoxicological problems connected to this method, an extraction with mustard was preferred (Filser et al., 1993a, b).

From these data it becomes clear that *Aporrectodea caliginosa* and *L. terrestris* are the dominant species which were found in 80% of the investigation plots, whereas *L. rubellus, L. castaneus* and *A. rosea* represent the subdominant species on the farm.

Similar to the Collembola, it was also found that the frequently observed heterogeneity of the earthworm distribution in the topsoil is directly correlated to the soil heterogeneity. This results in an especially complex distribution of earthworms in the farming system under consideration, as the Tertiary Hills represent a landscape of extremely diverse soils. The abundances of the earthworms are controlled by the clay contents, the C/N-ratios, the soil pH and the metabolic quotient of the site under consideration. However, the distribution patterns of the dominant species are equal for all investigated sampling points (Nagel and Beese, 1992).

After the change of the land use systems, the C/N turnover was markedly increased, and the mineralisation was enhanced. In the beginning, earthworms decrease the carbon mineralisation, obviously because they ingest plant residues and disperse them in the soil. Also the N-mineralisation is decreased, eventually due to immobilisation.

When two sites on the farms with comparable soil type but different tillage system were investigated for their megafauna, it was found that microbial biomass had become homogeneous. The distribution of Collembola and earthworms was strongly related to soil health. Soils polluted with heavy metals or synthetic chemicals contained significantly less animals (Filser et al., 1993a, b).

Earthworm activities had a large influence on the transport of water and solutes. This was especially true for compacted soils.

When ^{15}C labelled straw was fed to microcosms containing earthworm populations, it was observed that the straw degradation could be segregated into two independent processes. In the presence of earthworms, the CO_2 production was enhanced and the N leaching was strongly impeded after a short initial burst. The N-mineralisation, however, was highly enhanced. It is speculated that this immobilisation of N is due to the distribution of the straw residues in the earthworm ducts by the animals and due to a change in chemical composition of the material which would lead to a slower degradation rate. Microcosms with a mixed community of mesofauna and earthworms showed effects differing form those of one single animal group (Nagel et al., 1995). Especially, significant was the finding that the presence of Collembola increased soil respiration, whereas earthworms decreased it.

3.5.3 Indicator value

Both faunal groups, Collembola and earthworms, are responsible for recycling nutrients and they develop and preserve the soil structure. Due to their high abundances and the large number of different species with distinct roles in the food web it was hypothesised at the beginning of the FAM project that earthworms and Collembola above all would be well suited as indicators of certain soil properties, and in particular of sustainable farming systems.

In fact, recent literature indicates that some Collembola species would be extremely selective regarding their food sources (McLean et al., 1996). However, it is difficult to gain insight into the relationships between these micro arthropods and their habitat. Analyses about the factors controlling the species distribution and abundance are scarce. So far, they yielded temperature, soil type, soil moisture, acidity and microbial characteristics as the main parameters of influence. Species-specific differences in the nitrogen turnover of the soil was recorded and attributed to different feeding habits of the Collembola, their habitat specificities or mouthpart size and, importantly, the degree of food specialisation (Mebes and Filser, 1998). Specialist behaviour was especially pointed out for the species *I. notabilis* and *L. cyaneus*, which had previously been described as exclusive fungal feeders. The other species found in the FAM study seemed to be nutritional generalists, feeding mainly on soil algae, fungi and even on dead earthworms.

In two independent studies, the soil mesofauna was investigated in a total of eight hops fields. The soil types were characterised as colluvial cambisols, a calcari regosol and a gleysol. By far the lowest abundances and biomasses of Collembola were found in the cambisols that had been treated with sulphur fungicides (Filser, 1993). Interestingly in these fields the copper contents of topsoils were found to be highest. In the other investigated fields the number and abundance of Collembola was comparable to other reports. However, *Folsomia quadrioculata* and *Isotomia bipunctata*, two species that are frequently reported to live on arable land, were never found in hops fields. The author concludes that these species might serve as indicators for sustainability, as their occurrence might be restricted to alternative farming methodologies (Filser, 1993).

In the Scheyern study, defaunated soil was used to study effects of recolonisation, in order to evaluate possible alterations in the community structure after setting arable land aside (Mebes and Filser, 1998). Unfortunately this is the only available study on this topic, and the results are ambiguous. Indications are found that Collembolan communities vary under different arable management strategies, and that single species might influence the turnover of nitrogen. Whether this is also valid under field conditions remains to be elucidated.

Of course, the abundances of soil organisms are extremely variable in space and time. This must be taken into account to reveal reliable data sets.

Neuronal networks have been applied to estimate the numerical abundance of Collembola in varying environments. Due to difficulties in the sampling procedure, these approaches have not been successful (Kampichler, 1998), but they would certainly need to be followed further.

Mebes (1999) collected Collembola in special mini Barber traps and minicontainers and investigated differences in abundance and species dominance structure on arable land after land use changes. Here, clear-cut evidence was presented for a shift in species distribution after few years and Mebes (1999) concluded that they will be useful as indicators for land use changes. However, this argumentation disregards the fact that the land use change is already visible macroscopically on the landscape level. It would be more useful to recognise microarthropods as feeding specialists on fungi and bacteria of special interest, to indicate changes in biodiversity.

At least in the Scheyern farm, it was not possible to instrumentalise earthworms as indicators for soil quality or sustainability. Their abundances did not significantly differ between fields, fallows or between organic and integrated farm plots. Ploughing exerted some influence, but not on the species distribution (Filser et al., 1996).

3.5.4 Collembola and their influence on microbial diversity

Collembola graze on fungal mycelia and bacterial biofilms and keep their growth constantly in the logarithmic phase, which is positive for the soil turnover of nutrients (Parkinson et al., 1979). In their faeces, fungal spores are frequently found. This might be a reason for the spread of fungi within the soil (Lussenhop, 1996; Visser et al., 1987).

The direct influence of Collembola on the mineralisation of dead organic matter is underestimated in most studies. Ghilarov (1971) and Van Straalen (1998) calculated an amount of 6% of the total leaf litter in a forest ecosystem. With this, the animals liberate 40 times more nitrogen than present in the surrounding medium. However in agricultural fields, positive correlation between nitrogen and Collembola abundances were found to rely on their preference for fertilised sites (Filser et al., 1999). It has to be mentioned that a possible nitrogen mobilisation by this species might be important for many ecosystems, but is insignificant in a fertilised agricultural soil.

Species numbers and abundance were observed to increase after the transformation of the management systems from conventional farming to integrated or organic farming, respectively. Strong influences of the setting aside of land were also conserved. Cluster analysis performed to elucidate dominance structures within the community were shown to also reflect exactly the different management practices (Mebes, 1999).

A restricted number of experiments exist on the influence of Collembola on nitrate turnover. Mebes and Filser (1998) found that mixed populations of Collembola tended to mobilise nitrogen in the soil, whereas single species rather had immobilising effects on the available nitrogen. The role of *L. cyaneus* on the N-cycle was investigated by Mebes (1999) in laboratory and field experiments. In the laboratory, significant nitrogen immobilisation was observed as compared to defaunated soils when microcosms were inoculated with *L. cyaneus* in a density of 6500 individuals m^{-2}. Mebes interprets the high immobilisation rates as a consequence of a nutrient deficiency of the Collembola. The animals fed from the fungi in the soil, and died when this food source was consumed. In an animal free microcosm soil bacteria multiplied unhampered and immobilised the available nitrogen.

Soil decontamination of copper and other metals may occur using amino acid hydrolysates. This technique changes both chemical and physical soil properties (Fischer et al., 1997). The lowering of the metal concentrations was assumed to increase bacterial and fungal growth, thus attracting immigrating micro- and mesofauna.

When copper decontaminated soils (starting concentration 318 mg kg^{-1} dry mass, end concentration 121 mg kg^{-1}; Böckl et al., 1998) were investigated for the recolonisation by Collembola, it became clear that the biomass of the treated plots did not differ significantly from controls over 12 weeks. In comparison, the abundances of soil protozoa and nematodes had changed drastically during this time, with a huge increase of protozoa and an 80% loss of the nematode population.

3.5.5 Perspective

Collembola and earthworms are judged to be an essential part of the endogenous potential of agroecosystems. They are strongly influenced by agronomy and the given conditions of the soils, and they themselves influence N- and C-turnover in the field. With regards to sustainable land use, it will be necessary to find out which management practices would be the most beneficial to the mesofauna in order to keep soil fertility and stability high. To our present knowledge all events increasing the organic matter of the soil, all measures that keep the soil undisturbed and all applications that stabilise pH in a neutral range are important in this respect. Interestingly these factors are also beneficial for a good diversity and abundance of the soil mesofauna.

We are lacking conclusive experiments about the development of populations under landscape change, and we have to intercompare data on mesofauna with other food web partners and the nutrient turnover in the landscape. With view to sustainable management of landscapes, the functional biodiversity of the mesofauna has to be addressed. The added value of a large abundance of species is not to be underestimated. Reliability studies will have to point out the significance of small-scale lab studies for interpretations on the field.

Acknowledgements

The scientific activities of the *FAM Munich Research Network on Agroecosystems* were financially supported by the German Federal Ministry of Education and Research (BMBF 0339370). Overhead costs of the Research Station Scheyern are funded by the Bavarian State Ministry for Science, Research and the Arts.

References

Anderson TH, Domsch KH, 1989. Ratios of microbial biomass carbon to total organic carbon in arable soils. Soil Biology Biochemistry 21(4), 471–479.

Beare MH, Parmelee RW, Hendrix PF, Cheng W, Coleman DC, Crossley Jr DA, 1992. Microbial and faunal interactions and effects on litter nitrogen and decomposition in agroecosystems. Ecological Monographs 62, 569–591.

Bellinger PF, Christiansen KA, Janssens F, 1996–2006. Checklist of the Collembola of the World. http://www.collembola.org.

Böckl M, Blay K, Fischer K, Mommertz S, Filser J, 1998. Colonisation of a copper decontaminated soil by micro- and mesofauna. Applied Soil Ecology 9, 489–494.

Brussaard L, van Veen JA, Kooistra MJ, Lebbink G, 1988. The Dutch Programme on Soil Ecology of Arable Farming Systems. 1. Objectives, approach and some preliminary results., Ecological Bulletins (no. 39), 35–40.

Dunger W, 1983. Tiere im Boden. Die Neue Brehm-Bücherei 327. A. Ziemsen, Wittenberg.

Edwards CA, Lofty JR, 1977. The influence of invertrbrates on root growth of crops with minimal or zero cultivation. Ecological Bulletin 25, 348–356.

Filser J, 1993. Die Bodenmesofauna unter der landwirtschaftlichen Intensivkultur Hopfen: Anpassung an bewirtschaftungsbedingte Bodenbelastungen? Int. Natursch. Landschaftspfl. 6, 368–386.

Filser J, 1995a. The effect of green manure on the distribution of collembolan in permanent row crop. Biology and Fertility of Soils 19, 303–308.

Filser J, 1995b. Collembola as indicators for long-term effects of intensive management. Acta Zoologica Fennica 196, 326–328.

Filser J, Dette A, Fromm H, Lang A, Mebes KH, Munch JC, Nagel R, Winter K, Beese F, 1999. Reactions of soil organisms to site-specific management: The first long-term study at the landscape level. Ecosystem (Supplement) 28, 139–147.

Filser J, Fromm H, Mommertz S, Nagel RF, Wahl F, Winter K, Beese F, 1993a. Raum-zeitliche Muster von Zustandsgrößen der Tier- und Mikroogranismengesellschaften in Böden. FAM-Bericht 3, 61–75.

Filser J, Fromm H, Nagel RF, Winter K, 1993b. Bodentier- und Mikroorganismengesellschaft in einer heterogenen Agrarlandschaft. Mitteilungen der Deutschen Bodenkundlichen Gesellschaft 72, 507–510.

Filser J, Fromm H, Nagel RF, Winter K, 1995. Effects of previous intensive agricultural management on microorganisms and the biodiversity of soil fauna. Plant and Soil 170, 123–129.

Filser J, Lang A, Mebes KH, Mommertz S, Palojärvi A, Winter K, 1996. The effect of land use change on soil organisms – an experimental approach. Verhandlungen der Gesellschaft für Ökologie 26, 671–679.

Fischer K, Blay K, Kotalik J, Riemschneider P, Klotz D, Kettrup A, 1997. Extraktion eines kupferbelasteten Bodens unter Verwendung eines aminosäurehaltigen Reststoffhydrolysats. II. Elutionsdynamik und Bilanzierung der Aminosäuren. Z. Pflanzenernähr. Bodenkunde 160, 511–518.

Fromm H, Winter K, Filser J, Hantschel R, Beese F, 1992. The influence of soil type and cultivation system on the spatial distributions of the soil fauna and microorganisms and their interactions. Geoderma 60, 109–118.

Ghilarov MS, 1971. Invertebrates which destroy the forest litter and ways to increase their activity. In: Duvigreaud P. (Ed.), Productivity of forest ecosystems. Syrup. Proc. UNESCO, Brussels, 433–439.

Hole FD, 1981. Effects of animals on soil. Geoderma 25, 75–112.

Kampichler C, 1998. The potential of soil habitat features for the modelling of numerical abundance of Collembola: A neural network approach. Verh. Ges. Ökologie Bd 28, 151–159.

Kampichler C, Dzeroski S, Wieland R, 2000. Application of machine learning techniques to the analysis of soil ecological databases: Relationships between habitat features and Collembolan community characteristics. Soil Biology and Biochemistry 32, 197–209.

Lenz R, Eisenbeis G, 1998. An extraction method for nematodes in decomposition studies using the minicontainer-method. Plant and Soil 198, 109–116.

Lussenhop J, 1996. Collembola as mediators of microbial symbiont effects to soybean. Soil Biology and Biochemistry 28, 363–369.

McLean MA, Kaneko N, Parkinson D, 1996. Does selective grazing by mites and Collembola affect litter fungal community structure? Pedobiologia 40, 97–105.

Mebes KH, 1999. Collembolengemeinschaften in Agrarökosystemen: Steuerung durch Umweltfaktoren, Einfluss auf den Stoffumsatz. PhD thesis, TU München, ISBN 3-8265-4896-5, FAM-Bericht 33, pp. 179.

Mebes KH, Filser J, 1997. A method for estimating the significance of surface dispersal for population fluctuations of Collembola in arable land. Pedobiologia 41, 115–122.

Mebes KH, Filser J, 1998. Does the composition of Collembola affect nitrogen turnover? Applied Soil Ecology 9, 241–247.

Nagel RF, 1996. Die Bedeutung von Regernwürmern für den C- und N-Umsatz in einer heterogenen Agrarlandschaft. PhD thesis, LMU München, FAM-Bericht 11, ISBN 3-8265-1575-7, pp. 126.

Nagel RF, Beese F, 1992. Veränderung des Transportverhaltens gelöster Stoffe durch Regenwurmgänge. Mitteilungen der Deutschen Bodenkundlichen Gesellschaft 67, 107–110.

Nagel RF, Fromm H, Beese F, 1995. The influence of earthworms and soil mesofauna on the C and N mineralization in agricultural soils – a microcosm study. Acta Zoologica Fennica 196, 22–26.

Parkinson D, Visser S, Whittaker JB, 1979. Effects of collembolan grazing on fungal colonization of leaf litter. Soil Biology and Biochemistry 11(5), 529–535.

Petersen H, 1978. Some properties of two high-gradient extractors for soil microarthropods, and an attempt to evaluate their extraction efficiency. Natura Jutlandica 20, 95–122.

Ponge JF, 1993. Biocoenoses of Collembola in Atlantic temperate grass-woodland ecosystems. Pedobiologia 37, 223–244.

Raw F, 1959. Estimating earthworm populations by using formalin. Nature 184, 1661–1662.

Reddy MV, Reddy VR, Yule DF, Cogle AL, George PJ, 1994. Decomposition of straw in relation to tillage, moisture, and arthropod abundance in a semi-arid tropical alfisol. Biology and Fertility of Soils 17, 45–50.

Ruz Jerez E, Ball PR, Tillman RW, 1988. The role of earthworms in nitrogen release from herbage residues. In: Jenkinson DS, Smith KA (Eds.), Nitrogen Efficiency in Agricultural Soils. Elsevier Applied Science, London, pp. 355–370.

Seastedt TR, 1984. The role of microarthropods in decomposition and mineralization processes. Annual Review Entomology 29, 25–46.

Setala H, Huhta V, 1990. Evaluation of the soil fauna impact on decomposition in a simulated coniferous forest soil. Biology and Fertility of Soils 10, 163–169.

Tian G, Adejuyigbe CO, Adeoye GO, Kang BT, 1998. Role of soil microarthropods in leaf decomposition and N release under various land-use practices in the humid tropics. Pedobiologia 42, 33–42.

Van Straalen, NM, 1998. Evaluation of bioindicator systems derived from soil arthropod communities. Applied Soil Ecology 9, 429–437.

Visser S, Parkinson D, Hassall M, 1987. Fungi associated with Onychiurus subtenuis (Collembola) in an aspen woodland. Canadian Journal of Botany 65(4), 635–642.

Vreeken-Buijs MJ, Brussaard L, 1996. Soil mesofauna dynamics, wheat residue decomposition and nitrogen mineralization in buried litterbags. Biology and Fertility of Soils 23, 374–381.

Werner MR, 1990. Earthworm ecology and sustaining agriculture. Components, vol. 1, no.4 (Fall 1990), University of California Sustainable Agriculture Research & Education Program.

Yeates GW, 1981. Soil nematode populations depressed in the presence of earthworms. Pedobiologia 22, 191–195.

Chapter 3.6

Spiders (Araneae) in Arable Land: Species Community, Influence of Land Use on Diversity, and Biocontrol Significance

A. Lang and J. Barthel[*]

3.6.1 Introduction

Increased agricultural intensity over the last decades and the use of modern farming methods correspond to significant decreases in both faunal and floral biodiversity in arable land (Plachter, 1991). The prominent factors considered to be responsible for this development are application of pesticides and herbicides, use of drainage and inorganic fertilisers, changing patterns of crop rotation, increasing field sizes accompanied with a loss in spatial and structural heterogeneity, and decrease in and degradation of habitats and field boundaries (e.g. Knauer, 1980; Pfiffner and Niggli, 1996; Thomas and Jepson, 1997; Baines et al., 1998; Morris, 2000). Many species of agricultural weeds and their associated arthropod fauna are nowadays rare and restricted to field edges. However, uncultivated areas in European

[*] Dedicated to Jutta Barthel who died on 14th November 2002.

Perspectives for Agroecosystem Management
Edited by P. Schröder, J. Pfadenhauer and J.C. Munch

agricultural landscapes such as permanent fallows, field margins and hedgerows have been at the same time severely cut down (e.g. Kaule, 1991; Plachter, 1991). In recent years, an increasing number of studies have dealt with the significance of uncultivated areas with respect to wildlife habitats, refuge areas and, especially, pest control in neighbouring arable fields (Malfait and De Keer, 1990; Frei and Manhart, 1992; Boatman, 1994; Feber et al., 1996; Baines et al., 1998; Sunderland and Samu, 2000). Within the framework of the FAM research project, the spider fauna was extensively studied between 1991 and 1996 at Research Station Scheyern, as well as in six further agricultural areas in southern Bavaria. The focus of the research was the influence of habitat and landscape characteristics on foliage-dwelling spiders in uncultivated areas, and the predation impact of spiders within arable fields. Spiders were chosen as target organisms, since they fulfil several features relevant for 'indicator organisms' (Barthel, 1998): They represent a functional important predator group (also capable of limiting pest organisms), are a species and individual rich arthropod taxon, occur in (nearly) all terrestrial ecosystems, add a substantial contribution to the biodiversity of habitats and are relatively easy to collect using various sampling methods. Knowledge about their ecological demands has developed rapidly over the last years, and biotope relationships and preferences are known for many species in mid-Europe (e.g. Maurer and Hänggi, 1990; Martin, 1991; Hänggi et al., 1995; Platen et al., 1995; Platen, 1996). Furthermore, red data books and species checklists exist for Germany and most of its federal states, as well as standard and good-quality identification literature for Central Europe (e.g. Blick and Scheidler, 2003; Platen et al., 1991, 1995, 1998; Nentwig et al., 2004). The objectives of the FAM spider fauna study were to (i) conduct an inventory survey of the foliage-dwelling spider fauna in fields and their margins, (ii) determine the influence of land use on spiders in agricultural landscapes, (iii) clarify the variability of the spider community in dependence of environmental factors and (iv) study the function of spiders in agro-ecosystems with special reference to their role as biological control agents (Plachter et al., 1991, 1993; Agricola et al., 1996; Filser et al., 1997).

3.6.2 The community of foliage-dwelling spiders in arable land

The composition of foliage-dwelling spider fauna in 97 field margins was investigated in seven agricultural landscapes of southern Bavaria (Barthel and Plachter, 1996). Field margins (width up to 10 m) were considered grassy strips of banks between arable fields or meadows, generally not cultivated or ploughed and containing only single shrubs or trees. The foliage-dwelling spider fauna was recorded by standardised visual inspections in specified 1×50 m plots within the margins (Barthel, 1997a). A total of 75 spider species were observed between 1993 and 1995, with an average of 11–12 species per margin (range 3–26 species). Approximately 30% of all species were recorded in most of the studied landscapes, whereas over 50% were observed only in one or two areas. The most abundant species were *Argiope bruennichi*, *Theridion impressum*, *Larinioides folium*, *Pisaura mirabilis*, *Araneus quadratus* and *Evarcha arcuata* (spiders with a proportion >5% of the total catch and in decreasing order). Although less abundant (<5% of the total catch), *Aculepeira ceropegia*, *Mangora acalypha* and *Tetragnatha pinicola* were also very common and occurred in 80–90% of the sampling plots. Fourteen per cent of the total 75 spider species are listed in the German or Bavarian red data book or both (Platen et al., 1998; Blick and Scheidler, 2003), but these species were always rare and comprised less than 0.2% of the overall individual catch.

The occurrence of foliage-dwelling spiders at Research Station Scheyern in arable fields and permanent set-asides (left to natural regeneration in 1992/93) were recorded between 1992 and 1995 to study spider utilisation of different agricultural habitats. Cultivated areas were classified in fields with various crops and grasslands, and set-asides in former fields and abandoned grasslands (Barthel, 1997a). Because of the low sample size of five replicates per set-aside type, the total number of spider species recorded was lower than that of the above study of 97 field margins. In Scheyern, the total number of spider species was 28 for field set-asides and 41 for grassland set-asides. However, the mean number of spider species per set-aside was quite similar to that in the above-studied field margins (10 species in field set-asides, 14 species in grassland set-asides). The occurrence of the dominant species was almost identical to uncultivated areas of the other agro-ecosystems studied (see species mentioned above). Additionally, *Clubiona reclusa* and *A. quadratus* showed high abundances in uncultivated areas of Research Station Scheyern. The proportion of species listed in red data books (see above) comprised 8% of the total species in Scheyern. Although the most dominant species of spiders were also found in the managed fields, in particular *T. impressum* and *A. ceropegia*, individual and species numbers were significantly lower in fields and grasslands under management.

The results of the FAM project clearly demonstrated that managed fields and grassland are colonised by only few foliage-dwelling spiders, in terms of both species number and individual abundance. These findings have been confirmed by others (e.g. Nyffeler and Benz, 1979; Morris, 2000). Areas intensely cultivated were represented by a species-poor fraction of the spider fauna in adjacent uncultivated habitats. In grasslands, intense grazing by cattle or regular mowing reduces spider abundance in the herbaceous layer, either directly or by altering climatic conditions and vegetation structure (e.g. Duffey, 1966; Hatley and MacMahon, 1980; Robinson, 1981; Maelfait and De Keer, 1990; Nyffeler and Breene, 1990; Gerstmeier and Lang, 1996; Baines et al., 1998; Morris, 2000; Bell et al., 2001). Most cultivated fields provide a living space for foliage-dwelling spiders for only approximately 5 months, since harvest and subsequent tilling destroys many spiders and their cocoons. Therefore, spiders must re-colonise the area each season (Nyffeler and Benz, 1979; Samu et al., 1999). However, some foliage-dwelling spiders do cope indeed well with these adverse conditions. A review of regularly occurring spiders in higher densities in European fields has been provided by Barthel (1997a). For instance, *A. ceropegia* was a typical 'agrobiont' species in our study, as it was more abundant in fields than in adjacent margins during June and July (Plachter et al., 1991; Barthel, 1997a; Luczak, 1979). For this species, the key factor is its life history pattern, since juvenile *A. ceropegia* emerge from their cocoons and can migrate from the field before harvest time. Nevertheless, this and other species require field margins as evasion habitats and as sites for hibernation (Luczak, 1979; Sotherton, 1984; Maelfait and De Keer, 1990; Thomas et al., 1991; Barthel, 1997a; Pfiffner and Luka, 2000). It is also well known that due to the ephemeral nature of agricultural habitats, spider species with an adequate dispersal and colonisation power dominate the communities in arable land (e.g. Thomas et al., 1992; Duffey, 1993; Wise, 1993). Spiders can drift long distances by aerial dispersal (ballooning), which is especially true for the Linyphiidae as many of them are able, unlike numerous other species, to balloon also as adults due to their small body size and weight (Duffey, 1978; Bishop and Riechert, 1990; Weyman, 1993; Barthel, 1997a).

In the FAM study, the foliage-dwelling spider faunas in field margins, set-asides of former fields and abandoned grassland were very similar to each other. On the basis of the results of the FAM research project, a typical species spectrum has been obtained for the foliage-dwelling spider community in arable land with regard to the region studied (Table 1). This listing includes 29 species from 11 families, showing a broad distribution in our study area or occurrence in over 20% of the single sample plots. Comparison of our species list with European studies leads to the conclusion that those species listed in Table 1 can be characterised as typical foliage-dwelling spiders of agro-ecosystems in Central Europe and were not randomly collected from the local species pool (cf. Duffey, 1962; Geiler, 1963; Vilbaste, 1964, 1965; Buchar, 1968; Raatikainen and Huhta, 1968; Hempel et al., 1971; Kajak, 1971; Huhta and Raatikainen, 1974; Luczak, 1979, 1980; Nyffeler and Benz, 1979, 1987, 1989; Krause, 1987; Ingrisch et al., 1989; Nyffeler and Breene, 1990; Maelfait and De Keer, 1990; Scheidler, 1990; Alderweireldt, 1993; Samu et al., 1996; Ysnel and Canard, 2000; Ludy and Lang, 2004).

Table 1 Characteristic species spectrum of foliage-dwelling spiders in the agro-ecosystems of the FAM project (cf. Barthel, 1997a, 1998).

Family	Species
Agelenidae	*Agelena labyrinthica* (Clerck 1757)
Araneidae	*Aculepeira ceropegia* (Walckenaer 1802)
Araneidae	*Araneus diadematus* (Clerck 1757)
Araneidae	*Araneus quadratus* (Clerck 1757)
Araneidae	*Araniella cucurbitina* (Clerck 1757)
Araneidae	*Argiope bruennichi* (Scopoli 1772)
Araneidae	*Cyclosa oculata* (Walckenaer 1802)
Araneidae	*Larinioides folium* (Schrank 1803)
Araneidae	*Mangora acalypha* (Walckenaer 1802)
Clubionidae	*Clubiona neglecta* (O.P.-Cambridge 1862)
Clubionidae	*Clubiona reclusa* (O.P.-Cambridge 1863)
Linyphiidae	*Floronia bucculenta* (Clerck 1757)
Linyphiidae	*Linyphia triangularis* (Clerck 1757)
Linyphiidae	*Microlinyphia pusilla* (Sundevall 1829)
Philodromidae	*Philodromus cespitum* (Walckenaer 1802)
Pisauridae	*Pisaura mirabilis* (Clerck 1757)
Salticidae	*Evarcha arcuata* (Clerck 1757)
Salticidae	*Heliophanus flavipes* (Hahn 1832)
Tetragnathidae	*Metellina segmentata* (Clerck 1757)
Tetragnathidae	*Tetragnatha extensa* (Linné 1758)
Tetragnathidae	*Tetragnatha pinicola* (L. Koch 1870)
Theridiidae	*Enoplognatha latimana* (Hippa & Osala 1982)
Theridiidae	*Enoplognatha ovata* (Clerck 1757)
Theridiidae	*Theridion bimaculatum* (Linné 1767)
Theridiidae	*Theridion impressum* (L. Koch 1870)
Thomisidae	*Xysticus bifasciatus* (C.L. Koch 1837)
Thomisidae	*Xysticus cristatus* (Clerck 1757)
Thomisidae	*Xysticus kochi* (Thorell 1872)
Thomisidae	*Xysticus ulmi* (Hahn 1826)

3.6.3 The influence of land use on foliage-dwelling spiders in field margins

The foliage-dwelling spider community in field margins at Scheyern represents a case study on the substantial influence of general land use patterns on biodiversity. Together with the management reorganisation of the farm in 1992, actual field sizes were reduced, leading to a greater amount of field margins; these increased from 120 to 150 m ha^{-1}. Accompanying the implementation of new set-aside land, habitat space and the connectedness among uncultivated areas thus considerably improved. This led to a significant threefold increase in both spider species and individual numbers within the field margins of the station (e.g. Barthel, 1998); these findings were also evident on a landscape scale. The seven agricultural landscapes studied in southern Bavaria differed in their margin densities and ranged from 90 to 300 m margin length per hectare. Field size was negatively correlated with margin density, and there was a positive relationship between margin density and species number of spiders within an area. Margin density could explain up to 74% of the variation in total species number of an area (Barthel and Plachter, 1996; Barthel, 1997a, 1998). These results show that margin density plays a vital role in the community of foliage-dwelling spiders in agricultural land: supplying additional suitable living space for more species and higher individual abundance. Field margins also have a functional significance with respect to connectivity among habitats, which may facilitate migration between and colonisation of new biotopes (e.g. Topping and Sunderland, 1994a; Samu et al., 1999; Morris, 2000). The relationship between landscape and spider diversity, as well as spider abundance, has been also demonstrated in other studies (e.g. Asselin, 1988; Pfiffner and Luka, 2000). Further, recent modelling studies have shown that insertion of uncultivated areas within agricultural landscapes increases metapopulation persistence and population densities of spiders within these agro-ecosystems (e.g. Topping and Sunderland, 1994b; Halley et al., 1996).

Several additional habitat characteristics have been recorded in field margins of the studied agricultural landscapes, to analyse other factors influencing the composition of the foliage-dwelling spider community (Barthel and Plachter, 1995, 1996; Barthel, 1997a). Correlation and regression analyses revealed that the most distinctive factors influencing the number of spider species, either positively or negatively, were margin density of an area (+), margin width (+), percentage cover of herbaceous plants (+) and the number of mechanical treatments (−) (Anderlik-Wesinger et al., 1996; Barthel and Plachter, 1996; Barthel, 1997a, 1997b). In contrast to the other factors tested, margin width showed a non-linear relationship (with species numbers). Species density increased with increasing margin width but levelled off at a margin width of approximately 3–4 m. This pattern may be a likely result of single incidental mowing or ploughing events of the narrow field margins, thus completely destroying the area together with its spider community. At wider field margins (>3 m), the possibility that at least a narrow strip survives mechanical treatment is higher. Thus, there is a lower risk that spider communities are affected by frequent and regular disturbances (Barthel and Plachter, 1995). Important reasons for the general positive correlation between abundance of spider species and margin width include (i) a concurrent increase in plant diversity and cover of herbs, representing structural features commonly enhancing the densities of web-building spiders, and a diverse spectrum of microclimatic conditions and (ii) a lower input of agrochemicals in wider margins (Anderlik-Wesinger et al., 1996; Barthel and Plachter, 1996; Baines et al., 1998; Rypstra et al., 1999; Samu et al., 1999; Bell et al., 2001). Kampichler et al. (2000) translated the collected data into a fuzzy rule-based model. The predictive power of the model was high, and it showed the following: (i) margin width and disturbance determined habitat persistence; (ii) habitat persistence and margin density, in turn, determined colonisation potential of the field margin; and (iii) colonisation potential and herbaceous plant cover determined spider species numbers. Therefore, this model confirmed the actual observations of the FAM project of Anderlik-Wesinger et al. (1996), Barthel and Plachter (1996) and Barthel (1997a). Furthermore, the results are consistent with similar observational and modelling studies (e.g. Asselin, 1988; Topping and Sunderland, 1994a, 1994b; Halley et al., 1996; Topping, 1997; Topping and Lövei, 1997; Pfiffner and Luka, 2000). Moreover, the model demonstrated the hierarchical influence of habitat factors and pointed out their compensatory powers. For instance, a high margin density may compensate for low habitat persistence and vice versa (Kampichler et al., 2000). The suggestions of this model are also in accordance with the proposed conceptual framework of Samu et al. (1999), with regard to the distribution, dispersal and abundance of spiders in agricultural systems, since Samu et al. advocate to consider how factors operate and interact at three levels of spatial hierarchy, namely, micro-habitat, habitat and landscape.

3.6.4 Biocontrol significance of spiders

Arable land typically harbours communities of polyphagous invertebrate natural enemies, among them many predacious beetles and spiders (Ekschmitt et al., 1997). There has been much concern about their role as biological control agents (e.g. Luff, 1987; Wise, 1993; Nyffeler et al., 1994; Sunderland, 1999), and although these generalist predators lack prey specificity, they also prey on pest organisms and can exert a substantial predation impact on given prey groups or pest species in the field (Symondson et al., 2002; Maloney et al., 2003). However, as spiders and other omnivorous natural enemies prey on a wide range of different prey groups including other predatory arthropods and non-pest organisms, their direct and indirect effects within agro-ecosystems can be manifold and sometimes unpredictable (Wise, 1993; Wise et al., 1999; Symondson et al., 2002). Spiders owe their prominent position within the invertebrate community in arable land not only to their predacious nature and their multiple food web links but also to their sheer abundance. Among the invertebrate natural enemies of agro-ecosystems, they belong to the most dominant groups (e.g. Wise, 1993; Ekschmitt et al., 1997; Nyffeler, 2000), contributing up to 35–50% to the ground-dwelling invertebrate predators at Research Station Scheyern farm in terms of both individual numbers and biomass (Filser et al., 1996; Mommertz et al., 1996; Lang et al., 1997; Lang, 1998, 2000). Within the framework of the FAM research project the effects of the arthropod predator community, especially of spiders, were studied in experimental enclosures in arable fields with special reference to biological control (cf. Lang et al., 1995). In contrast to the previous sections of this contribution, the focus of the predation studies was put on soil-dwelling spiders rather than on spiders on plants. Epigeal spiders are generally more abundant in fields than foliage-dwelling spiders; thus, their substantial predation effects seemed more likely. On the experimental farm in Scheyern, peak abundances of 20–90 ground-dwelling spiders per square metre were found in fields (Lang, 1998), whereas Barthel (1997a) recorded only minor numbers of foliage-dwelling spiders in the Scheyern fields. However, it must be acknowledged that published studies about the densities of spiders of higher strata in arable fields are generally too sparse to depict a general picture. For example, foliage-dwelling spiders can reach densities of up to 25 spiders per square metre in maize fields (Ludy and Lang, 2004), and even relatively small numbers of web spiders on crop plants may nonetheless decrease prey numbers significantly (e.g. Carter and Rypstra, 1995).

In a maize field at Research Station Scheyern, Lang (1996) manipulated numbers of wolf spiders (Araneae, Lycosidae) and ground beetles (Coleoptera, Carabidae) within closed field cages and recorded the subsequent effects on potential prey groups. By directly testing predator-enriched versus predator-reduced cages, it could be shown that wolf spiders significantly contribute to the reduction of some prey groups (Lang, 1998; Lang et al., 1999). Although neither wolf spiders nor ground beetles alone reduced numbers of thrips (Thysanoptera) and aphids (Homoptera, Aphididae), their combined predation pressure depressed these herbivorous prey groups. In the case of aphids, control was evident only in mid-season when aphid abundance was lower than later in the season at high aphid densities, which confirms the importance of achieving an early control of pest numbers (Ekbom et al., 1992; Östman et al., 2001). In a similar experiment with open enclosures in a winter wheat field, ground-dwelling spiders (Lycosidae and Linyphiidae) did not limit aphid numbers, whereas the ground beetle assemblage reduced this pest group (Lang, 2003). The carabid predation impact on aphids was most evident earlier in the season and disappeared later. However, at the beginning of July, the joint impact of spiders and ground beetles was responsible for a reduction in aphid numbers. Thus, our results clearly show that spiders can cause, or rather contribute to, a reduction of pest organisms in arable fields, which is in accordance with studies from other countries and those conducted on other crops (e.g. Oraze and Grigarick, 1989; Riechert and Bishop, 1990; De Barro, 1992; Provencher and Riechert, 1994; Carter and Rypstra, 1995; Riechert and Lawrence, 1997; Snyder and Wise, 1999, 2001; Maloney et al., 2003; Schmidt et al., 2003; Snyder and Ives, 2003). In our studies, the spider effects were significant only together with the impact of other generalist predators.

We interpret this as a simple biomass effect: The more predators are present, the more prey is reduced (Lang, 2003). As generalist predators in arable land seem to have a very low prey capture rate (e.g. Nyffeler and Benz, 1988; Bilde and Toft, 1998), cutback of prey organisms is possibly negligible if only few numbers of predators are present. With respect to pest organisms, this pattern may even be amplified if a pest is of low quality or even toxic and the predators actively avoid it as it seems to be the case for some aphids (Bilde and Toft, 1994; Toft, 1995; Lang and Gsödl, 2001, 2003). However, besides the pure biomass effect (more predators eat more prey), there exist indications that a higher diversity of invertebrate predators per se contributes to a more efficient pest control (e.g. Marc and Canard, 1997; Riechert and Lawrence, 1997; Sunderland et al., 1997; Sunderland, 1999; Riechert, 1999; Tscharntke and Kruess, 1999; Maloney et al., 2003). Some potential reasons for this positive effect of biodiversity are as follows: (i) Different predator species occupy different and complementary niches; thus, the prey may not find spatial refuges. (ii) Pest organisms are exposed to natural enemies with differing phenologies throughout their life cycles and can therefore not exploit temporal refuges. (iii) Predator species exert a synergistic impact on pests, i.e. the joint effect of the predators is greater than the sum of their individual effects, if one predator facilitates the killing of the prey by another predator (e.g. Ferguson and Stiling, 1996; Losey and Denno, 1998; Sunderland et al., 1997; Sunderland, 1999). On the contrary, invertebrate natural enemies in agro-ecosystems do also prey on, or interfere with, other arthropod predators, and this intraguild interference may be antagonistic to pest control, and even lead to an enhancement of the concerned pest population (e.g. Rosenheim et al., 1993; Raymond et al., 2000; Snyder and Ives, 2001; Snyder and Wise, 2001). In our studies we also found evidence for intraguild interference. In a winter wheat field, the presence of ground beetles led to a decrease in numbers of wolf spiders. Nevertheless, the overall impact of the spider–ground beetle assemblage was still strong enough to limit aphid numbers in mid-season (Lang, 2003). Therefore, the results of the FAM study confirm the important role of spiders and other invertebrate predators for biological control, thereby validating other studies (cf. reviews of Symondson et al., 2002; Maloney et al., 2003), additionally showing that the demonstrated intraguild interference within the generalist predator community did not negate its effectiveness in reducing pest numbers (Lang, 2003).

3.6.5 Concluding remarks

The FAM project identified a species spectrum of foliage-dwelling spiders characteristic for agro-ecosystems in Europe (Barthel, 1997a, 1998). Typically for arable land, only few species were dominating the community, and these species could be recorded in all studied habitats (arable fields, grasslands, set-asides, field margins), though in differing densities. Nevertheless, the spider fauna is also influenced by the local and regional species pool (cf. Sunderland and Samu, 2000). It could be shown that spider species composition and richness is affected on different spatial scales: on a micro-habitat, a habitat and a landscape level (e.g. Barthel and Plachter, 1996; Barthel, 1997a, 1997b; Kampichler et al., 2000). The prominent factors supporting a richer spider community were structural and floral diversity of habitats, field margin width, reduction of mechanic disturbances, and margin density within the landscape (cf. Anderlik-Wesinger et al., 1996; Kampichler et al., 2000). In conclusion, recommendations for the promotion of the spider fauna in agro-ecosystems include smaller field sizes, a reduction of mowing and other disturbances, the implementation of undisturbed habitats, a minimum margin width of 3–5 m, and a higher margin density (>200 m ha^{-1}) (e.g. Barthel and Plachter, 1995; Plachter et al., 1995; Barthel, 1997a). Ideally, a mosaic of set-asides of various life spans should be achieved within the agro-ecosystem, since set-asides of different ages contain different communities depending on the species concerned and the local circumstances (Gibson et al., 1992; Topping and Sunderland, 1994a, 1994b; Barthel, 1997a; Baines et al., 1998). Higher abundance and species richness of spiders in arable land will, for instance, pay off in terms of an improved pest control in fields, as was shown by the predation studies of the FAM project (e.g. Lang, 1998, 2003; Lang et al., 1999). One crucial point is, however, the immigration of spiders from adjacent 'source' habitats to the field. In intensely managed crops, field conditions are often too adverse for spiders, resulting in only low immigration rates and population densities (e.g. Barthel, 1997a). Organic farming practices have been shown to provide a more suitable within-field environment for spiders and other predators, and consequently spider numbers and diversity in organic fields are much higher compared to conventionally managed fields (e.g. Filser et al., 1995; Pfiffner and Niggli, 1996; Feber et al., 1998; Pfiffner and Luka, 2000, 2003). The combination of external field management (e.g. set-asides, field margins, hedgerows) and internal field management (e.g. intercropping, cover cropping, mulching) is therefore a mandatory strategy to fully exploit the beneficial effects of spiders. As the observed species compositions of foliage-dwelling spiders reflected biodiversity and structural heterogeneity of agricultural landscapes in general, and were also associated with a higher diversity of herbaceous plants, grasshoppers and birds (Anderlik-Wesinger et al., 1996; Janßen et al., 1997; Laußmann and Plachter, 1998), a tentative group of spider species was suggested to indicate the (bio)diversity of agro-ecosystems on different spatial scales (Barthel, 1997a, 1998).

Many management practices have been explored and developed to increase arthropod predator density and diversity in agro-ecosystems with the aim to utilise the beneficial predation impact of natural enemies such as spiders, a research goal also pursued in the FAM project. Suggested successful approaches range from the scale of micro-habitat manipulations to the habitat and the landscape (e.g. Nentwig, 1988; Frank and Nentwig, 1995; Barthel, 1997a; Rypstra et al., 1999; Sunderland and Samu, 2000; Denys and Tscharntke, 2002; Symondson et al., 2002). However, it is surprising that only a limited number of studies seem to exist trying to link these augmentation methods and habitat characteristics to pest control effects, i.e. to directly and rigorously prove that certain habitat manipulations or features, respectively, promote a more efficient biocontrol (e.g. Riechert, 1990; Riechert and Bishop, 1990; Carter and Rypstra, 1995; Jmhasly and Nentwig, 1995; Wyss et al., 1995; Östman et al., 2001). Because of long-distance dispersal of spiders (ballooning), the overall heterogeneity of arable landscapes may have a stronger influence on densities and species numbers of spiders than small scale changes on a field level would have (Halley et al., 1996; Topping, 1997; Samu et al., 1999; Sunderland and Samu, 2000). Therefore, there is an urgent need of studies exploring the relationship between landscape characteristics and biological pest control efficiency (cf. Kruess and Tscharntke, 1994; Thies and Tscharntke, 1999; Östman et al., 2001).

Acknowledgements

The scientific activities of the FAM Munich Research Network on Agroecosystems were financially supported by the German Federal Ministry of Education and Research (BMBF 0339370). Overhead costs of Research Station Scheyern were funded by the Bavarian State Ministry for Science, Research and the Arts. Thanks to Juliane Filser, Claudia Ludy and Ursula Olazábal for reading and commenting an earlier draft of the chapter.

References

Agricola U, Barthel J, Laussmann H, Plachter H, 1996. Struktur und Dynamik der Fauna einer süddeutschen Agrarlandschaft nach Nutzungsumstellung auf ökologischen und integrierten Landbau. Verh. Ges. Ökol. 26, 681–692.

Alderweireldt M, 1993. A five year study of the invertebrate fauna of crop fields and their edges. Part 2. General characteristics of the spidertaxocoenosis. Bulletin et Annales de la Société royale belge d'Entomologie 129, 63–68.

Anderlik-Wesinger G, Barthel J, Pfadenhauer J, Plachter H, 1996. Einfluß struktureller und floristischer Ausprägungen von Rainen in der

Agrarlandschaft auf die Spinnen (Araneae) der Krautschicht. Verh. Ges. Ökol. 26, 711–720.

Asselin A, 1988. Changes in grassland use consequences on landscape patterns and spider distribution. Münstersche Geographische Arbeiten 29, 85–88.

Baines M, Hambler C, Johnson PJ, Macdonald DW, Smith H, 1998. The effects of arable field margin management on the abundance and species richness of Araneae (spiders). Ecography 21, 74–86.

Barthel J, 1997a. Einfluß von Nutzungsmuster und Habitatkonfiguration auf die Spinnenfauna der Krautschicht (Araneae) in einer süd-deutschen Agrarlandschaft. PhD Thesis, Phillips-Universität Marburg, Verlag Agrarökologie, Bern, Hannover, ISBN 3-909192-02-5, Agrarökologie Band 25, 175 pp.

Barthel J, 1997b. Habitat preferences of *Enoplognatha latimana* Hippa *et* Oksala, 1982, and *Enoplognatha ovata* (Clerck, 1757) (Araneae: Theridiidae) in agricultural landscapes in Southern Bavaria (Germany). Proceedings of the 16th European Colloquium of Arachnology, Siedlce, 13–25.

Barthel J, 1998. Entwicklung von Indikationsverfahren durch Langzeit-beobachtungen und deren Eignung für den Naturschutz am Beispiel von Spinnen (Araneae). Schr.-R. f. Landschaftspfl. u. Natursch. 58, 161–190.

Barthel J, Plachter H, 1995. Distribution of foliage-dwelling spiders in uncultivated areas of agricultural landscapes (Southern Bavaria, Germany) (Arachnida, Araneae). In: Ruzicka V (Ed.), Proceedings of the 15th European Colloquium of Arachnology, pp. 11–21.

Barthel J, Plachter H, 1996. Significance of field margins for foliage-dwelling spiders (Arachnida, Araneae) in an agricultural landscape of Germany. Revue Suisse de Zoologie, vol. hors série, 45–49.

Bell JR, Wheather CP, Cullen WR, 2001. The implications of grassland and heathland management for the conservation of spider communities: A review. Journal of Zoology, London 255, 377–387.

Bilde T, Toft S, 1994. Prey preference and egg production of the carabid beetle *Agonum dorsale*. Entomologia Experimentalis et Applicata 73, 151–156.

Bilde T, Toft S, 1998. Quantifying food limitation of arthropod predators in the field. Oecologia 115, 54–58.

Bishop L, Riechert SE, 1990. Spider colonization of agroecosystems: Mode and source. Environmental Entomology 19, 1738–1745.

Blick T, Scheidler M, 2003. Rote Liste gefährdeter Spinnen (Araneae) Bayerns. Schr.-R. Bayer. Landesamt f. Umweltsch. 166, 308–321.

Boatman N, 1994. Field Margins: Integrating agriculture and conservation. British Crop Protection Council, BCPC Monograph, 58.

Buchar J, 1968. Analyse der Wiesenarachnofauna. Acta Univ. Carolinae – Biologica 1967, Prague, pp. 289–318.

Carter PE, Rypstra AL, 1995. Top-down effects in soybean agroecosystems: Spider density affects herbivore damage. Oikos 72, 433–439.

De Barro PJ, 1992. The impact of spiders and high temperatures on cereal aphid (*Rhopalosiphum padi*) numbers in an irrigated perennial grass pasture in South Australia. Annals of Applied Biology 121, 19–26.

Denys C, Tscharntke T, 2002. Plant-insect communities and predator-prey ratios in field margin strips, adjacent crop fields, and fallows. Oecologia 130, 315–324.

Duffey E, 1962. A population study of spiders in limestone grassland. Description of study area, sampling methods and population characteristics. Journal of Animal Ecology 31, 571–599.

Duffey E, 1966. Spider ecology and habitat structure (Arachnida, Araneae). Senckenbergiana Biology 47, 45–49.

Duffey E, 1978. Ecological strategies in spiders including some characteristics of species in pioneer and mature habitats. Symposia of the Zoological Society of London 42, 109–123.

Duffey E, 1993. A review of factors influencing the distribution of spiders with special reference to Britain. Memoirs of the Queensland Museum 33, 497–502.

Ekbom BS, Wiktelius S, Chiverton AP, 1992. Can polyphagous predators control the bird cherry-oat aphid (Rhopalosiphum padi) in spring cereals? Entomologia Experimentalis et Applicata 65, 215–223.

Ekschmitt K, Weber M, Wolters V, 1997. Spiders, Carabids, and Staphylinids: The ecological potential of predatory macro-arthropods. In: Benckiser G (Ed.), Fauna in Soil Ecosystems. Marcel Dekker, New York, pp. 307–362.

Feber RE, Bell J, Johnson PJ, Firbank LG, MacDonald DW, 1998. The effects of organic farming on surface-active spider (Araneae) assemblages in wheat in Southern England, UK. Journal of Arachnology 26, 190–202.

Feber RE, Smith H, MacDonald DW, 1996. The effects on butterfly abundance of the management of uncropped edges of arable fields. Journal of Applied Ecology 33, 1191–1205.

Ferguson KI, Stiling P, 1996. Non-additive effects of multiple natural enemies on aphid populations. Oecologia 108, 375–379.

Filser J, Dörsch P, Fromm H, Lang A, Mebes KH, Mommertz S, Nagel R, Palojärvi A, Pöckl B, Röver J, Schauer C, Winter K, Beese F, 1995. Steuerung biotischer Stoffumsetzungen im Habitat Boden – Beitrag zur Minimierung der Belastung von Wässern und der Atmosphäre durch C- und N-Verbindungen. FAM-Bericht 5, 151–168.

Filser J, Lang A, Mebes KH, Mommertz S, Palojärvi A, Winter K, 1996. The effect of land use change on soil organisms – An experimental approach. Verh. Ges. Ökol. 26, 671–679.

Filser J, Mommertz S, Angermayr L, Jell B, Lang A, Mebes KH, 1997. Steuerung von Nährstoffumsetzungen durch Bodentiere – Beitrag zur Minimierung der Belastung von Wässern und der Atmosphäre durch C- und N-Verbindungen. FAM-Bericht 13, 83–94.

Frank T, Nentwig W, 1995. Ground dwelling spiders (Araneae) in sown weed strips and adjacent fields. Acta Oecologica 16, 179–193.

Frei G, Manhart C, 1992. Nützlinge und Schädlinge an künstlich angelegten Ackerkrautstreifen in Getreidefeldern. Verlag Agrarökologie, Bern, Hannover, Agrarökologie 4, 1–140.

Geiler H, 1963. Die Spinnen- und Weberknechtfauna nordwestsächsischer Felder (Die Evertebratenfauna mitteldeutscher Feldkulturen V). Z Ang Zool 50, 257–272.

Gerstmeier R, Lang C, 1996. Beitrag zur Auswirkungen der Mahd auf Arthropoden. Z Ökologie u Naturschutz 5, 1–14.

Gibson CWD, Hambler C, Brown VK, 1992. Changes in spider (Araneae) assemblages in relation to succession and grazing management. Journal of Applied Ecology 29, 132–142.

Hänggi A, Stöckli E, Nentwig W, 1995. Lebensräume mitteleuropäischer Spinnen. Miscellanea Faunistica Helvetia 4, 1–460.

Halley JM, Thomas CFG, Jepson PC, 1996. A model for the spatial dynamics of linyphiid spiders in farmland. Journal of Applied Ecology 33, 471–492.

Hatley CA, MacMahon JA, 1980. Spider community organization: Seasonal variation and the role of vegetation architecture. Environmental Entomology 9, 632–639.

Hempel W, Hiebsch H, Schiemenz H, 1971. Zum Enfluß der Weidewirtschaft auf die Arthropoden-Fauna im Mittelgebirge. Faun. Abh. Mus. Tierk. Dresden 3, 235–281.

Huhta V, Raatikainen M, 1974. Spider communities of leys and winter cereal fields in Finland. Annales Zoologici Fennici 11, 97–104.

Ingrisch S, Wasner U, Glück E, 1989. Vergleichende Untersuchungen der Ackerfauna auf alternativ und konventionell bewirtschafteten Flächen. In: König W et al.: Alternativer und konventioneller Landbau. Schriftenr Landesanstalt f Ökologie, Landschaftsentwicklung u Forstplanung Nordrhein-Westfalen 11, 113–271.

Janssen B, Homes V, Plachter H, 1997. Struktur der Fauna der Agrarlandschaft in Abhängigkeit von landwirtschaftlichen Nutzungsformen und Biotopneuschaffung. FAM-Bericht 13, 197–208.

Jmhasly P, Nenwtig W, 1995. Habitat management in winter wheat and evaluation of subsequent spider predation on insect pests. Acta Oecologica 16, 389–403.

Kajak A, 1971. Productivity investigations of two types of meadows in the Vistula Valley. IX. Production and consumption of field layer spiders. Ekologia Polska 19, 197–211.

Kampichler C, Barthel J, Wieland R, 2000. Species density of foliage-dwelling spiders in field margins: A simple, fuzzy rule-based model. Ecological Modelling 129, 87–99.

Kaule K, 1991. Arten- und Biotopschutz. UTB, Stuttgart.

Knauer N, 1980. Möglichkeiten und Schwierigkeiten bei der Schaffung funktionsfähiger Naturschutzgebiete in der Agrarlandschaft. Landwirtsch. Forsch., Sonderheft 37, 105–116.

Krause A, 1987. Untersuchungen zur Rolle von Spinnen in Agrarbiotopen. Dissertation, Rheinische Friedrich-Wilhelms-Universität, Hohe landwirtschaftliche Fakultät, Bonn, 306 pp.

Kruess A, Tscharntke T, 1994. Habitat fragmentation, species loss, and biological control. Science 264, 1581–1584.

Lang A, 1996. Einfluss von Wolfspinnen und Laufkäfern auf Beutepopulationen: Ein Freilandexperiment auf einer landwirtschaftlichen Fläche. DGaaE Nachr. 10, 43–44.

Lang A, 1998. Invertebrate epigeal predators in arable land: Population densities, biomass and predator-prey interactions in the field with special reference to ground beetles and wolf spiders. PhD Thesis, LMU München, Shaker Verlag, Aachen, ISBN3-8265-3449-2, FAM-Bericht 23, 139 pp.

Lang A, 2000. The pitfalls of pitfalls: A comparison of pitfall trap catches and absolute density estimates of epigeal invertebrate predators in arable land. Journal of Pest Science 73, 99–106.

Lang A, 2003. Intraguild interference and biocontrol effects of generalist predators in a winter wheat field. Oecologia 134, 144–153.

Lang A, Filser J, Henschel JR, 1999. Predation by ground beetles and wolf spiders on herbivorous insects in a maize crop. Agriculture, Ecosystems & Environment 72, 189–199.

Lang A, Filser J, Mommertz S, Gigglinger S, 1995. Relationships between lycosid spiders and their prey in an agroecosystem. An outline of a project. Proceedings of the 15th European Colloquium of Arachnology, Ceské Budějovice, 107–110.

Lang A, Gsödl S, 2001. Prey vulnerability and active predator choice: A carabid beetle and its aphid prey. Journal of Applied Entomology 125, 53–61.

Lang A, Gsödl S. 2003. "Superfluous killing" of aphids: A potentially beneficial behaviour of the predator *Poecilus cupreus* (L.) (Coleoptera: Carabidae)? Journal of Plant Diseases and Protection 110, 583–590.

Lang A, Krooß S, Stumpf H, 1997. Mass-length relationships of epigeal arthropod predators in arable land (Araneae, Chilopoda, Coleoptera). Pedobiologia 41, 233–239.

Laußmann H, Plachter H, 1998. Der Einfluss der Umstrukturierung eines Landwirtschaftsbetriebes auf die Vogelfauna: Ein Fallbeispiel aus Süddeutschland. Die Vogelwelt 119, 7–19.

Losey JE, Denno RF, 1998. Positive predator-predator interactions: Enhanced predation rates and synergistic suppression of aphid populations. Ecology 79, 2143–2152.

Luczak J, 1979. Spiders in agrocoenoses. Polish Ecological Studies 5, 151–200.

Luczak J, 1980. Spider communities in crop fields and forests of different landscapes of Poland. Polish Ecological Studies 6, 735–762.

Ludy C, Lang A, 2004. How to catch foliage-dwelling spiders (Araneae) in maize fields and their margins: A comparison of two sampling methods. Journal of Applied Entomology 128, 501–509 .

Luff ML, 1987. Biology of polyphagous ground beetles in agriculture. Agricultural and Zoological Review 2, 237–278.

Maelfait JP, De Keer R, 1990. The border zone of an intensively grazed pasture as a corridor for spiders Araneae. Biological Conservation 54, 223–238.

Maloney D, Drummond FA, Alford R, 2003. Spider predation in agroecosystems: Can spiders effectively control pest populations? Maine Agricultural and Forest Experiment Station, University of Maine, Technical Bulletin 190, 1–32.

Marc P, Canard A, 1997. Maintaining spider diversity in agroecosystems as a tool in pest control. Agriculture, Ecosystems & Environment 62, 229–235.

Martin D, 1991. Zur Autökologie der Spinnen (Arachnida: Araneae). I. Charakteristik der Habitatausstattung und Präferenzverhalten epigäischer Spinnenarten. Arachnologische Mitteilungen 1, 5–26.

Maurer R, Hänggi A, 1990. Katalog der schweizerischen Spinnen. Documenta Faunistica Helvetiae 12.

Mommertz S, Schauer C, Kösters N, Lang A, Filser J, 1996. A comparison of D-Vac suction, fenced and unfenced pitfall trap sampling of epigeal arthropods in agro-ecosystems. Annales Zoologici Fennici 33, 117–124.

Morris MG, 2000. The effects of structure and its dynamics on the ecology and conservation of arthropods in British grasslands. Biological Conservation 95, 129–142.

Nentwig W, 1988. Augmentation of beneficial arthropods by strip-management. 1. Succession of predacious arthropods and long-term change in the ratio of phytophagous and predacious arthropods in a meadow. Oecologia 76, 597–606.

Nentwig W, Hänggi A, Kropf Ch, BlickT, 2004. Spinnen Mitteleuropas/ Central European Spiders. An internet identification key. http://www.araneae.unibe.ch, version 8.12.2003, obtained from the Internet on July 1, 2004.

Nyffeler M, 2000. Ecological impact of spider predation: A critical assessment of Bristowe's and Turnbull's estimates. Bulletin of British Arachnological Society 11 (2000) 367–372.

Nyffeler M, Benz G, 1979. Zur ökologischen Bedeutung der Spinnen der Vegetationsschicht von Getreide- und Rapsfeldern bei Zürich. Journal of Applied Entomology 87, 348–376.

Nyffeler M, Benz G, 1987. The foliage-dwelling spider community of an abandoned grassland ecosystem in Eastern Switzerland assessed by sweep sampling. Bull. Soc. Entomol. Suisse 60, 383–389.

Nyffeler M, Benz G, 1988. Feeding ecology and predatory importance of wolf spiders (*Pardosa* spp.) (Araneae, Lycosidae) in winter wheat fields. Journal of Applied Entomology 106, 123–134.

Nyffeler M, Benz G, 1989. Foraging ecology and predatory importance of a guild of orb-weaving spiders in a grassland habitat. Journal of Applied Entomology 107, 166–184.

Nyffeler M, Breene RG, 1990. Spiders associated with selected European hay meadows, and the effects of habitat disturbance, with the predation ecology of the crab spiders, *Xysticus* spp. (Araneae, Thomisidae). Journal of Applied Entomology 110, 149–159.

Nyffeler M, Sterling WL, Dean DA, 1994. Insectivorous activities of spiders in United States field crops. Journal of Applied Entomology 118, 113–128.

Östman O, Ekbom B, Bengtsson J, 2001. Landscape heterogeneity and farming practice influence biological control. Basic and Applied Ecology 2, 365–371.

Oraze M, Grigarick AA, 1989. Biological control of aster leafhopper (Homoptera: Cicadellidae) and midges (Diptera: Chironomidae) by *Pardosa ramulosa* (Araneae: Lycosidae) in California rice fields. Journal of Economic Entomology 82, 745–749.

Pfiffner L, Luka H, 2000. Overwintering of arthropods in soils of arable fields and adjacent semi-natural habitats. Agriculture, Ecosystems & Environment 78, 215–222.

Pfiffner L, Luka H, 2003. Effects of low-input farming systems on carabids and epigeal spiders – A paired farm approach. Basic and Applied Ecology 4, 117–127.

Pfiffner L, Niggli U, 1996. Effects of bio-dynamic, organic and conventional farming on ground beetles (Col. Carabidae) and other epigeic arthropods in winter wheat. Biological Agriculture and Horticulture 12, 353–364.

Plachter H, 1991. Naturschutz. Fischer Verlag, Stuttgart.

Plachter H, Agricola U, Barthel J, Laussmann H, 1993. Inventarisierung der Tierwelt im Hinblick auf naturschutzbezogene Wirkungen unterschiedlicher Landbewirtschaftung – Abschlussbericht Aufbauphase 1990–1992. FAM-Bericht 3, 93–111.

Plachter H, Agricola U, Barthel J, Janssen B, Laussmann H, 1995. Untersuchungen zur Tierwelt im Hinblick auf naturschutzbezogene Wirkungen unterschiedlicher Landbewirtschaftung. FAM-Bericht 5, 339–357.

Plachter H, Kühn I, Laussmann H, Barthel J, 1991. Inventarisierung der Tierwelt im Hinblick auf naturschutzbezogene Wirkungen unterschiedlicher Landbewirtschaftung. FAM-Bericht 1, 75–90.

Platen R, 1996. Spinnengemeinschaften mitteleuropäischer Kulturbiotope. Arachnologische Mitteilungen 12, 1–45.

Platen R, Blick T, Bliss P, Drogla R, Malten A, Martens J, Sacher P, Wunderlich J, 1995. Checklist of the arachnids (excl. Acarida) of Germany (Arachnida: Araneida, Opilionida, Pseudoscorpionida). Arachnologische Mitteilungen Sonderband 1, 1–55.

Platen R, Blick T, Sacher P, Malten A, 1998. Rote Liste der Webspinnen (Arachnida: Araneae). In: Bundesamt für Naturschutz (Ed.), Rote Liste gefährdeter Tiere Deutschlands, Schr.-R. Landschaftspfl. Naturschutz 55, 268–277.

Platen R, Moritz M, v. Broen B, 1991. Rote Liste der Webspinnen- und Weberknechtarten (Arach.: Araneidae, Opilionida) des Berliner Raumes und ihre Auswertung für Naturschutzzwecke (Rote Liste). In: Aughagen A, Platen R, Sukopp H (Eds.), Rote Listen der gefährdeten Pflanzen und Tiere in Berlin. Landschaftsentwicklung und Umweltforschung 6, 169–205.

Provencher L, Riechert SE, 1994. Model and field test of prey control effects by spider assemblages. Environmental Entomology 23, 1–17.

Raatikainen M, Huhta V, 1968. On the spider fauna of Finnish oat fields. Annales Zoologici Fennici 5, 254–261.

Raymond B, Darby AC, Douglas AE, 2000. Intraguild predators and the spatial distribution of a parasitoid. Oecologia 124, 367–372.

Riechert SE, 1990. Habitat manipulations augment spider control of insect pests. Acta Zoologica Fennica 190, 321–325.

Riechert SE, 1999. The hows and whys of successful pest suppression by spiders: Insights from case studies. Journal of Arachnology 27, 387–396.

Riechert SE, Bishop L, 1990. Prey control by an assemblage of generalist predators: Spiders in garden test systems. Ecology 71, 1441–1450.

Riechert SE, Lawrence K, 1997. Test for predation effects of single versus multiple species of generalist predators: Spiders and their insect prey. Entomologia Experimentalis et Applicata 84, 147–155.

Robinson JV, 1981. The effect of architectural variation in habitat on a spider community: An experimental field study. Ecology 62, 73–80.

Rosenheim JA, Wilhoit LR, Armer CA, 1993. Influence of intraguild predation among generalist insect predators on the suppression of an herbivore population. Oecologia 96, 439–449.

Rypstra AL, Carter PE, Balfour RA, Marshall SD, 1999. Architectural features of agricultural habitats and their impact on the spider inhabitants. Journal of Arachnology 27, 371–377.

Samu F, Sunderland KD, Szinetár C, 1999. Scale-dependent dispersal and distribution patterns of spiders in agricultural systems: A review. Journal of Arachnology 27, 325–332.

Samu F, Vörös G, Botos E, 1996. Diversity and community structure of spiders of alfalfa fields and grassy field margins in south Hungary. Acta Phytopathologica et Entomologica Hungarica 31, 253–266.

Scheidler M, 1990. Influence of habitat structure and vegetation architecture on spiders. Zoologischer Anzeiger 225, 333–340.

Schmidt MH, Lauer A, Purtauf T, Thies C, Schaefer M, Tscharntke T, 2003. Relative importance of predators and parasitoids for cereal aphid control. Proceedings of the Royal Society of London B 270, 1905–1909.

Snyder WE, Ives AR, 2001. Generalist predators disrupt biological control by a specialist parasitoid. Ecology 82, 705–716.

Snyder WE, Ives AR, 2003. Interactions between specialist and generalist natural enemies: Parasitoids, predators, and pea aphid biocontrol. Ecology 84, 91–107.

Snyder WE, Wise DH, 1999. Predator interference and the establishment of generalist predator populations for biocontrol. Biological Control 15, 283–292.

Snyder WE, Wise DH, 2001. Contrasting trophic cascades generated by a community of generalist predators. Ecology 82, 1571–1583.

Sotherton NW, 1984. The distribution and abundance of predatory arthropods overwintering on farmland. Annals of Applied Biology 105, 423–429.

Sunderland KD, 1999. Mechanisms underlying the effects of spiders on pest populations. Journal of Arachnology 27, 308–316.

Sunderland KD, Axelsen JA, Dromph K, Freier B, Hemptinne JL, Holst NH, Mols PJM, Petersen MK, Powell W, Ruggle P, Triltsch H, Winder L, 1997. Pest control by a community of natural enemies. Acta Jutlandica 72, 271–326.

Sunderland KD, Samu F, 2000. Effects of agricultural diversification on the abundance, distribution, and pest control potential of spiders: A review. Entomologia Experimentalis et Applicata 95, 1–13.

Symondson WOC, Sunderland KD, Greenstone MH, 2002. Can generalist predators be effective biocontrol agents? Annual Review of Entomology 47, 561–594.

Thies C, Tscharntke T, 1999. Landscape structure and biological control in agroecosystems. Science 285, 893–895.

Thomas CFG, Jepson PC, 1997. Field-scale effects of farming practices on linyphiid spider populations in grass and cereals. Entomologia Experimentalis et Applicata 84, 59–69.

Thomas MB, Wratten SD, Sotherton NW, 1991. Creation of 'island' habitats in farmland to manipulate populations of beneficial arthropods: Predator densities and emigration. Journal of Applied Ecology 28, 906–917.

Thomas MB, Wratten SD, Sotherton NW, 1992. Creation of 'island' habitats in farmland to manipulate populations of beneficial arthropods: Predator densities and species composition. Journal of Applied Ecology 29, 524–531.

Toft S, 1995. Value of the aphid *Rhopalosiphum padi* as food for cereal spiders. Journal of Applied Ecology 32, 552–560.

Topping CJ, 1997. Predicting the effect of landscape heterogeneity on the distribution of spiders in agroecosystems using a population dynamics driven landscape-scale simulation model. Entomol. Entomological Research in Organic Agriculture 15, 325–336.

Topping CJ, Lövei GL, 1997. Spider density and diversity in relation to disturbance in agroecosystems in New Zealand, with a comparison to England. New Zealand Journal of Ecology 21, 121–128.

Topping CJ, Sunderland KD, 1994a. A spatial population dynamics model for *Lepthyphantes tenuis* (Araneae: Linyphiidae) with some simulations of the spatial and temporal effects of farming operations and land-use. Agriculture, Ecosystems & Environment 48, 203–217.

Topping CJ, Sunderland KD, 1994b. The potential influence of set-aside on populations of *Lepthyphantes tenuis* (Araneae: Linyphiidae) in the agroecosystem. Aspects of Applied Biology 40, 225–228.

Tscharntke T, Kruess A, 1999. Habitat fragmentation and biological control. In: Hawkins BA, Cornell HV (Eds.), Theoretical Approaches to Biological Control. Cambridge University Press, Cambridge (UK), pp. 190–205.

Vilbaste A, 1964. Über die Fauna und Dynamik der Spinnen auf den Auwiesen Estlands. Eesti NSV Teaduste Akad Toimetised (Biol.) 13, 284–301.

Vilbaste A, 1965. Suveaspekti amblikefaunast kultuurniitudel. (Über den Sommeraspekt der Spinnenfauna auf den Kulturwiesen). Eesti NSV Teaduste Akad Toimetised (Biol.) 14, 329–337.

Weyman GS, 1993. A review of the possible causative factors and significance of ballooning in spiders. Ethology Ecology and Evolution 5, 279–291.

Wise DH, 1993. Spiders in Ecological Webs. Cambridge University Press, Cambridge.

Wise DH, Snyder WE, Tuntibunpakul P, 1999. Spiders in decomposition food webs of agroecosystems: Theory and evidence. Journal of Arachnology 27, 363–370.

Wyss E, Niggli U, Nentwig W, 1995. The impact of spiders on aphid populations in a strip-managed apple orchard. Journal of Applied Entomology 119, 473–478.

Ysnel F, Canard A, 2000. Spider biodiversity in connection with the vegetation structure and the foliage orientation of hedges. Journal of Arachnology 28, 107–114.

Part IV

Influence of the Land Use Changes
on the Abiotic Environment

Chapter 4.1

Development and Application of Agro-Ecosystem Models

E. Priesack, A. Berkenkamp, S. Gayler,
H.P. Hartmann and C. Loos

4.1.1 Introduction

   Increased crop production by modern agricultural land use systems most often was achieved at the cost of increased stress on the environment, resulting in contamination of water resources by fertilizers and pesticides, increased trace gas emissions or decreased soil fertility (Kimbrell, 2002; Smil, 1997). Therefore, strategies to better protect the subsurface environment have to be developed based on an improved understanding of the dynamics and interaction of physical, chemical and biological processes in soil–plant and groundwater systems. Numerical simulation models that describe agro-ecosystems are necessary tools to integrate and extrapolate available knowledge on these different interacting processes. They can also be used in scenario studies to develop and evaluate alternative methods for a sustainable agricultural management and thus can support decision-making processes in rural land use planning (Priesack and Beese, 1995).

Perspectives for Agroecosystem Management
Edited by P. Schröder, J. Pfadenhauer and J.C. Munch
Copyright © 2008 Elsevier B.V. All rights reserved.

Mainly for purposes of evaluation and extrapolation, from the beginning of the FAM Munich Research Network on Agroecosystems it was planned not only to monitor nitrogen pools and fluxes of the agro-ecosystems but also to simulate the nitrogen balances and nitrogen dynamics at pedon and field scale of different sites of the research farm (Schröder et al., 2002; Stenger and Hantschel, 2001; Stenger et al., 2002). Moreover, the FAM research project provided complete experimental data sets to test, improve and further develop available simulation models that had already found widespread use in N-turnover and N-leaching studies (de Willigen, 1991; Diekkrüger et al., 1995; Engel et al., 1993). In particular, effects due to new and more site-specific management practices became the focus of model development and effects due to enhanced or restricted soil water flow caused by preferential and lateral flow or by soil freezing and thawing had to be considered for the simulation of soil nitrogen cycling and transport.

4.1.2 Modular model development

In the scientific literature the term model is used in different ways. Often by the term model an executable computer program is understood, which is based on a mathematical formulation of logical rules and equations and can describe or represent a natural system in a simplified form using input data and parameter values (Refsgaard and Henriksen, 2004). In the following we will understand as a model a finite system of equations and algorithms that represents a dynamical system describing certain aspects of the development of a natural process. However, a dynamical system is defined not only in the strict sense by a system of partial and ordinary differential equations but also in a wider sense by a finite series of differential equation systems augmented by a finite set of algorithmic rules and including the solution algorithms that solve the differential equations. Models that are defined in this way are considered as deterministic or mechanistic models (Addiscott and Wagenet, 1985). The finite series of differential equations and algorithms defining the model can be build up in a way that single components as given by certain subseries describe a single natural process, e.g. transport of a chemical or growth of a plant organ. This componentwise composition of the model system based on equations describing single processes defines the modularity of the model, the single process models providing the elementary modules from which larger submodels, e.g. the water flow model or the crop growth model, and finally the total model can be constructed.

This modularity of the model allows a thorough model system analysis starting with the validation of the single process model and ending by the comparison of simulation results with experimental data sets to test the mutual couplings of processes and related feed back loops that determine the total model. Furthermore, model modularity facilitates model extension, since the model can be easily expanded by adding further components. Moreover, the complete model can be designed as an open model, which allows the user to insert his self-defined and self-programmed submodels.

Such an open and modular model concept was realised by the development of the model system EXPERT-N, resulting in one of the first soil–plant system models with an open and modular model architecture (Abrahamsen and Hansen, 2000). Because of consequently implementing several different submodels that describe the same single process given by different soil–plant system models such as CERES (Jones and Kiniry, 1986; Ritchie, 1991) or LEACHN (Hutson and Wagenet, 1992), from the beginning of the EXPERT-N model development, attention had to be paid to the exchangeability of single process models. Basis of the model development was documentation and review of known models and modelling approaches (Engel et al., 1993) that lead to the model structure and the partitioning into modular model groups of water flow, heat transfer, solute transport, plant growth and agricultural management and their further division into single process components.

The current version EXPERT-N 3.0 is a model system that provides different approaches to simulate vertical one-dimensional soil water flow, soil heat transfer, solute transport, soil carbon and nitrogen turnover, crop growth and soil management. The subroutines currently available have either been taken from published models such as LEACHN (Hutson and Wagenet, 1992), CERES (Hutson and Wagenet, 1992; Jones and Kiniry, 1986; Ritchie and Godwin, 1989), HYDRUS (Simunek et al., 1998), SUCROS (van Laar et al., 1992), NCSOIL (Nicolardot and Molina, 1994), SOILN (Johnsson et al., 1987) and DAISY (Hansen et al., 1991; Svendsen et al., 1995) or been newly developed (Berkenkamp et al., 2002; Stenger et al., 1999) including the nitrogen model N-SIM (Schaaf et al., 1995) and the crop growth model SPASS (Gayler et al., 2002; Wang and Engel, 1998, 2000).

Furthermore, EXPERT-N comprises a user-friendly data input system to enter data on soil, crop, management, climate and weather conditions (Priesack and Bauer, 2003).

4.1.3 Retrospective modelling of soil water flow

Given the basic soil properties of the field site soil profile and given the time series data on weather conditions and on agricultural management for the crop vegetation period, an agro-ecosystem model such as EXPERT-N can be applied to simulate soil water flow. The model can also be tested by comparing measured and simulated data if time series of measured data on soil water contents, soil matric potentials or water flow of the considered soil profile are additionally provided. Since this test of a predictive model is based on retrospective data sets it is considered as a retrospective analysis or retrospective modelling approach. In the following we will summarise results obtained by different retrospective approaches to simulate soil water flow at specific field sites of the research farm Scheyern. Water flow simulations describing the non-steady-state vertical movement of water through the soil were based on a numerical solution of the one-dimensional Richards equation similar to the finite-element solution of HYDRUS 6.0 (Simunek et al., 1998):

$$\frac{\partial \theta}{\partial t} = \frac{\partial}{\partial z}\left[K(h) \cdot \left(\frac{\partial h}{\partial z} - 1\right)\right] - S(t, z, h) \tag{1}$$

where t denotes time [d], z soil depth (positive downward) [mm], θ volumetric water content [mm^3 mm^{-3}], $h = h(t, z)$ soil matric potential [kPa] converted to [mm] water head, $K(h)$ unsaturated hydraulic conductivity [mm d^{-1}] and $S = S(t, z, h)$ the water sink term for root water uptake [mm mm^{-1} d^{-1}].

To obtain the required information on soil hydraulic properties of the soil profile, not only forward approaches based on data of soil texture, bulk density and pedotransfer functions (ptfs) were applied (Frolking et al., 1998; Priesack et al., 1999), but also inverse modelling methods based on retrospective data of the spatial and temporal variation of state variables were used (Priesack and Achatz, 1999; Priesack et al., 2001).

In case of the forward approach we compared results of water balance estimation obtained by applying different ptfs that were partly derived for the area of the research farm (Priesack et al., 1999; Scheinost et al., 1997; Sinowski et al., 1997). Results show that only a careful selection of ptfs and the following quality examination of the estimated water retention and hydraulic conductivity curves lead to accurate simulations of soil water flow and water balances at selected field sites. In particular, to improve the simulations of infiltration during heavy rainfall at several field sites we had to include a preferential soil water flow model based on a bi-modal representation of water retention curves (Priesack, 2006; Priesack and Durner, 2006). Besides its non-equilibrium nature 'an important characteristic of preferential flow is that during wetting, part of the moisture front can propagate quickly to significant depths while bypassing a large part of the matrix pore space', see Simunek et al. (2003), where an overview on currently available models for describing preferential flow and transport in the vadose zone is given.

In case of inverse modelling of water flow or solute transport, related model parameters are optimised by utilising data from transient flow or transport experiments. This is achieved by exploiting the model's ability to reproduce the transient data sets in terms of the dependence of the model on its parameters. Despite some well-known problems mainly due to non-uniqueness and instability of the optimised parameter sets (Kool and Parker, 1988), inverse methods are an increasingly attractive method to estimate model parameters related to water flow (Lambot et al., 2004; Priesack et al., 2001; Simunek and van Genuchten, 1996), root water uptake (Hupet et al., 2003; Musters and Bouten, 1999; Vrugt et al., 2001) and even included solute transport (Inoue et al., 2000; Priesack et al., 2001; Roulier and Jarvis, 2003; Sonnleitner et al., 2003). By applying such inverse parameter estimation techniques it was possible to identify adequate soil hydraulic properties to simulate water flow and water balances at different field sites of the FAM research station (Priesack et al., 1999, 2001).

4.1.4 Simulation of organic matter turnover and nitrate leaching

The intensification of agricultural land use during the past five decades has led to a strong increase in crop production. This higher productivity was most often achieved by increased application of nitrogen (N) fertilizers at the risk of soil and groundwater contamination due to nitrogen leaching below the root zone (Smil, 1997). To decrease nitrogen leaching N fertilizer application has to match the dynamics of crop N demand. Therefore, not only the periods of highest N use of the crop have to be identified but also the N mineralisation-immobilisation-turnover dynamics of soil organic matter have to be known to adequately size the amount of N fertilizers.

Application of soil–plant system simulation models can help to calculate adequate N supply for both optimal crop growth and minimal N losses. Often these models correctly describe crop N demand and N uptake, but simulation of soil N turnover remains difficult (de Willigen, 1991; Diekkrüger et al., 1995; Shaffer et al., 2001). In particular, data sets on quality and quantity of organic N fertilizers and remaining crop residues including decaying roots after crop harvest are often incomplete. As a consequence, the decomposition of remaining crop residues at the soil surface, of crop residues incorporated into the soil and of dead roots is often neglected or inadequately described by N simulation models. Most models that describe the turnover of soil organic matter (SOM) distinguish several conceptual SOM pools of different SOM amounts and different decomposition rates that may not have any physical meaning but represent empirical relationships (Cabrera et al., 2005; Falloon and Smith, 2000; Gijsman et al., 2002; Petersen et al., 2005; Plante et al., 2006; Smith et al., 2002).

The vertical transport of nitrogen in soils is simulated by the numerical solution of the one-dimensional advection–dispersion transport equation:

$$\frac{\partial}{\partial t}[(\theta + \rho \cdot k_d) \cdot c] = \frac{\partial}{\partial z}\left[\theta \cdot D(\theta,q)\frac{\partial c}{\partial z} - q \cdot c\right] + T(t,z,c) \tag{2}$$

where t denotes time [d], z soil depth [mm], θ the volumetric water content [$mm^3\ mm^{-3}$], ρ [$kg\ dm^{-3}$] the soil bulk density, $c = c(t,z)$ the nitrogen species concentration of either urea, ammonium or nitrate [$kg\ dm^{-3}$], $D = D(\theta,q)$ the dispersion coefficient [$mm^2\ d^{-1}$], $q = q(t,z)$ the Darcy water flow velocity [$mm\ d^{-1}$], $T = T(t,z,c)$ the sink resp. source term including plant N uptake [$kg\ dm^{-3}\ d^{-1}$], mineralised N etc.; finally $k_d = k_d(z)$ is the adsorption constant for urea resp. ammonium [$dm^3\ kg^{-1}$]. We solved this equation by applying the procedure based on the numerical scheme following the LEACHN model (Hutson and Wagenet, 1992).

In addition to the model for vertical N movement, a complete N-model includes submodels for N mineralisation, urea hydrolysis, nitrification, denitrification and NH_3 volatilisation. N turnover described by the LEACHN model follows the concept of Johnsson et al. (1987), who assume N mineralisation results from the decomposition of three different organic matter pools, namely, litter, manure and humus.

For each pool N mineralisation resp. N immobilisation is determined from the carbon (C) decomposition and the C/N ratio of the pool by using the following equations:

$$\frac{dN_{lit}}{dt} = \left[-\frac{N_{lit}}{C_{lit}} + \frac{f_e}{r_0}(1 - f_h)\right] \cdot k_{lit} \cdot e_\theta \cdot e_T \cdot C_{lit} \tag{3}$$

$$\frac{dN_{man}}{dt} = \left[-\frac{N_{man}}{C_{man}} + \frac{f_e}{r_0}(1 - f_h)\right] \cdot k_{man} \cdot e_\theta \cdot e_T \cdot C_{man} \tag{4}$$

$$\frac{dN_{hum}}{dt} = \frac{f_e \cdot f_h}{r_0}[(k_{lit} \cdot C_{lit} + k_{man} \cdot C_{man}) - k_{hum} \cdot N_{hum}] \cdot e_\theta \cdot e_T \tag{5}$$

where the subscripts lit (litter), man (manure) and hum (humus) refer to the corresponding SOM pools and C resp. N denotes their amount of C resp. N [$kg\ ha^{-1}$]. k is the reaction rate [d^{-1}] of C mineralisation for the three soil organic matter pools, the efficiency factor f_e denotes the fraction of the decomposed organic carbon converted to humus or to litter via incorporation into microbial biomass and not respired as CO_2, the humification factor f_h defines the relative humus fraction which is produced, and r_0 is the C/N ratio of microbial biomass and humus.

Similar to the organic matter turnover rates, urea hydrolysis and nitrification rates are also described by first-order kinetics with the same correction functions e_θ and e_T accounting for the impact of water content and temperature (Johnsson et al., 1987).

Following a careful determination of soil hydraulic properties for several sites of the research farm Scheyern and by use of global optimisation methods, a parameterisation of the C and N turnover model of the SOILN agro-ecosystem model (Johnsson et al., 1987) was found that allows to reproduce the observed general course of soil ammonium and nitrate contents (Priesack et al., 2001).

Except when only a part of the plant residues is ploughed in and considerable amounts of residues remain on the soil surface the turnover model overestimates N immobilisation. Furthermore, after frost periods during thawing, the observed N mineralisation is often higher than predicted by the model simulation (Priesack et al., 2001; Stenger, 1996). To improve turnover simulations of plant residues incorporated into the soil, we included a submodel for the simulation of crop residue decomposition at the soil surface into our agro-ecosystem model EXPERT-N (Engel and Priesack, 1993; Priesack, 2006; Priesack and Bauer, 2003; Stenger et al., 1999). It was shown that the submodel describing the mineralisation of crop residues on the soil surface could simulate the decrease in soil cover observed at two field sites of the Scheyern research farm (Berkenkamp et al., 2002). Simulated CO_2 production and N mineralisation from the crop residues on the soil surface result in a simulated decrease in C/N ratio of these residues, leading to a reduced N immobilisation after incorporation into the soil. This is in correspondence with the observed lower N immobilisation after incorporation of crop residues that already had remained on the soil surface for a longer time. It also corresponds to the occurrence of higher N immobilisation following incorporation directly after harvest (Berkenkamp et al., 2002). By use of this refined C and N mineralisation model which accounts for the specific agricultural practice aiming to reduce soil erosion by preserving a soil cover of crop residues, it was possible to simulate N transport and N balances of different field sites.

This was achieved based on calculated soil water fluxes and water balances by application of the convection–dispersion equation describing solute transport in soils and by prescribing root N uptake as a sink term (Priesack et al., 2001). The needed dispersion parameters were estimated from the evaluation of lysimeter tracer experiments that were carried out with undisturbed soil monoliths taken from field sites of the Scheyern research farm. Table 1 gives an example of calculated N balances for two different years of a 5 ha field (Huber et al., 2005). The calculated N balance of the whole agricultural field resulted from weighted averages of simulated N balances at grid points within the field where basic soil

Table 1 Simulated nitrogen balance [kg N ha^{-1} a^{-1}] of a field site in the years 1994 and 2000.

		Crop N uptake	N leaching	N storage	Net N mineralisation	Gaseous N losses
1994	Mean value	167	55	9	58	17
	Standard deviation	14	14	18	5	10
2000	Mean value	128	37	103	111	50
	Standard deviation	12	13	38	26	26

profile data including soil texture, bulk density and organic matter content of the soil horizons have been measured.

Overall, the observed reduction of nitrate concentrations in groundwater after changing the agricultural practice and reducing the fertilizer N input corresponds to the reduction of N leaching out of the rooted soil zone which could be retrospectively simulated by the N turnover and N transport model.

4.1.5 Crop growth modelling

Because the composition of organic matter takes place by the process of photosynthesis, modelling of crop biomass growth determines how the internal cycling of organic matter is introduced into the model system and how the most important part of C input into the agro-ecosystem is described. Moreover, because of the high water and nitrogen demand of the growing crop, the crop growth simulation also strongly determines simulation of soil water flow and soil nitrogen transport by the calculation of the sink due to root uptake of available water and nitrogen. In particular, if complete crop rotations are considered, differences in crop growth simulations can lead to not only differences in root uptake of soil solution but also distinct differences in simulated additions to the soil organic matter pools from dead root biomass as well as from incorporated vegetative aboveground biomass after harvest (Priesack et al., 2006).

Therefore, one of the most crucial steps in the development of agro-ecosystem models is the choice, integration and parameterisation of the crop growth model. In the EXPERT-N model system not only the well-known generic crop growth models CERES (Jones and Kiniry, 1986; Ritchie and Godwin, 1989) and SUCROS (van Laar et al., 1997) were modularised and implemented, but also a new crop growth model, the SPASS model, was developed trying to combine strengths and to avoid weaknesses of both the CERES and SUCROS models (Wang, 1997; Wang and Engel, 1998, 2000, 2002; Wang et al., 2002). Furthermore, based on the new SPASS model, a potato crop model was developed and tested using experimental data from the fields of the Scheyern research farm (Gayler et al., 2002).

First, by applying only simple uptake functions and prescribing observed N-uptake rates, the parameterisation of the C- and N-turnover model was achieved completely independent from crop growth models. Then, all three generic crop growth models (CERES, SUCROS, SPASS) were calibrated using observations of germination, emergence, tillering, anthesis and maturity as well as measured biomass data of vegetative aboveground parts, storage organs and roots of cereals from 1991 until 1998 (Priesack et al., 2006). Using this site-specific model parameterisation, simulation of crop development, biomass growth and yield correlated very well with observations, except for the SUCROS model of barley showing an overestimation of aboveground biomass growth rates in the early vegetation states (Priesack et al., 2006). In this way, crop biomass growth and yields at the field scale could be fairly well simulated in a retrospective way by all three crop growth models. Only by comparing simulation results of leaf area index and dry matter biomass production at single sites of a uniformly managed field with results derived from remote sensing data, rather low correlations were found. One reason of this discrepancy can be seen in the rather low sensitivity of the root growth model to the heterogeneous soil properties, since predicted rooting depths were all similar in contrast to observed significant differences in measured rooting depths at the different field sites. However, the overall mean of leaf area index and dry matter production of the total field could be well reproduced by the simulations.

4.1.6 Assessment of nitrous oxide emissions

Increased nitrous oxide (N_2O) levels in the atmosphere contribute to the greenhouse effect and to the depletion of the stratospheric ozone layer. Levels appear to be rising at a rate of 0.8 ppb a^{-1}. On the basis of a 100 year horizon, N_2O has a global warming potential 296 times that of carbon dioxide (CO_2) (IPPC, 2001). If recent estimates are correct, N_2O emissions account for 6% of the global warming potential ascribed to anthropogenic sources (FAO and IFA, 2001). Of these human initiated N_2O emissions the main part, i.e. ca. 78%, is considered to result from crop and livestock production comprising a large share (ca. 35%) of the current global annual N_2O emissions (FAO and IFA, 2001). Therefore, agricultural management practices need to be evaluated in order to identify options for reducing N_2O release to the atmosphere.

Nitrification and denitrification are considered as major sources of N_2O production in soils (Firestone and Davidson, 1989); hence, N_2O emissions rates from soils depend on nitrification and denitrification rates. Thus, they are strongly influenced by several interacting soil attributes such as temperature, contents of water, oxygen, nitrate ammonium and organic matter. Hence, observed temporal and spatial N_2O emission patterns are site-specific and N_2O emission estimates derived from measurements are confined to sites similar in climate conditions, soils and agricultural management. However, process-based mathematical modelling and simulation of soil N_2O production including modelling of subsequent N_2O transport to the soil surface often helps to extrapolate N_2O emission estimates in space and time even without the need to measure (or to continuously measure) at every specific site of interest.

On the basis of the models of water flow, heat transfer, nitrogen transport and nitrogen turnover including their adequate site-specific parameterisation, we implemented the N_2O transport model by assuming local equilibrium between gas and liquid phase. Additionally, we extended the heat transfer to simulate freezing and thawing following the approaches of the models SOILN, SHAW and DAISY (Flerchinger and Saxton, 1989; Hansen et al., 1990; Jansson, 1999; Jansson and Halldin, 1980).

The transport of N_2O was simulated using the following transport equation for the gaseous N_2O-N concentration c_{N_2O} [mg cm^{-3}] in the soil air (Priesack, 2006):

$$\frac{\partial}{\partial t}[(\varepsilon + \theta K_H)c_{N_2O}] = \frac{\partial}{\partial z}\left(D\frac{\partial c_{N_2O}}{\partial z} - qK_H c_{N_2O}\right) + \phi \tag{6}$$

$$\phi = k_{nit, N_2O}\,\theta c_{NH_4} + k_{den, N_2O} - k_{red, N_2}c_{N_2O} \tag{7}$$

where the variables are defined as follows: ε [cm^3 cm^{-3}] denotes the volumetric content of gas filled soil porosity, $K_H(T)$ [1] the Henry constant representing the gas–liquid partition coefficient assuming the validity of Henry's law, D [cm^2 d^{-1}] the diffusion–dispersion coefficient, q [cm d^{-1}] the average water flow velocity of Darcy flow, c_{NH_4} [mg cm^{-3}] the ammonium-N concentration in the soil solution, θ [cm^3 cm^{-3}] the volumetric soil water content, t [s] time and z [cm] depth. Furthermore, k_{nit, N_2O} represents the N_2O production rate during nitrification [d^{-1}], k_{den, N_2O} the N_2O production rate during denitrification [mg cm^{-3} d^{-1}] and k_{red, N_2} the reduction rate of N_2O to N_2 [d^{-1}].

The N_2O production rate during denitrification k_{den, N_2O} is estimated by

$$k_{den, N_2O} = k_{den, N_2O, max}f_\theta f_{NO_3}\,f_T \tag{8}$$

where the maximal rate $k_{den, N_2O, max}$ [mg cm^{-3} d^{-1}] gives the rate value for optimal conditions. This optimal rate is reduced under non-optimal conditions by different reduction factors accounting for soil moisture, nitrate availability and soil temperature (Johnsson et al., 1987). Freezing and thawing have strong impact on microbial activity. One possible explanation for enhanced microbial activity during thawing is the release of dissolved organic carbon, e.g. due to disruption of the soil. During thawing we consider this effect by a higher $k_{den, N_2O, max}$ proportional to the thawing rate:

$$k_{den, N_2O, max, Thaw} = k_{den, N_2O, max}\left(1 - C_f \frac{\partial Ice}{\partial t}\right) \qquad (9)$$

with $k_{den, N_2O, max, Thaw}$ [mg cm^{-3} d^{-1}] the increased maximal denitrification rate, C_f [cm^3 cm^{-3} d^{-1}] constant and $\partial Ice/\partial t$ change of ice content (<0 during thawing). Moreover, during thawing we suspend the limitation due to low NO$_3$, i.e. $f_{NO_3} = 1$, supposing that NO$_3$ is also released.

A rewetting factor f_{rew} is used to simulate increased N$_2$O production during denitrification (Priesack et al., 1998):

$$k_{den, N_2O} = k_{den, N_2O, max}\,\tilde{f}_\theta\,\tilde{f}_{NO_3}\,f_T, \qquad (10)$$

by lowering the limit water content θ_d for the occurrence of denitrification to account for the higher O$_2$-consumption due to higher microbial activity after rewetting

$$\tilde{\theta}_d = \left(1 - \frac{f_{rew}}{3}\right)\theta_d \qquad (11)$$

giving a modified moisture reduction factor $\tilde{f}_\theta$ and by reducing the limitation of denitrification caused by low soil nitrate contents assuming an additional nitrate release during rewetting directly available for denitrification

$$\tilde{f}_{NO_3} = \max(f_{NO_3}, f_{rew}) \qquad (12)$$

Reduction of N$_2$O to N$_2$ by denitrification is modelled by a first-order kinetic. Thus change of N$_2$/N$_2$O ratio depends solely on length of the period in which N$_2$O remains in the soil. Consequently, if gas diffusion is severely hindered at high water contents or in the frozen soil, the simulated N$_2$/N$_2$O ratio increases.

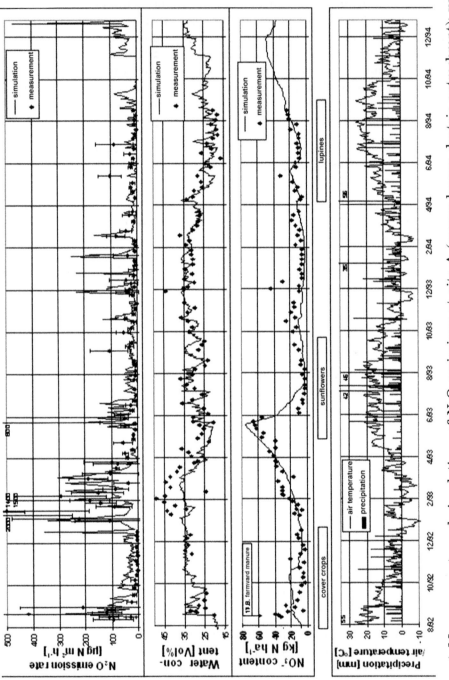

Figure 1 Measurements and simulations of N_2O emissions at site A (coarse-loamy, dystric eutrochrept); volumetric water content in 0–30 cm depth; NO_3 content in 0–30 cm depth; mean daily air temperature and precipitation.

The observed dynamics of N_2O release rates show a low background emission for most sampling days and extremely high flux rates only on a few days (Figures 1 and 3).

The highest N_2O emissions occurred during the first freeze-thaw and dry-rewetting cycles. Subsequent thawing or rewetting events had significantly smaller effects in triggering N_2O production.

Comparing measured data with simulation results of N_2O emissions, of soil water and nitrate contents, it can be concluded that the model is able to describe the observed seasonal dynamics if appropriate maximal N_2O production rates were chosen. The frost model improved simulations of water dynamics. However, the observed ponding after thawing could not be simulated in an adequate way.

By observing maximal N_2O emission rates we could estimate maximal N_2O production rates and in this way the N_2O release during the vegetation period of potato could be described including the rewetting event after a period of low precipitation during July 1995 (Figure 2). This was possible only because water flow as well as nitrate transport could

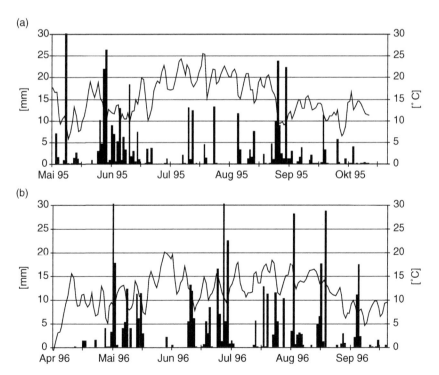

Figure 2 (a) Precipitation and daily average air temperature 1995.
(b) Precipitation and daily average air temperature 1996.

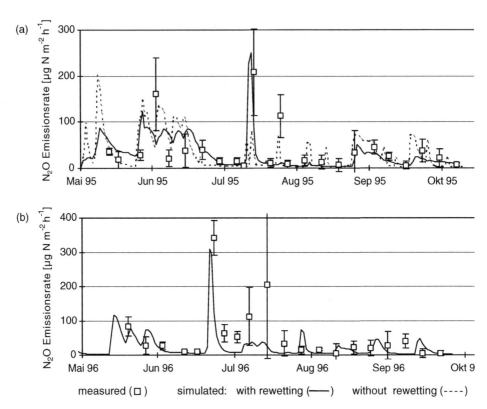

Figure 3 (a) Simulation of N₂O-emission rates from potato ridge 1995.
           (b) Prognosis of N₂O-emission rates from potato ridge 1996.

be accurately simulated. Without changing the model parameterisation, the N_2O emissions from potato ridges were fairly well predicted for the next year using only additional climate and management data of 1996 (Figure 3). This was not possible for the inter-row area between the potato ridges. Here additional runoff water from an upslope area infiltrated into the soil. This led to an underestimation of water contents by the simulation and as a consequence to an underestimation of denitrification including N_2O production and N_2O release from the soil for both vegetation periods 1995 and 1996.

In conclusion, the model estimation of N_2O emissions from agricultural soils needs a good simulation of water flow and nitrate transport. Also maximal N_2O production rates during denitrification as well as during nitrification should be approximately known. These optimal production rates vary between different soil horizons and soil types and are often highest within the surface layer. They reflect different chemical (accumulation of organic nutrients) and biological (increased microbial populations) soil properties and therefore also reflect the history and long-term impact of former soil use management.

The model has been further tested using data from different field sites in Europe and North America and provided in most cases good predictions of N_2O emissions from soils (Frolking et al., 1998; Kaharabata et al., 2003).

4.1.7 Conclusions

Modelling approaches are indispensable tools to integrate current knowledge and available data necessary to estimate the magnitudes of water, carbon and nitrogen flows within an agricultural landscape section of the type represented by the Klostergut Scheyern. Since we cannot monitor all relevant state variables and fluxes with a high enough spatial and temporal resolution, process-based simulation models help to extrapolate available experimental data to understand the sensitivities of our complex agricultural production systems, especially in the context of achieving sustainable land use.

By extending our agro-ecosystem models to include descriptions of soil-borne N_2O production, in particular during winter, and to consider actual management practices such as erosion protection measures (e.g. using soil cover by plant residues), we were able to better quantify risks of atmospheric pollution by N_2O emissions and risks of groundwater pollution by nitrate leaching. Water flow and crop growth modelling provided water balances of the root zone (actual evaporation, actual transpiration, water fluxes out of the root zone), thus providing necessary input data for the hydrological modelling of the watershed Schnatterbach, which includes the Klostergut Scheyern. The applied crop growth models seem to need more sensitivity to soil properties as the simulated crop biomass, leaf area indices and yield did not adequately reflect the inner-field variability of these values as observed by or derived from remote sensing, albeit the observed and simulated average field values of dry matter biomass and grain yield were in rather good agreement. In particular, root growth models need to be improved (Priesack et al., 2006).

Overall, the development and improvement of agro-ecosystem and N-turnover models like Expert-N provide operational integrated knowledge on the dynamics of actual crop production systems, which can be directly applied in decision support systems helping the farmer and his consultant to improve agricultural management and land use practice both ecologically and economically.

Acknowledgements

The scientific activities of the FAM Munich Research Network on Agroecosystems were financially supported by the German Federal Ministry of Education and Research (BMBF 0339370). Overhead costs of the Research Station Scheyern were funded by the Bavarian State Ministry for Science, Research and the Arts.

References

Abrahamsen P, Hansen S, 2000. Daisy: an open soil-crop-atmosphere system model. Environmental Modelling and Software 15, 313–330.

Addiscott TM, Wagenet RJ, 1985. Concepts of solute leaching in soils: A review of modelling approaches. Journal of Soil Science 36, 411–424.

Berkenkamp A, Priesack E, Munch JC, 2002. Modelling the mineralisation of plant residues on the soil surface. Agronomie 22, 711–722.

Cabrera ML, Kissel DE, Vigil MF, 2005. Nitrogen Mineralization from Organic Residues: Research Opportunities. Journal of Environmental Quality 34(1), 75–79.

de Willigen P, 1991. Nitrogen turnover in the soil-crop system: comparison of fourteen simulation models. Fertilizer Research 27, 141–149.

Diekkrüger B, Söndgerath D, Kersebaum KC, McVoy CW, 1995. Validity of agroecosystem models a comparison of results of different models applied to the same data set. Ecological Modelling 81, 3–29.

Engel T, Klöcking B, Priesack E, Schaaf T, 1993. Simulationsmodelle zur Stickstoffdynamik. Agrarinformatik, 25. Verlag Eugen Ulmer, Stuttgart, 484 pp.

Engel T, Priesack E, 1993. Expert-N, a building block system of nitrogen models as a resource for advice, research, water management and policy. In: Hamers T (Ed.), Integrated Soil and Sediment Research: A Basis for Proper Protection. Kluwer Academic Publishers, Dordrecht, pp. 503–507.

Falloon PD, Smith P, 2000. Modelling refractory soil organic matter. Biology and Fertility of Soils 30, 388–398.

FAO, IFA, 2001. Global estimates of gaseous emissions of NH_3, NO and N_2O from agricultural land. International Fertilizer Industry Association and Food and Agriculture Organization of the United Nations, http://www.gm-unccd.org/FIELD/Multi/FAO/FAO3.pdf, Rome, 106 pp.

Firestone MK, Davidson EA, 1989. Microbial basis of NO and N_2O production and consumption in soil. In: Andreae MO, Schimel DS (Eds.), Exchange of Trace Gases Between Terrestrial Ecosystems and the Atmosphere. Wiley, Chichester, pp. 7–21.

Flerchinger GN, Saxton KE, 1989. Simultaneous heat and water model of a freezing snow-residue-soil system I. Theory and development. Transactions of the ASAE 32, 565–571.

Frolking SE, Mosier AR, Ojima DS, Li C, Parton WJ, Potter CS, Priesack E, Stenger R, Haberbosch C, Dörsch P, Flessa H, Smith KA, 1998. Comparison of N_2O emissions from soils at three temperate agricultural sites: simulations of year-round measurements by four models. Nutrient Cycling in Agroecosystems 52, 77–105.

Gayler S, Wang E, Priesack E, Schaaf T, Maidl FX, 2002. Modelling biomass growth, N-uptake and phenological development of potato crop. Geoderma 105, 367–383.

Gijsman AJ, Hoogenboom G, Parton WJ, Kerridge PC, 2002. Modifying DSSAT Crop Models for Low-Input Agricultural Systems Using a Soil Organic Matter-Residue Module from CENTURY. Agronomy Journal 94(3), 462–474.

Hansen S, Jensen HE, Nielsen NE, Svendsen H, 1990. DAISY – Soil Plant Atmosphere System Model. The Royal Veterinary and Agricultural University, Copenhagen, 273 pp.

Hansen S, Jensen HE, Nielsen NE, Svendsen H, 1991. Simulation of nitrogen dynamics and biomass production in winter wheat using the Danish simulation model DAISY. Fertilizer Research 27, 245–259.

Huber B, Winterhalter M, Mallén G, Hartmann HP, Gerl G, Auerswald K, Priesack E, Seiler KP, 2005. Wasserflüsse und wassergetragene Stoffflüsse in Agrarökosystemen. In: Osinski E, Meyer-Aurich A, Huber B, Rühling I, Gerl G, Schröder P (Eds.), Landwirtschaft und Umwelt – ein Spannungsfeld. Ergebnisse des Forschungsverbunds Agrarökosysteme München (FAM). oekom Verlag, München, pp. 57–98.

Hupet F, Lambot S, Feddes A, van Dam JC, Vanclooster M, 2003. Estimation of root water uptake parameters by inverse modeling with soil water content data. Water Resources Research 39, 1312, doi:10.1029/2003WR002046.

Hutson JL, Wagenet RJ, 1992. LEACHM: Leaching Estimation And Chemistry Model: A process-based model of water and solute movement, transformations, plant uptake and chemical reactions in the unsaturated zone. Version 3.0. Research Series No. 93–3. Cornell University, Ithaca, NY.

Inoue M, Simunek J, Shiozawa S, Hopmans JW, 2000. Simultaneous estimation of soil hydraulic and solute transport parameters from transient infiltration experiments. Advances Water Resources 23, 677–688.

IPPC, 2001. Climate change 2001: The scientific basis. Contribution of Working Group I to the Third Assessment Report of the Intergovernmental Panel on Climate Change. Cambridge University Press, http://www.grida.no/climate/ipcc_tar/wg1/index.htm, Cambridge, UK, 881 pp.

Jansson PE, 1999. Simulation model for soil water and heat conditions. Description of the SOIL model. Swed. Univ. Agric. Sci., Dept. Soil Sci., Uppsala, Sweden.

Jansson PE, Halldin S, 1980. Soil water and heat model. Technical description. 26, Swedish Coniferous Forest Project, Dept. of Ecology and Environmental Research, Swedish University of Agricultural Sciences, Uppsala, Sweden.

Johnsson H, Bergström L, Jansson PE, Paustian K, 1987. Simulated nitrogen dynamics and losses in a layered agricultural soil. Agriculture, Ecosystems and Environment 18, 333–356.

Jones CA, Kiniry JR, 1986. CERES-Maize: A Simulation Model of Maize Growth and Development. Texas A&M University Press, Temple, TX.

Kaharabata SK, et al., 2003. Comparing measured and Expert-N predicted N_2O emissions from conventional till and no till corn treatments. Nutrient Cycling in Agroecosystems 66(2), 107–118.

Kimbrell A. (Ed.), 2002. Fatal Harvest: The Tragedy of Industrial Agriculture. Island Press, Washington, DC.

Kool JB, Parker JC, 1988. Analysis of the inverse problem for transient unsaturated flow. Water Resources Research 24, 817–830.

Lambot S, Hupet F, Javaux M, Vanclooster M, 2004. Laboratory evaluation of a hydrodynamic inverse modeling method based on water content data. Water Resources Research 40, W03506, doi:10.1029/2003WR002641.

Musters PAD, Bouten W, 1999. Assessing rooting depths of an Austrian pine stand by inverse modeling soil water content maps. Water Resources Research 35, 3041–3048.

Nicolardot B, Molina JAE, 1994. C and N fluxes between pools of soil organic matter: model calibration with long-term field experimental data. Soil Biology & Biochemistry 26, 245–251.

Petersen BM, Berntsen J, Hansen S, Jensen LS, 2005. CN-SIM—a model for the turnover of soil organic matter. I. Long-term carbon and radiocarbon development. Soil Biology & Biochemistry 37(2), 359–374.

Plante AF, Conant RT, Stewart CE, Paustian K, Six J, 2006. Impact of Soil Texture on the Distribution of Soil Organic Matter in Physical and Chemical Fractions. Soil Science Society of America Journal 70(1), 287–296.

Priesack E, 2006. Expert-N Dokumentation der Modell-Bibliothek. FAM Bericht 60. Hieronymus, München, pp 296.

Priesack E, Achatz S, 1999. Inverse modelling of soil nitrogen transport. In: Feyen J, Wiyo K (Eds.), Modelling of Transport Processes in Soils. Wageningen Pers, Leuven, Belgium, pp. 641–649.

Priesack E, Achatz S, Stenger R, 2001. Parametrization of soil nitrogen transport models by use of laboratory and field data. In: Hansen S (Ed.), Modeling Carbon and Nitrogen Dynamics for Soil Management. Lewis publishers, Boca Raton, pp. 459–481.

Priesack E, Bauer C, 2003. Expert-N Datenmanagement. FAM-Bericht 59, Hieronymus, München, 114 pp.

Priesack E, Beese F, 1995. Changing modelling concepts and their relation to scenario studies. In: Schoute JFT, Finke PA, Veeneklaas FR, Wolfert HP (Eds.), Scenario Studies for the Rural Environment. Kluwer Academic Publishers, Dordrecht, pp. 131–140.

Priesack E, Durner W, 2006. Closed-form expression for the multi-modal unsaturated conductivity function. Vadose Zone Journal 5, 121–124.

Priesack E, Gayler S, Hartmann HP, 2006. The impact of crop growth sub-model choice on simulated water and nitrogen balances. Nutrient Cycling in Agroecosystems 75, 1–13.

Priesack E, Haberbosch C, Stenger R, 1998. Modellierung der N_2O-Emission mit Expert-N. In: Lay JP (Ed.), Freisetzung und Verbrauch der klimarelevanten Spurengase N_2O und CH_4 beim Anbau nachwachsender Rohstoffe. Inititativen zum Umweltschutz 11. Zeller Verlag, Osnabrück.

Priesack E, Sinowski W, Stenger R, 1999. Estimation of soil property functions and their application in transport modelling. In: Wu L (Ed.), International workshop on the characterization and measurement of the hydraulic properties of unsaturated porous media, October 22–24, 1997, Riverside, CA. Department of Environmental Sciences, University of California, Riverside, CA, pp. 1121–1129.

Refsgaard JC, Henriksen HJ, 2004. Modelling guidelines – terminology and guiding principles. Advances in Water Resources 27, 71–82.

Ritchie JT, 1991. Wheat phasic development. In: Ritchie JT (Ed.), Modeling plant and soil systems. Agronomy 31. ASA, CSSA, SSSA, Madison, WI., pp. 31–54.

Ritchie JT, Godwin DC, 1989. CERES Wheat 2.0 – Documentation for version 2 of the CERES wheat model. http://nowlin.css.msu.edu/wheat_book/.

Roulier S, Jarvis N, 2003. Analysis of inverse procedures for estimating parameters controlling macropore flow and solute transport in the dual-permeability model MACRO. Vadose Zone Journal 2, 349–357.

Schaaf T, Priesack E, Engel T, 1995. Comparing field data from north Germany with simulations of the nitrogen model N-SIM. Ecological Modelling 81, 223–232.

Scheinost AC, Sinowski W, Auerswald K, 1997. Regionalization of soil water retention curves in a highly variable soilscape, I. Developing a new pedotransfer function. Geoderma 78, 129–143.

Schröder P, Huber B, Olazabal U, Kämmerer A, Munch JC, 2002. Land use and sustainability: FAM Research Network on Agroecosystems. Geoderma 105, 155–166.

Shaffer MJ, Ma L, Hansen S (Eds.), 2001. Modeling Carbon and Nitrogen Dynamics for Soil Management. Lewis Publishers, Boca Raton.

Simunek J, Huang K, van Genuchten MT, 1998. The HYDRUS code for simulating the one-dimensional movement of water, heat, and multiple solutes in variably-saturated media.Version 6.0. 144, U.S. Salinity Laboratory, USDA, ARS, Riverside, CA.

Simunek J, Jarvis N, van Genuchten MT, Gärdenäs A, 2003. Review and comparison of models for describing non-equilibrium and preferential flow and transport in the vadose zone. Journal of Hydrology 272, 14–35.

Simunek J, van Genuchten MT, 1996. Estimating unsaturated soil hydraulic properties from tension disc infiltrometer data by numerical inversion. Water Resources Research 32, 2683–2696.

Sinowski W, Scheinost AC, Auerswald K, 1997. Regionalization of soil water retention curves in a highly variable soilscape, II. Comparison of regionalization procedures using a pedotransfer function. Geoderma 78, 145–159.

Smil V, 1997. Global population and the nitrogen cycle. Scientific American 277, 58–63.

Smith JU, Smith P, Monaghan R, MacDonald AJ, 2002. When is a measured soil organic matter fraction equivalent to a model pool? European Journal of Soil Science 53(3), 405–416.

Sonnleitner MA, Abbaspour KC, Schulin R, 2003. Hydraulic and transport properties of the plant-soil system estimated by inverse modeling. European Journal of Soil Science 54, 127–138.

Stenger R, 1996. Dynamik des mineralischen Stickstoffs in einer Agrarlandschaft. FAM-Bericht 10. Shaker Verlag, Aachen, 202 pp.

Stenger R, Hantschel R, 2001. Site effects on the variability of nitrogen turnover at the Scheyern experimental farm. In: Tenhunen JD, Lenz R, Hantschel R (Eds.), Ecosystem Approaches to Landscape Management in Central Europe. Ecological Studies 147. Springer Verlag, Berlin, pp. 229–247.

Stenger R, Priesack E, Barkle G, Sperr C, 1999. Expert-N A tool for simulating nitrogen and carbon dynamics in the soil-plant-atmosphere system. In: Gielen G (Ed.), NZ Land Treatment Collective Proceedings Technical Session 20: Modelling of Land Treatment Systems, New Plymouth, New Zealand, pp. 19–28.

Stenger R, Priesack E, Beese F, 2002. Spatial variation of nitrate-N and related properties at the plot scale. Geoderma 105, 259–275.

Svendsen H, Hansen S, Jensen HE, 1995. Simulation of crop production, water and nitrogen balances in two German agro-ecosystems using the DAISY model. Ecological Modelling 81(1–3), 197–212.

van Laar HH, Goudriaan J, van Keulen H, 1992. Simulation of crop growth for potential and water-limited production situations (as applied to spring wheat). Simulation Report CABO-TT no. 27. Centre for

Agrobiological Research and Department of Theoretical Production Ecology, Wageningen Agricultural University, Wageningen, 72 pp.

van Laar HH, Goudriaan J, van Keulen H, 1997. SUCROS97: Simulation of crop growth for potential and water-limited production situations. Quantitative Approaches in System Analysis, 14. C.T. de Wit Graduate School for Production Ecology and Resource Conservation, Wageningen, pp. 52 + appendices.

Vrugt JA, Hopmans JW, Simunek J, 2001. Calibration of a two-dimensional root water uptake model. Soil Science Society of America Journal 65, 1027–1037.

Wang E, 1997. Development of a Generic Process-oriented Model for Simulation of Crop Growth. Herbert Utz Verlag, München, 195 pp.

Wang E, Engel T, 1998. Simulation of phenological development of wheat crops. Agricultural Systems 58, 1–24.

Wang E, Engel T, 2000. SPASS: a generic process-oriented crop model with versatile windows interfaces. Environmental Modelling and Software 15, 179–188.

Wang E, Engel T, 2002. Simulation of growth, water and nitrogen uptake of a wheat crop using the SPASS model. Environmental Modelling and Software 17(4), 387–402.

Wang E, Robertson MJ, Hammer GL, Carberry PS, Holzworth D, Meinke H, Chapman SC, Hargreaves JNG, Huth NI, McLean G, 2002. Development of a generic crop model template in the cropping system model APSIM. European Journal of Agriculture 18, 121–140.

Chapter 4.2

Assessment of Soil Landscape Variability

M. Sommer, M. Wehrhan, M. Zipprich and U. Weller

4.2.1 Introduction

Soils are four-dimensional natural bodies (Schlichting, 1986) with the key characteristic of variation in time and space. Different approaches had been developed to handle spatial variability of soils from field to regional scale (overviews in Heuvelink and Webster, 2001; McBratney et al., 2000). At landscape scale, quantitative analysis still is rather empirical and descriptive by using geostatistics (e.g. Mausbach and Wilding, 1991), structural analysis (e.g. Fridland, 1976; Hole and Campbell, 1985) or hybrid techniques (reviewed in McBratney et al., 2000). Currently, we are far away from any thorough understanding of the underlying processes in time and space. Simple mechanistic and stochastic process models were published during the last decade (Heimsath et al., 1997, 2001, 2002; Minasny and McBratney, 1999, 2001, 2006). On the contrary, soil scientists face an increasing demand for soil information at the landscape scale, e.g. for environmental purpose or new production methods such as precision agriculture (Sommer, 2006). A new framework for handling soil variability

Perspectives for Agroecosystem Management
Edited by P. Schröder, J. Pfadenhauer and J.C. Munch

was given by Vogel and Roth (2003). According to the so-called 'scaleway', soil variability can be separated at every scale into a scale-typical, predictable part (structure) and a random part (texture), which becomes structure at the following subscale level. For a full phenomenology, structure has to be imaged explicitly. On field and landscape scale, structure results from pedogenesis, which itself is controlled by the soil forming factors, as it was demonstrated by numerous studies (reviewed in Birkeland, 1999; Sommer and Schlichting, 1997).

Recent developments of methods such as remote sensing, terrain analysis on digital elevation models (DEM) and geophysical measures (e.g. ground-penetrating radar, electromagnetic induction [EMI]) allow prediction of soil variability from field to landscape scale, so-called 'digital soil mapping' (review in McBratney et al., 2003). In part, these non-invasive methods image soil forming factors (terrain analysis – relief); in part they image soil properties (Grunwald, 2006). Nevertheless, most methods are only proxies for relevant soil properties and soil types. Additional pedotransfer functions (PTF) or soil inference systems (McBratney et al., 2002) are needed for data interpretation. In addition, most of the studies using non-invasive methods focus on single soil properties, such as water content, clay content or soil organic carbon. An integrative multi-purpose method is missing up to now.

At our research farm, the 'Versuchsgut Scheyern', soil database at pedon scale was excellent (50 m raster, Figure 1, Sinowski, 1995). Nevertheless, it was insufficient for detecting soil pattern at field and landscape scale due to correlation length of soil types and properties well below 50 m. Therefore, it was the task of our group to apply and advance new instruments for the assessment of soil landscape variability. Here, we present the results of different non-invasive methods, i.e. terrain analysis, EMI and remote sensing. The chapter will focus on the methodological aspects. Interpretation of the data has been published elsewhere (e.g. Sommer et al., 2003, 2004; Weller et al., 2007).

4.2.2 Terrain analysis

Terrain is controlling the superficial water and sediment transport through differences in gravitation potentials. Water flow and transport capacity is accelerated at convex terrain elements due to slope increase in flow direction. This leads to enhanced erosion. On the contrary, the transport of water–sediment suspensions slows down in concave parts of the landscape and leads to sedimentation (a process called 'colluviation'). Colluviated sites receive additional water and nutrient inputs and, therefore, are sites of high plant biomass. The terrain-related processes act through time, which means they influence soil pattern (soil forming factor 'relief'; Jenny, 1941) as well as actual water flow and matter transport.

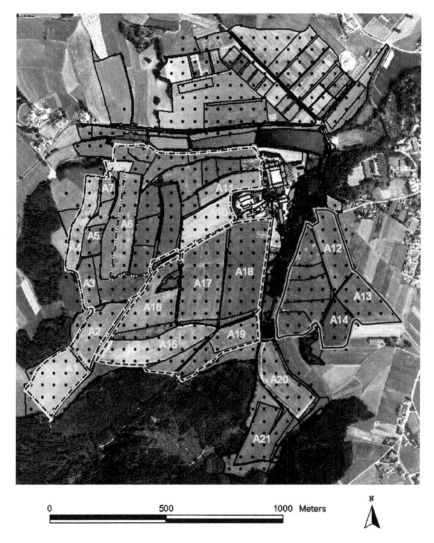

0   500   1000 Meters

N

Figure 1 Research farm "Klostergut Scheyern": field boundaries (arable land as A1, A2, etc.) and soil raster (for EM38: -·-- boundary = test area; -·--· boundary = unit 1, --- boundary = unit 2, —— = unit 3). (For colour version of this figure, please see page 426 in colour plate section.)

This is the reason why terrain analysis based on DEM is one of the intensively used non-invasive methods to predict both soil types and properties (Moore et al., 1991, 1993).

Terrain information for the 'Versuchsgut Scheyern' was obtained by laser altimetry. The scattered data with a mean distance between measurement points of 3.7 m were resampled into a regular 5 m × 5 m grid yielding a high-precision digital elevation model (DEM 5). Precision of the Laser-DEM 5 in height were measured at a plain control surface and yielded a maximum overestimation of the height by +35 cm (mean deviation = −0.3 cm, standard deviation = 7 cm, n = 496 points). On the basis of the DEM 5, different terrain parameters were calculated with the software package 'System for Automated Geoscientific Analyses' (SAGA-Gis, compare http://www.saga-gis.uni-goettingen.de/html/index.php): (i) Morphographic terrain units (summit and bottom areas), (ii) local terrain parameters (slope, convergence/divergence; Figures 2 and 3) and (iii) complex terrain parameters (local catchment area, topographic wetness index; Figures 4 and 5). During our research work, it turned out that only a few parameters could be used as proxies for soil type distribution. On the basis of the raster points, we can predict only 60% of all colluvial soils using a unique set of thresholds, i.e. local concavity and local catchment area (>1250 m^2). We had to learn that colluvial soils could be found at almost every relief position, even in flat areas near watersheds. The main reason for the low explanatory power of terrain analysis was a extreme spatial variability of parent materials. For this we focussed on geophysical methods during the project. Nevertheless, terrain attributes might be used for future spatially distributed modelling, e.g. for surface flow and erosion processes.

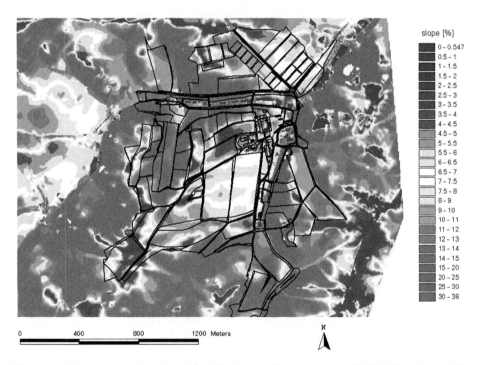

Figure 2 Terrain analysis with SAGA on basis of the DEM 5; slope (%). (For colour version of this figure, please see page 427 in colour plate section.)

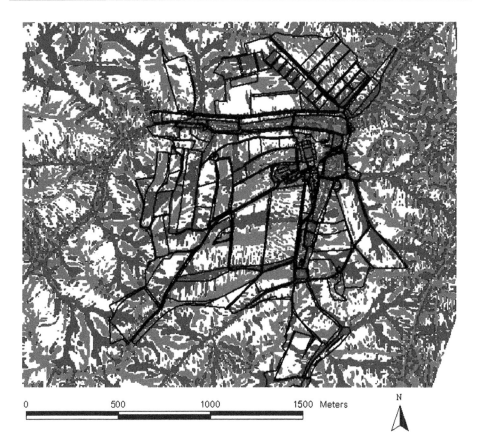

0   500   1000   1500 Meters

N

Figure 3 Terrain analysis with SAGA on basis of the DEM 5; divergence (gray) and convergence areas (dark gray), linear elements (white). (For colour version of this figure, please see page 428 in colour plate section.)

4.2.3 Electromagnetic induction

Parent materials of the study sites are from tertiary flurial deposits (gravels to clays) partly covered by pleistocene aeolian deposits (loess). Clay contents of these sediments vary tremendously. The clay content of a soil – part of which is lithogenic, part is pedogenic – decisively influences many soil functions such as water balance, erosion and fertility. On the contrary, the clay minerals dominate the electrical conductivity (EC) for soils with free drainage under a humid climate. Here, we use the EMI technique to assess the clay content via EC measurements. In contrast to direct current measurements (Tabbagh et al., 2000), it is contact-free. This enables fast measurements of integral clay content and extends applicability to meadows and fallow land. Several investigations successfully apply EMI to map the clay content of individual fields (e.g. Dalgaard et al., 2001). At the catchment scale, however, units of different sampling dates

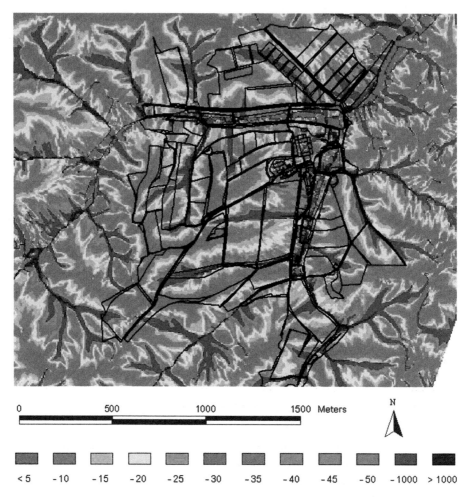

0   500   1000   1500 Meters   N

< 5   - 10   - 15   - 20   - 25   - 30   - 35   - 40   - 45   - 50   - 1000   > 1000

Figure 4 Terrain analysis with SAGA on basis of the DEM 5; local catchment area (no. grid cells of 25 m^2). (For colour version of this
figure, please see page 429 in colour plate section.)

and land management add variation to EC and constrain the mapping
across field boundaries. This results in imaginary discontinuities of predicted clay content at field borders. To apply EMI technique as a tool in
landscape analysis, we developed a method to enable reliable clay content
mapping at the landscape scale across the boundaries of individual fields
and over different sampling dates.

EC measurement and soil calibration data

The EC measurements were made using Geonics device EM38 in vertical operation mode (McNeill, 1980) along sloping tracks. The location
was determined by differential GPS. Aside from our own measurements,

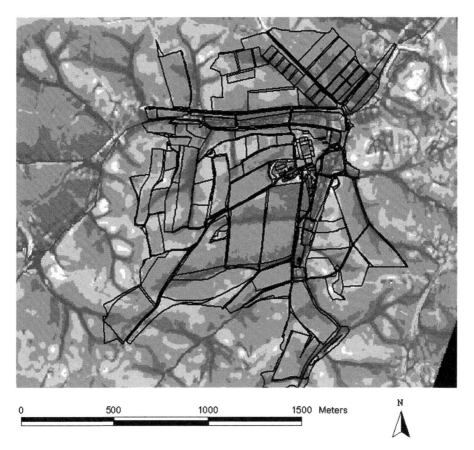

Figure 5 Terrain analysis with SAGA on basis of the DEM 5; topographic wetness index, defined as TWI = ln [local catchment area/tan (slope)], values as gradient (dark gray) from 4.3 to 23.5. (For colour version of this figure, please see page 430 in colour plate section.)

our investigations are based on measurements of Durlesser and Sperl (Durlesser, 1999). Additional measurements were carried out at different dates in several years. The soils were near field capacity on all dates. To correlate EC with the horizon-wise measured clay content at the raster points, a weighted sum was calculated from the density function of the measuring signal (The regression was restricted to points that showed neither influence of ground water nor stagnic properties, and which were located on the same field and in a distance of less than 5 m to the nearest EC measurement. For correlating soil clay content C with interpolated EC measurements, we assume a linear relationship:

$$C = \alpha_0 + \alpha_1 EC \tag{1}$$

Modelling approach

To eliminate the influence of land use and sampling date on EC and to obtain a reliable clay prediction across field boundaries, we developed the 'nearest neighbours EC correction' technique (Weller et al., 2007, Figure 6). This method takes advantage of the spatial autocorrelation of soil properties: We assume continuity of clay content at field boundaries. Considering Eq. (1), we define a spatially autocorrelated variable, $EC_{fit}(x)$, which is linearly correlated to the time and field-dependent $EC(x)$ values at measurement positions x:

$$EC_{fit}(x) = \beta_0 + \beta_1 EC(x) \tag{2}$$

For each pair (A, B) of neighbouring fields, we choose pairs of points of EC measurements (i, j) close to each other on either side of the field boundary (so-called nearest neighbours). Then, for each pair of measurements, a linear transform can be written as

$$\beta_{A0} + \beta_{A1} EC(i) = \beta_{B0} + \beta_{B1} EC(j) + \varepsilon_{ij} \tag{3}$$

where $EC(i)$ is measured in field A and $EC(j)$ at a point close by in field B. ε_{ij} is a random variable with mean 0. Eq. (3) leads to a linear equation system with many pairs of neighbouring points of adjacent EC data sets. The result is a pair of linear transformation parameters (β_0, β_1) for each field and date, which are used to recalculate the EC values per field so that the steps between adjacent fields are minimised. Subsequently, we use ordinary kriging to interpolate these EC_{fit} values spatially and at the locations of the analysed soil profiles. To evaluate its quality, we compare the 'nearest neighbours EC correction' to the raw EC_{25} measurement, merely corrected for the influence of soil temperature (cf. Sheets and Hendrickx, 1995). The 'nearest neighbours EC correction' method is developed on a test area with dense measurements near the field boundaries (Figure 1). It is subsequently applied to four contiguously measured units of the

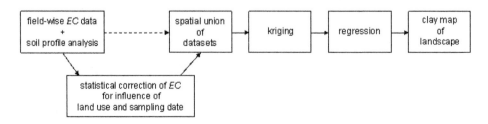

Figure 6 Workflow for prediction of clay content from EC at landscape scale.

'Klostergut Scheyern' (Figure 1), which are spatially separated disabling the selection of pairs of 'nearest neighbours'.

Results

Let us first confirm the validity of our assumptions. We assume EC to be linearly related to clay content (Eq. (1)) and clay to be continuous at boundaries between fields of different land use and sampling dates. First, this is tantamount to the assumption of a linear influence of land use and date on EC. Figure 7a shows a strong relation between measurements taken on two fields at different dates and under different cultivation. Second, our assumptions imply continuity of EC_{fit} at field boundaries. The semivariogram of EC_{fit} strongly confirms the spatial autocorrelation (Figure 7b).

If we disregard the influence of land use and sampling date, the correlation between EC_{25} and clay content only gives a relatively small coefficient of determination ($R^2 = 0.66$) for the test area (Figure 8a). The method 'nearest neighbours EC correction', which accounts for the influence of land use and sampling date, lowers the variance of EC by more than 50% (Figure 7b). For the prediction of weighted clay content with EC_{fit}, a coefficient of determination of $R^2 = 0.85$ was achieved for the test area (Figure 8b). The regression functions between EC and clay content are built only once for all fields and sampling dates together, which is essential for the application of the method in the framework of mapping at landscape scales. A significantly smaller set of calibration points is sufficiently compared to a 'field-wise' calibration of the relation clay to EC separately for each sampling date and land use – or even management history. For the test area, a reduction of the number of calibration points from 46 to 8,

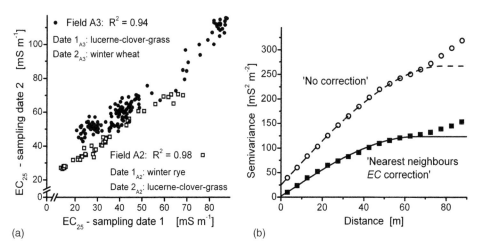

(a) (b)

Figure 7 (a) Influence of sampling date and land use on EC, (b) semivariograms for EC.

i.e. one per field, only increases the median root mean square error (RMSE) from 4.3 to 4.7% clay. Besides improving clay prediction at point support, our method enables us to continuously map patterns of clay content across the boundaries of individual fields.

Finally, we apply the 'nearest neighbours EC correction' method at the farm scale. The results for each of the contiguous units of the research farm 'Klostergut Scheyern' are summarised in Table 1. The EC correction improves the prediction of clay content for all units. As the 'nearest neighbours' method was applied 'a posterior' to existent data, here, often a large distance between 'nearest neighbours' pairs of points limited its efficiency.

The predicted weighted clay content for the research farm 'Klostergut Scheyern' is shown in Figure 9. The clay content exhibits strong heterogeneity even inside single fields. The knowledge of this pattern may aid

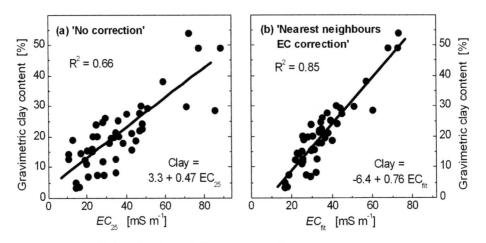

Figure 8 (a), (b) Prediction of clay content for test area.

Table 1 Correlation between weighted clay content and EC signal.

	Number of raster points	No EC correction	Nearest neighbours EC correction		
		R^2	R^2	Error of clay prediction	Percentage of variance explained
Test area	46	0.66	0.85	4.3	77
Unit 1	109	0.49	0.72	5.3	50
Unit 2	65	0.52	0.60	4.9	16
Unit 3	47	0.73	0.76	6.0	7

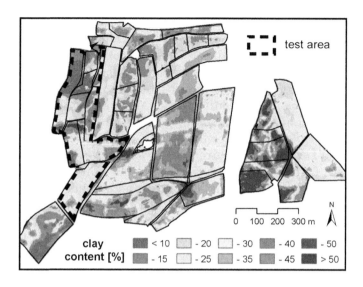

Figure 9 Prediction of clay content for research farm 'Klostergut Scheyern'.

precision agriculture to increase efficiency of production and decrease groundwater input of fertiliser residues. In the eastern part of the farm, the high and heterogeneous clay contents delineate numerous landslides indicating potential risks. Looking at the management systems (Figure 4 of Chapter 1.1), one has to notice that the clay content of the organic fields in the north-eastern part of the farm is much lower compared to the conventional fields in the centre. This indicates a fertility drawback, which has to be accounted for when the efficiency of the two management systems is compared. Because of this systematic 'bias', a sound comparison between both systems is only possible at selected sites (pedon scale), not in total (landscape scale).

4.2.4 Remote sensing

A general aim of remote sensing applications in agricultural landscapes is the estimation of vegetation parameters related to crop status and its variability in space and time from characteristic spectral signatures. Optical air- or spaceborne sensors register radiation over a wide range of the electromagnetic spectrum that is reflected from surface materials. Vegetation reflects incoming solar radiation in a characteristic manner (Figure 10a). In the visible region (VIS), reflectance is low because of absorption by chlorophyll pigments especially in the red domain. Transmittance is weak and almost no radiation penetrates inside the canopy after interception by the uppermost leaf layer. Therefore, the visible range is useful for estimating soil cover as soil reflectance is relatively high compared to vegetation (Baret, 1991). In contrast, transmittance and reflectance increase rapidly in the near-infrared (NIR) domain (red edge) (Horler et al., 1983). Absorption is absent and canopy layers (and soil) underneath the upper layer contribute significantly to the total measured reflectance. This tends towards a limit termed 'infinite reflectance' (Baret, 1991). This multiple reflectance in the NIR appears to be a suitable estimator of green LAI (LAI_g) (Figure 10b).

Since LAI_g (area of green leaves per ground unit [$m^2 m^{-2}$]) is directly related to photosynthetic activity of plants and at least to dry matter production and yield, a number of quantitative approaches were developed to derive LAI_g from spectral signatures (Kurz, 2003). The complexity ranges from simple (or multivariate) regression analysis to radiative transfer models (Moran et al., 1997). Since we wished to use a robust, operational estimator suitable for field trials, a simplified reflectance model developed by Clevers (1986) was applied. The semi-empirical model corrects for soil background reflectance and estimates LAI_g from only few input-parameters, but considers the physical relation between NIR-reflectance and LAI_g.

The spatial pattern of crop variables on the field scale indicates variability of subsoil properties, which, for example, control rooting depth and plant available water capacity. Crop growth integrates over all site effects

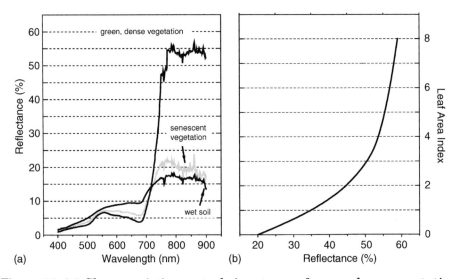

Figure 10 (a) Characteristic spectral signatures of green, dense vegetation, senescent vegetation and a wet soil surface, (b) NIR reflectance of a vegetation canopy (+soil) as a function of LAI_g.

(Moran et al., 1997). Therefore, plant canopies can be used as indicators for soil properties if climatic environment and local relief position are known (Auerswald et al., 1997; Maidl et al. 1999), This indicator function is mostly effective at the end of the growing season, when site-specific stress factors such as water and nutrient deficits lead to premature senescence of plant components. In this context, most flight missions (except those for grassland studies) were performed at the end of the growing season. In this stage of development, the loss of chlorophyll results in similar spectral signatures of vegetation and bare soils (Leeuwen and Huete, 1996). To overcome this problem, the model was modified by introducing an additional variable, derived from the differences in optical properties of senescent vegetation in the NIR and the technique of non-invasive LAI determination in the field.

Since dry matter (DM) weight is the sum of all assimilation processes during the vegetation period (Geisler, 1983) and thus a time-integrating indicator for the relation to soil properties, the estimation of this crop parameter is preferable to LAI. Thus, the relationship of estimated LAI to DM production was analysed with the aim to minimise or avoid time-consuming and destructive Ground truth measurements of these crop parameters are independent of annual changes of the climatic environment, different crop varieties and management influences.

Methods

Multispectral imagery was acquired with a DAEDALUS AADS 1268 scanner on a DO 228 (1994–1999) and a Cessna 208 B (2000–2002) platform under optimum flight conditions (clear sky, near-maximum sun inclination). Flight altitude was 450 m above ground resulting in a ground resolution (pixel size) of about $1\,m^2$. Rectification was applied involving the DEM 5. The SOLSPEC model (Bird, 1984) was used for corrections of atmospheric influences. Additional measurements of soil and canopy reflectance at eight representative sites were performed by the GTCO (Ground Truth Center Oberbayern) with a Mikropack SD 2000 radiospectrometer on 7th of July in 2002. Measurement conditions were comparable to those of the previous years according to crop status and soil moisture content. Simultaneously to flight missions, crop parameters were determined at representative sites mainly in winter wheat fields. Because of crop rotation, the ground truth of each year consists of samples from different fields. Measured crop parameters of winter wheat included LAI, fresh matter (FM) weight, DM weight and plant height (PH). LAI was measured with an LAI 2000 plant canopy analyser (LI-COR Inc., 1991). FM was ascertained by the sum of four samples of $0.25\,m^2$ each within an area of $25\,m^2$. Regression analysis was applied to ascertain the spatial distribution of DM weight from remotely sensed LAI values. An overview of flight missions from the years 1994 until 2001 is given in Table 2.

Table 2 Overview of available data sets of DAEDALUS images, studied fields and measured ground truth parameters from 1994 to 2001.

Date of mission	Covered area (%)	Target	Fields under study	[a]	Ground truth	n[b]
29.04.1994	100	WW	A01/A04/A07/A15/A17/A19/A20	+	Moisture content of soil surface	160
04.07.1994	100	WW	A01/A04/A07/A15/A17/A19/A20	+	LAI; FM; DM; PH	20
03.05.1995	100	WW	A03/A05/A12/A13/A16/A18/A21	+	LAI; FM; DM; PH	35
29.05.1995	100	WW	A03/A05/A12/A13/A16/A18/A21	+	LAI; FM; DM; PH	30
11.07.1995	100	WW	A03/A05/A12/A13/A16/A18/A21	+	LAI; FM; DM; PH	30
11.06.1996	100	WW	A02/A06/A15/A17/A19/A20	−	LAI; FM; DM; PH	34
05.08.1996	100	WW	A02/A06/A15/A17/A19/A20	−	LAI; FM; DM; PH	34
19.07.1999	100	WW	A02/A05/A05/A11–A13/A16/A18	−	FM; DM; PH	22
02.08.1999	75-W	GL/M	W02/A18	+	LAI; FM; DM; PH	12
03.05.2000	60-W	GL	W02–W04/W06/F18/W12/ W18–W20	+	FM; DM (different vegetation units)	36
28.06.2000	100	WW	A08/A09/A14	+	LAI; FM; DM; PH	26
02.05.2001	60-E	GL	W11/W12/W14/W16/W21–W25	+	FM; DM (different vegetation units)	38
27.06.2001	100	WW	A04/A07/A10	+	LAI; FM; DM; PH	17

[a] Coincidence of flight mission and maximum spatial differentiation of target on the field scale.
[b] Used for modelling.

Modelling approach

The physical relations describing the interception of solar radiation by a vegetation canopy and the subsequent scattering of radiation towards a sensor incorporate both structural and optical properties of leaf layers and illumination and observation geometry. The simplified model used in this study describes the non-linear relationship between NIR reflectance and

LAI_g using a Mitscherlich curve with two unknown parameters:

$$\rho_{NIR,cor} = \rho_{NIR,\infty} \times (1 - e^{-\alpha \times LAI_g}) \tag{4}$$

where $\rho_{NIR,cor}$ is the corrected NIR reflectance, $\rho_{NIR,\infty}$ is the infinite reflectance value in the infrared region and α is a shape parameter, representing the combination of extinction and scattering coefficients.

Before estimating LAI from Eq. (4), a correction for soil background reflectance has to be applied to a NIR passband (wavelength depends on type of sensor and bandwidth of spectral channels). In general, soil reflectance is relatively low in the blue domain and increases monotonically with wavelength from the VIS to the NIR region. These optical properties vary with different soils and their complex composition and are mainly related to colour, roughness and water content, but experimental studies indicate that for a given soil variability, the reflectance at one wavelength is functionally related to the reflectance in another wavelength. This linear relationship is referred to as the 'soil line' concept (Huete, 1989; Baret, 1991; Rondeaux et al., 1996) and is widely accepted and implemented in vegetation reflectance observations. Consequently, the reflectance of bare soil in different wavelength is independent of soil moisture content. For the correction of ρ_{NIR}, the following equation can be used:

$$\rho_{NIR,cor} = \rho_{NIR} - \frac{C_2 \times \left(\rho_G \times \rho_{R,Veg} - \rho_R \times \rho_{G,Veg} \right)}{C_1 \times \rho_{R,Veg} - \rho_{G,Veg}} \tag{5}$$

where C_1 and C_2 are the ratios of soil reflectance in the green/red and infrared/red domain, respectively, ρ_G and ρ_R the detected reflectances in the green and the red domain, respectively and $\rho_{G,Veg}$ and $\rho_{R,Veg}$ the reference reflectances of a dense, healthy vegetation canopy. After soil background correction, LAI_g is calculated by the inverse of Eq. (4). The curve in Figure 10b now runs through the origin. In the case of living, but non-photosynthetic plant material, the correction for soil background will result in values of LAI_g near 0, although biomass is still present. On the contrary, the LAI 2000 plant canopy analyser determines total LAI (LAI_t), including all light-blocking objects in the field of view of the instruments optics. Consequently, LAI_t of senescent vegetation is underestimated by the semi-empirical model.

To overcome this problem, an additional parameter termed β is introduced in the inverse of Eq. (4) after soil background correction (Eq. (6)). The resulting curve fitted to measured LAI_t is then the product of a two-step correction (soil background and senescent vegetation):

$$LAI_t = -\frac{1}{\alpha} \times \ln\left(\frac{\rho_{NIR,cor} - \rho_{NIR,\infty}}{-\beta} \right) \tag{6}$$

This modified model requires three parameters, which are unknown. In general, the infinite reflectance $\rho_{NIR,\infty}$ varies with different crops, stage of development and crop management. The evaluation of the experimental data showed that NIR reflectance of the same LAI_t is in most cases significantly higher in the integrated than in the ecologically managed fields. Assuming values between 52 and 62% for a closed, green winter wheat canopy as reported in literature (Clevers, 1986; Price and Bausch, 1995) for $\rho_{NIR,\infty}$ in all fields leads to a drastic underestimation of LAI_t in the latter cases. For this reason, the maximum NIR reflectances $\rho_{NIR,max}$ as an approximation for $\rho_{NIR,\infty}$ for individual fields were used in Eq. (6). This appears to be more realistic in the studied stage of development and could be improved by the radiospectrometer measurements of winter wheat canopies performed in 2002 above varying soil backgrounds. Another advantage is the reduction of unknown parameters in Eq. (6). The remaining parameters α and β were determined by a least-squares fit to the samples for individual fields using Monte Carlo simulation. Means of α and β from 100 trials were used to estimate LAI_t from NIR reflectance. As an example, the original and the modified model were applied to measure LAI_t in field A9 on 28th of June 2000. The result is illustrated in Figure 11. RMSE for the relationship between measured and estimated LAI_t decreases significantly, when LAI_t is predicted by the modified model (RMSE = 0.54 and 0.17, respectively).

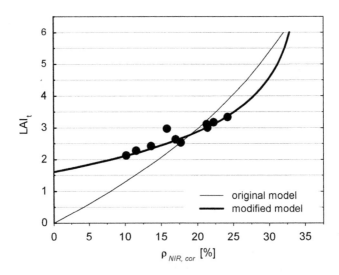

Figure 11 Comparison of modelling results applying the original model and the modified model to field measurements of LAI_t in field A14 (28.06.2000).

Spatial distribution of DM production

The main objective of the most remote sensing missions was the determination of spatial distribution of DM production assuming this vegetation parameter to be a suitable indicator for the pattern of soil properties on the field scale. The correlations between estimated LAI_t and DM production were significantly high at the end of the growing season of all years. A non-linear power function was found to be the best fit to the scattered data. With the exception of 1996 (late flight mission due to weather conditions in July, complete maturity of winter wheat) and field A14 in 2000, the slope of the functions varies little (Table 2). The lowest correlation was found on 4th of July 1994 ($R^2 = 0.72$), when defective field measurements of LAI_t caused corresponding errors in LAI_t estimations. Calculated RMSE values indicate that DM weight can be predicted with a sufficient accuracy (± 5.6–15.5% of mean measured DM weight) from remote sensing data. Compared with the variance of DM weight in the single fields observed during the study, the error is relative small. Ground truth data showed coefficients of variation (CV) of about 20 to 35%, which correspond to factors from 1.6 to 2.0 in DM production in single fields.

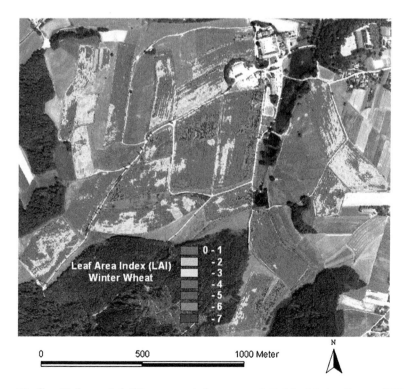

Figure 12 Spatial variability remotely sensed LAI_t (data from different years).

   The distinct relationships were the used to calculate DM weight from remotely sensed LAI_t. Figure 12 presents the results for the years 1994–2000. It is obvious that variability of DM weight, found on the basis of point samples, is spatially structured. Although the pattern of DM production on the field scale shows slight variations of the shape and expansion in different years, the general pattern remains equal. This indicates a strong influence of time-invariant soil properties since external growth conditions such as climate and management do not vary for the individual fields. This demonstrates that the presented method is very effective in distinguishing management units on the field scale for either precision farming or the determination of soil units with specific properties controlling crop growth.

   Nevertheless, field determinations of FM and DM weight are time-consuming and destructive. Thus, the reduction of field samples to a minimum is desirable. For this purpose, values from all campaigns performed in a comparable phenological stage of winter wheat were pooled in one data set. Figure 13 presents the regression of estimated LAI_t on DM weight using 92 samples. As expected, scattering increases compared with the regressions obtained from the single years caused by different climatic growth conditions and management influences (compare Table 3). Nevertheless, a high correlation exists between the two vegetation parameters ($R^2 = 0.72$). The calculated RMSE of $153.8\ \mathrm{g\,m^{-2}}$ corresponds to a mean deviation of the overall mean of measured DM weight of $\pm 16\%$. The result indicates that this 'global' relationship between estimated LAI_t and DM weight on basis of 4 years on different fields can be used in future research for the derivation of DM production and its spatial variability on the field scale in highly variable agricultural landscapes.

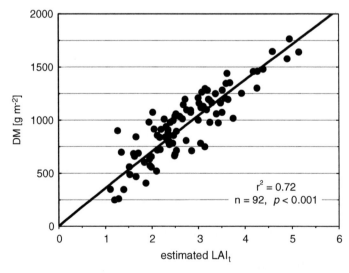

Figure 13 Non-linear regression of estimated LAI_t on DM weight for the pooled data set; values are from 1994, 1995, 2000 and 2001.

Table 3 Regressions and estimates of errors for the relationship of estimated LAI_t on measured DM weight from 1994 to 2001.

Date	N (fields)	Regression	R^2 $(p < 0.05)$	RMSE	Deviation from mean DM (%)
04.07.1994	20	DM = 395.89 $LAI_t$$^{0.8107}$	0.72	143.4	15.5
11.07.1995	30	DM = 477.02 $LAI_t$$^{0.7841}$	0.85	94.8	8.4
05.08.1996	6 (A15)	DM = 626.33 $LAI_t$$^{0.8387}$	0.90	90.4	6.6
	26 (others)	DM = 656.51 $LAI_t$$^{1.0001}$	0.75	155.7	9.6
28.06.2000	14 (A08–A09)	DM = 392.74 $LAI_t$$^{0.9759}$	0.92	58.6	6.2
	11 (A14)	DM = 817.14 $LAI_t$$^{0.3203}$	0.80	64.5	5.6
27.06.2001	17	DM = 248.73 $LAI_t$$^{1.1353}$	0.84	81.2	14
Total	92 (ex. 1996)	DM = 358.50 $LAI_t$$^{0.9714}$	0.72	153.8	16.0

Remote sensing data and soil properties

Spatial differences in DM indicate soil-dependent site qualities if other reasons, such as pests, weeds or management failure, can be excluded. In general, plant growth depends on (i) rooting depth, where dense plough layers, acid subsoils and reductive horizons will restrict it and (ii) the quality of that rooting volume, such as available water capacity (AWC), O_2 supply, nutrient contents, etc. In our case study, the analysis of the soilscape led us to the hypothesis of plant available water to be the most prominent influencing factor on plant growth. Therefore, we plotted AWC versus DM of winter wheat for both years. As can be seen from Figure 14, there is a non-linear relationship between both variables (compare Selige, 1997). In a dry year (1994), no differentiation between soils could be stated, whereas in a wetter year (2000), stagnant water soils showed lower DM values. Temporal O_2 deficiency for roots was the next important factor of site quality in our study area.

4.2.5 Conclusions

Interpretation of remote sensing data seems to be the most promising way to get information on soil (properties) because of its high spatial resolution and (potential) high areal coverage of agricultural landscapes (Sommer, 2006). Nevertheless, the major problem of using plants as indicators for soil (areas) results from convergence phenomena (Schumm, 1991) – different (soil-related) causes may have the same effect. As an example, plant growth may be inhibited either by a smaller rooting depth or by a lower quality of that rooting volume. A functional analysis of the soilscape is needed to decipher the exact site-specific causes for differences in plant growth first before any inference from

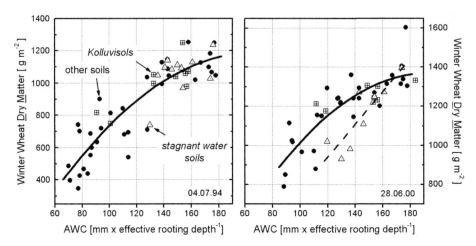

Figure 14 Relationship between available water capacity (pF1.8–pF4.2) and winter wheat dry matter in 1994 (left) and 2000 (right); data from all 50 m raster points of all fields in a single year (Sommer et al., 2003).

canopy data to soils can be made. To our experience, a hierarchy of soil forming factors generally explains most of the variability observed at landscape scale (structure in Vogel and Roth, 2003). Hierarchy in this sense means that a change in a factor on higher hierarchical level has a greater impact on soil pattern, regardless of differences in low-level factors. We have already developed a soil inference model on basis of a local factor hierarchy, details of which had been previously published (Sommer et al., 2003). There will be no general hierarchy for all landscapes, but local ones, which have to be developed by expert knowledge of soil scientists.

Acknowledgements

The scientific activities of the FAM Munich Research Network on Agroecosystems were financially supported by the German Federal Ministry of Education and Research (BMBF 0339370). Overhead costs of the Research Station Scheyern were funded by the Bavarian State Ministry for Science, Research and the Arts.

We are grateful for the help of Dr. Wolfgang zu Castell (IBB, GSF), Dr. Sven Ehrich (IBB, GSF), Dr. Danuta Kaczorek (SGGW, Warsaw) and Rainer Gryschko. Special thanks to Prof. Dr. Karl Auerswald (TU, Munich) who provided all data of former FAM research.

References

Auerswald K, Sippel R, Kainz M, Demmel M, Scheinost A, Sinowski W, Maidl FX, 1997. The crop response to soil variability in an agro-ecosystem. Advances in Geoecology 30, 39–53.

Baret F, 1991. Vegetation canopy reflectance: Factors of variation and application for agriculture. In: Belward A, Valenzuela R (Eds.), Rem. Sens. and Geographical Information Systems for Resource Management in Developing Countries. ECSE, EEC, EAEC, Brussels, pp. 145–167.

Bird RE, 1984. A simple spectral model for direct normal and diffuse horizontal irradiance. Solar Energy 32, 461–471.

Birkeland P, 1999. Soils and Geomorphology. Oxford University Press, Oxford.

Clevers JGPW, 1986. Application of remote sensing to agricultural field trials. Vol. 86 of Agricultural University Wageningen Papers. Agricultural University Wageningen.

Dalgaard M et al., 2001. Soil clay mapping by measurement of electromagnetic conductivity. In: Blackmore S, Grenier G (Eds.), Third European Conference on Precision Agriculture 2001. Full Paper CD. Agro Montpellier, Ecole Nationale Superieur Agronomique, Montpellier, pp. 367–372.

Durlesser HP, 1999. Determination of the Variation of Soil Physical Parameters Through Time and Space by Electromagnetic Induction (in German). PhD Thesis, TU München, Shaker Verlag, Aachen, ISBN 3-8265-6180-5, FAM-Bericht 35, 120 pp.

Fridland VM, 1976. Pattern of Soil Cover. Israel Program for Scientific Translation, Jerusalem.

Geisler G, 1983. Yield Physiology of Cultivated Plants of the Temperate Climate (in German). Verlag Paul Parey, Berlin.

Grunwald S (Ed.), 2006. Environmental Soil-Landscape Modeling – Geographic Information Technologies and Pedometrics. CRC Press, New York.

Heimsath AM, Chappell J, Spooner NA, Questiaux DG, 2002. Creeping soil. Geology 30, 111–114.

Heimsath AM, Dietrich WE, Nishiizumi K, Finkel RC, 1997. The soil production function and landscape equilibrium. Nature 388, 358–361.

Heimsath AM, Dietrich WE, Nishiizumi K, Finkel RC, 2001. Stochastic processes of soil production and transport: Erosion rates, topographic variation and cosmogenic nuclides in the Oregon Coast Range. Earth Surface Processes and Landforms 26, 531–552.

Heuvelink GBM, Webster R, 2001. Modelling soil variation: Past, present and future. Geoderma 100, 269–301.

Hole FD, Campbell J, 1985. Soil Landscape Analysis. Routledge and Kegan Paul, London.

Horler DNH, Dockray M, Barber J, 1983. The red edge of plant leaf reflectance. International Journal of Remote Sensing 4, 273–288.

Huete AR, 1989. Soil Influences in Remotely Sensed Vegetation-Canopy Spectra. In Asrar, G. (Eds.), Theory and Application of Optical Remote Sensing. John Wiley & Sons, New York.

Jenny H, 1941. Factors of Soil Formation. McGraw-Hill, New York.

Kurz F, 2003. Schätzung von Vegetationsparametern aus multispektralen Fernerkundungs-daten. PhD Thesis, TU München, 123 pp.

Leeuwen van WJD, Huete AR, 1996. Effects of Standing Litter on the Biophysical Interpretation of Plant Canopies with Spectral Indices. Remote Sensing of Environment 55, 123–138.

LI-COR Inc., 1991. LAI-2000 Plant Canopy Analyser Operating Manual. LI-COR Inc., Lincoln, NE, pp. 1–90.

Maidl FX, Brunner R, Sticksel E, Fischbeck G, 1999. Ursachen kleinräumiger Ertragsschwan-kungen im bayerischen Tertiärhügelland und Folgerungen für eine teilschlagbezogene Düngung. Journal of Plant Nutrition & Soil Science 162, 337–342.

Mausbach MJ, Wilding L (Eds.), 1991. Spatial Variabilities of Soils and Landforms. SSSA Spec. Publ., vol. 28. SSSA, Madison, WI.

McBratney AB, Mendonça Santos ML, Minasny B, 2003. On digital soil mapping. Geoderma 117, 3–52.

McBratney AB, Minasny B, Cattle SR, Vervoort RW, 2002. From pedo-transfer functions to soil inference systems. Geoderma 109, 293–327.

McBratney AB, Odeh IOA, Bishop TFA, Dunbar MS, Shatar TM, 2000. An overview of pedometric techniques for use in soil survey. Geoderma 97, 293–327.

McNeill JD, 1980. Electromagnetic Terrain Conductivity Measurement at Low Induction Numbers. Technical Note 6. Geonics Ltd., Mississauga, Ontario.

Minasny B, McBratney AB, 1999. A rudimentary mechanistic model for soil production and landscape development. Geoderma 90, 3–21.

Minasny B, McBratney AB, 2001. A rudimentary mechanistic model for soil production and landscape development: II. A two-dimensional model incorporating chemical weathering. Geoderma 103, 161–180.

Minasny B, McBratney AB, 2006. Mechanistic soil-landscape modelling as an approach to developing pedogenetic classification. Geoderma 133, 138–149.

Moore ID, Gessler PE, Nielson GA, 1993. Soil attribute prediction using terrain analysis. Soil Science Society of America Journal 57, 443–452.

Moore ID, Ladson AR, Grayson R, 1991. Digital terrain modelling: A review of hydrological, geomorphological, and biological applications. Hydrology Processes 5, 3–30.

Moran MS, Inoue Y, Barnes EM (1997): Opportunities and limitation for image-based remote sensing in precision crop management. Remote Sensing of Environment 61, 319–346.

Price JC, Bausch WC, 1995. Leaf area index estimation from visible and near-infrared reflectance data. Remote Sensing of Environment 52, 55–65.

Rondeaux G, Steven M, Baret F, 1996. Optimization of soil-adjusted vegetation indices. Remote Sensing of Environment 55, 95–107.

Schlichting E, 1986. Einführung in die Bodenkunde. Paul Parey, Hamburg, Berlin. 131 pp.

Schumm SA, 1991. To Interpret the Earth – Ten Ways to be Wrong. Cambridge University Press, Cambridge, 1–133.

Selige T, 1997. Remote sensing and relief analysis as tools for agricultural site assessment (in German). In: Felix-Henningsen P, Wegener HR (Eds.), Boden Landsch. vol. 17. University of Giessen, pp. 121–138.

Sheets KR, Hendrickx JMH, 1995. Non-invasive soil water content measurement using electromagnetic induction. Water Resources Research 31, 2401–2409.

Sinowski W, 1995. The three-dimensional variability of soil properties. PhD thesis. Technical University of Munich. Shaker, Aachen, ISBN 3-8265-0994-3. FAM-Bericht 7, 158 pp.

Sommer M, 2006. Influence of soil pattern on matter transport in and from terrestrial biogeosystems – A new concept for landscape pedology. Geoderma 133, 107–123.

Sommer M, Schlichting E, 1997. Archetypes of catenas in respect to matter – A concept for structuring and grouping catenas. Geoderma 76, 1–33.

Sommer M, Wehrhan M, Zipprich M, Weller U, 2004. Multidata fusion of remote sensing geophysical and relief data – A new tool for soil survey in precision agriculture, In: Stamatiadis S et al. (Eds.), Remote Sensing for Agriculture and the Environment. GAIA Center & OECD, Larissa, Greece, pp. 65–75.

Sommer M, Wehrhan M, Zipprich M, Weller U, zu Castell W, Ehrich S, Tandler B, Selige T, 2003. Hierarchical data fusion for mapping soil units at field scale. Geoderma 112, 179–196.

Tabbagh A, Dabas M, Hesse A, Panissod C, 2000. Soil resistivity: A non-invasive tool to map soil structure horizontation. Geoderma 97, 393–404.

Vogel HJ, Roth K, 2003. Moving through scales of flow and transport in soil. Journal of Hydrology 272, 95–106.

Weller U, Zipprich M, Sommer M, Wehrhan M, zu Castell W, 2007. Mapping clay-content across boundaries at landscape scale with electromagnetic induction. Soil Science Society of America Journal 71, 1740–1747.

Chapter 4.3

Changes in Nutrient Status on the Experimental Station Klostergut Scheyern from 1991 to 2001 – Statistical and Geostatistical Analysis

K. Weinfurtner

4.3.1 Introduction

In the discussion about sustainability of agriculture fertilisation plays an important role. On the one hand, fertilisation has to be high enough to get optimal yield and quality taking into consideration local conditions. On the other hand, the loss of nutrients and the nuisance for neighbouring ecosystems have to be avoided.

The first soil inventory in 1991 showed that the supply with phosphorus and potassium on the fields of the experimental station ranged from optimal to very high. Therefore, it was decide to resign of mineral P and K fertiliser to reduce the reserves, only organic manures supplied with P + K.

Perspectives for Agroecosystem Management
Edited by P. Schröder, J. Pfadenhauer and J.C. Munch

The character and level of N supply differed greatly at both farms. Since in the organic farming (OF) the use of mineral N fertilisation is prohibited, the N supply resulted from farmyard manure and N fixation by legume feeding crops and undersown legumes. In the farm with integrated crop cultivation (ICC) mineral N fertilisers were used in addition to farmyard manure.

The soil inventory of 2001 was carried out to determine which changes in soil reserves of the most important nutrients (C, N, P and K) caused by the changed management systems were observed in both farms. This was of special interest because long-term nutrient balances indicated changes in soil reserves. As an example for N a balance surplus of 80 kg N ha^{-1} yr^{-1} occurred between 1993 and 2001 in the ICC system (Weinfurtner et al., 2002). Investigations of Matthes et al. (2002) showed that only 50% of the N surplus could account for the investigated losses (NH$_3$, N$_2$O into the atmosphere, NO$_3^-$ to ground and surface water). As a possible sink for N an enrichment of nitrogen in the organic soil matter was assumed. For P and K especially negative balances were determined which indicated a reduction of soil reserves in the topsoil.

Therefore, the soil inventory of 2001 was statistically and geostatistically analysed to transfer the punctual data to spatial distributions of examined soil properties. The results then were compared with the results of the soil inventory of 1991 and with the nutrient balances given by REPRO.

4.3.2 Material and methods

Sample strategy

To ensure the comparability of both soil inventories the methodical concept of the first soil inventory in 1991 (Sinowski, 1995) was extensively assumed for the 2001 soil inventory. At every grid point of the 50 × 50 m grid pooled samples of the A$_p$ horizon of the arable land in the ICC system and the first 10 cm of the subsoil horizon were taken by cut-ins using a Pürckhauer sampler and undisturbed samples for determination of bulk density. At the grid points of the arable land in the ICC system sampling was divided in the first 10 cm of topsoil and the remaining A$_p$ horizon. This approach was used because partitioning of the 'old' A$_p$ horizon was expected caused by reduced soil tillage in the ICC system. The forecasted accumulation of C and N was estimated in the upper part of the A$_p$ horizon.

In addition to the grid points about 100 points in spatial neighbourhood to the grid point (distance 5–20 m) were sampled considering the relief and former cultivation. This expansion delivered important information about similarity of neighboured samples and improved the analysis of variogram. Hence the ascertainable autocorrelation of neighboured samples is the basis for an accurate spatial interpolation with the use of geostatistical procedures.

Laboratory methods

The sampled soil material was air dried and passed through a 2 mm sieve. All chemical analysis was done with the air dried material. Total and organic carbon and total nitrogen were analysed with a C/N-autoanalyser. Plant available P and K were determined with the CAL method (calcium acetate-lactacte). Soil pH was measured at a soil suspension ratio of 1:2.5 with 0.01 M $CaCl_2$. The bulk density was determined at the undisturbed samples using the core method (Hartge and Horn, 1989) after drying at 105°C. After drying the material from the cores was sieved through a 2 mm sieve to get the amounts of the soil material (<2 mm) and coarse material (>2 mm). All analysis was performed under consideration of the VDLUFA-guidelines (VDLUFA, 1991).

Data interpretation

Statistical analysis was carried out with SPSS. The significance of difference of averages was tested by using the parameter free U-Test according to Mann and Whitney (Bortz et al., 1990) because the tested soil parameters did not have a normal distribution.

All statistical calculations with the following levels of significance:

Significant	$p = 0,05$
high significant	$p = 0,01$
Extreme significant	$p = 0,001$

Geostatistics

Geostatistical software (GSwin) was used to analyse the spatial structure of the data and to define the semivariograms. The interpolation of measured punctual data to spatial information was performed in two steps. Assuming the intrinsic stationarity of the data, the degree of spatial correlation of a random variable $z(x_i)$ over a certain distance can be described by the semivariogram function (Eq. (1))

$$\gamma(\vec{h}) = \frac{1}{2 \cdot n} \cdot \sum_{i=1}^{n} (Z(\vec{x}_i) - Z(\vec{x}_i + \vec{h}))^2 \tag{1}$$

where $\gamma(\vec{h})$ is the semivariance, and n is the number of pairs of $Z(\vec{x}_i)$ at a separate lag $\vec{h}$.

The semivariograms were fit with a spherical (Eq. (2)), an exponential (Eq. (3)) or a Gaussian (Eq. (4)) model according to Dutter (1985).

$$\text{spherical model:} \quad \gamma(\vec{h}) = \left\{ C_0 + C_1 \left(\frac{3|\vec{h}|}{2A} - \frac{1}{2} \left(\frac{|\vec{h}|}{A} \right)^3 \right) \quad \text{for} \quad |\vec{h}| \leq A; \right. \tag{2}$$

$$C_1 \text{ for } |\vec{h}| > A$$

$$\text{exponential model:} \quad \gamma(\vec{h}) = C_0 + C_1(1 - e^{-(|\vec{h}|/A)}) \tag{3}$$

$$\text{Gaussian model:} \quad \gamma(\vec{h}) = C_0 + C_1\left(1 - e^{(-h^2/A^2)}\right) \tag{4}$$

$$C_1 \text{ for } |\vec{h}| > A$$

where $\gamma(\vec{h})$: semivariance depending on the range $\vec{h}$; C_0: nugget variance; C_1: sill; A: range.

A semivariogram is described by three attributes (Figure 1):

- Nugget variance (C_0): The nugget variance is the extrapolated inter-section of the semivariogram model with the ordinate. Ideally it is the value 0, but small-scale variability or measurement errors cause values greater than 0.
- Sill (C_1): The sill is attained when the semivariance does not increase in spite of increasing spatial distance. The bigger the sill is in comparison to the nugget variance the bigger is the spatial variability.
- Range (A): The range is the distance at which the sill is reached. Parameter values are spatial interdependent within the range only.

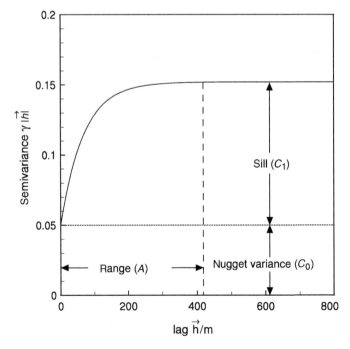

Figure 1 Example of an exponential semivariogram.

The variogram does not reach the sill by fitting with the exponential and Gaussian model but it approximates asymptotic for $\bar{h} \Rightarrow \infty$. The calculated range in these models is the intersection of the tangent in the origin with the sill of the variogram. The 'effective' range is the value $A \times 3$, which accords 95% of the maximal variance (Webster and Oliver, 1990).

For the calculation of the variograms the following settings were carried out:

- The maximal range $\bar{h}$ for the different types of utilisation (arable land, ICC; arable land, OF; grassland, fallow land) was determined with 250–300 m, because only a few pairs of variates could be generated for larger distances.
- For the calculation of semivariances lag classes has to be defined. Within these classes all distances between pairs of variates were averaged to one lag. If the lag classes are too small, the number of pairs within one class gets too small and the semivariance shows high range. Entz and Chang (1991) postulate 30 to 50 pairs of variates for every lag class. The calculations were carried out for lag classes of 16 and 30 m.
- Basically an isotropic allocation was assumed, that means that the search for pairs of variates was allowed in all directions.
- All valid pairs of variates for the particular utilisation were used for calculation of variograms to get as much pairs of variates as possible.

The spatial interpolation was carried out by kriging. The necessary semivariances for the krige matrix were calculated with the help of the semivariogram model that fits the experimental semivariogram best of all. For the kriging a block size of 12.5×12.5 m was defined. The estimation of every block was calculated from the 24 nearest points. The maximal distance within points were used for the calculation was the range of the semivariogram.

The disadvantage of spatial interpolation for every particular form of utilisation is that for border areas or small areas only a few measuring points are available and therefore the error of estimation becomes greater. Because of small differences between both soil inventories a precise spatial interpolation is necessary in order to detect changes.

To avoid this problem residual variograms were calculated according to Sinowski (1995). The following assumptions were defined:

The mean value of every form of utilisation (arable land, grassland, fallow land) represents the value of a soil characteristic adjusted by management.

The spatial variability which remains for every form of utilisation after generation of mean value is either stochastic or is caused by pedogenic influence.

$$\gamma_r(\vec{h}) = \frac{1}{2 \cdot n} \sum_{i=1}^{n} ((m_{j(\vec{x}_i)} - z(\vec{x}_i)) - (m_{j(\vec{x}_i + \vec{h})} - z(\vec{x}_i + \vec{h})))^2 \qquad (5)$$

where $\gamma_r(\vec{h})$: semivariance of residues with lag $\vec{h}$; n: amount of measuring points; $z(\vec{x}_i)$: specification of attribute at location $\vec{x}_i$; $z(\vec{x}_i + \vec{h})$: specification of attribute at location $\vec{x}_i + \vec{h}$; $m_{j(\vec{x}_i)}$: mean value of spatial item at location $\vec{x}_i$.

With Eq. (5) a residual variogram was calculated which was independent from borders of utilisation. For the calculation of variograms only the residual variance was used which remained for every measuring point after subtraction of the mean value of a plot. After the spatial interpolation the mean value of the plot was added to every krige block (12.5 × 12.5 m). The following maps were all based on calculation with residual variograms.

4.3.3 Results

Statistical analysis

All statistical analysis was carried out only for the topsoil. Apart from the different utilisations (arable land, OF; arable land, ICC; grassland and fallow land) in the arable land the single fields were examined. Mostly the small plots in the grassland were pooled to larger units. All meadow plots (W01–W12, W50/51) are the unit 'meadow'. The pasture plots W13–W16 represent the 'pasture north', W18–W20 the 'pasture west' and W21–W25 the 'pasture east'. The allocation of fields is given in Figure 2.

Comparison with the soil inventory of 1991 – Arable land

The trends were comparable for the arable areas in both farms (Table 1). The mean values for the concentrations of C, N, P and K decreased on the arable areas. But the decrease of N concentration was smaller as that of the C concentration and therefore the C/N ratio got narrower. This was in high gear in the farm with ICC where the mean value of the C/N ratio decreased from 9.7 to 9.0.

The bulk density was significantly higher in 1991 in both farms. However, the sampling 1991 was carried out in the middle of July before the harvest while the sampling 2001 took place after harvest and the use of harvesters. Furthermore investigations of Kaemmerer (2000) showed that an increase of bulk density was observed during the summer months even without the use of machines.

The mean value for pH did not change but this was not unexpected because the pH value is easy to control by liming and the arable areas were in 1991 within the target values. Therefore, only preservation liming was performed.

Both farms differed in relative to the thickness of the A_p horizon. Whereas the thickness in the farm with ICC did not change because of the reduced tillage a significant increase was observed in the farm with OF which was asserted by analysis of the plot card index.

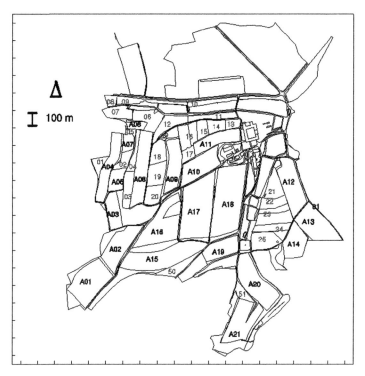

Figure 2 Allocation of plots of the experimental station Klostergut Scheyern: A01–A21: arable land; 01–25, 50/51: grassland; without description: fallow land.

Table 1 Soil chemical and soil physical parameters in the arable land for 1991 and 2001 (P/K_{CAL}, C_{org} and N_t in g kg^{-1}, bulk density in kg dm^{-3}, thickness of A_p horizon in cm).

	Integrated crop cultivation		Organic farming	
	1991	2001	1991	2001
pH (CaCl$_2$)	5.91	5.93	6.02	5.96
P_{CAL}	0.098	0.071***	0.101	0.079***
K_{CAL}	0.206	0.148***	0.213	0.175***
C_{org}	14.2	12.4***	15.6	13.9**
N_t	1.46	1.38**	1.60	1.50
C/N ratio	9.7	9.0***	9.8	9.3**
Bulk density	1.43	1.46**	1.43	1.50***
Thickness of A_p	23	23	22	25***

The levels of significance of 95%, 99% and 99.9% for changes between 1991 and 2001 are marked as *, ** and ***.

The reduction of nutrient concentration was significant at different levels of significance except for the N concentration in the OF system. The reduction of nutrient concentrations was less in the OF system than in the ICC system. For instance a reduction of K concentration at 29% was observed in the ICC system and of 16% in the OF system (Table 1). For K, a negative balance of about 20 kg ha^{-1} yr^{-1} was calculated within the years 1993–2001 for the ICC system, but a stable balance was calculated in the OF system. If the changes of concentrations were calculated as changes in supply taking into consideration A$_p$ thickness and bulk density the values agreed well with the balances (Table 2). Thereby it has to be taken into account that the years 1992 and 2001 were not included in the balance calculations with REPRO and therefore the balances for 1991 to 2001 had to be less than those calculated from 1993 to 2000. Similar observations were made for phosphorus.

Organic C and total N decreased little in the ICC system taking into consideration bulk density and A$_p$ thickness. The C and N reserves lowered from 46 300 kg ha^{-1} to 41 600 kg ha^{-1} for C and 4800 kg ha^{-1} to 4630 kg ha^{-1} for N, respectively. In contrast to the decrease in the ICC system a light increase of C reserves from 49 000 kg ha^{-1} to 51 125 kg ha^{-1} and of N reserves from 5030 kg ha^{-1} to 5210 kg ha^{-1} was observed in the OF system. Whereas the changes in the OF system and the decrease of organic C in the ICC system is confirmed by the results of the balances, the decrease of N in the ICC system is in conflict with the balances. A yearly surplus of about 40 kg ha^{-1} was calculated considering the recorded losses (N$_2$O, NH$_3$ and NO$_3$). Hence, an accumulation of 300 to 400 kg ha^{-1} could be assessed for the time between both inventories. However, the results of the soil inventory indicated a reduction of N reserves of about 170 kg ha^{-1} and therefore a difference between balance and inventory of about 500 kg ha^{-1}. Even if there were some uncertainness in the calculations the difference could not be explained by calculation errors.

The calculations for the fields in arable land in the ICC system showed mostly a decrease of N concentration (Table 3). Only on A15 and A19 a

Table 2 Balances for K and P in kg ha^{-1} calculated with REPRO (1993–2000) and calculated by changes of nutrient supply (1991–2001) (in kg ha^{-1}).

	K balance (REPRO)	K balance changes of nutrient supply	P balance (REPRO)	P balance changes of nutrient supply
Arable land, ICC	−173	−180	−85	−84
Arable land, OF	+33	−14	+20	−22

Table 3 Some soil chemical and soil physical parameters for 1991 and 2001 for A_p horizon on fields of the integrated crop cultivation (P/K_{CAL}, C_{org} and N_t in g kg^{-1}, bulk density in kg dm^{-3}, thickness of A_p horizon in cm).

	pH		P_{CAL}		K_{CAL}		C_{org}	
	1991	2001	1991	2001	1991	2001	1991	2001
A15	5.83	5.94	0.066	0.052	0.208	0.171[*]	12.5	11.4[*]
A16	6.26	6.30	0.088	0.063[**]	0.228	0.159[***]	14.9	11.6[***]
A17	5.75	5.96[***]	0.074	0.063[**]	0.221	0.149[***]	13.0	11.8[***]
A18	6.11	5.96[*]	0.123	0.088	0.218	0.155[***]	14.9	12.5[***]
A19	6.09	6.15	0.232	0.191	0.305	0.182[***]	15.7	14.3
A20	5.89	5.69[*]	0.117	0.072[***]	0.190	0.135[***]	13.2	12.8[*]
A21	5.38	5.55	0.026	0.021	0.062	0.081	17.4	14.1[*]

	N_t		C/N ratio		Bulk density		A_p thickness	
	1991	2001	1991	2001	1991	2001	1991	2001
A15	1.21	1.24	10.5	9.2[***]	1.46	1.47	24	22
A16	1.49	1.33[**]	10.0	8.8[***]	1.46	1.56[***]	24	24
A17	1.39	1.36	9.7	8.6[***]	1.42	1.46[**]	24	22
A18	1.55	1.39[**]	9.6	9.0[***]	1.39	1.47[***]	26	24[*]
A19	1.53	1.56	10.4	9.2[***]	1.34	1.42[**]	22	19
A20	1.38	1.38	9.5	9.3[**]	1.47	1.47	26	23[*]
A21	1.80	1.53	9.6	9.2	1.44	1.35[*]	20	20

The levels of significance of 95%, 99% and 99.9% for changes between 1991 and 2001 are marked as [*], [**] and [***].

small increase of N could be observed. The larger decrease of organic C and therefore a closer C/N ratio was observed for all fields. The decrease of C concentration was especially signified on the fields A16 and A21 and the closer C/N ratio on the fields A15–17 and A21. The decrease of P and K reserves can be noticed in almost all fields, the changes were mostly significant. The K concentration decreased especially on the fields A16–A20. The exception was A21 where the K concentration increased between 1991 and 2001. This could be explained by the surplus of 25 kg K ha^{-1} yr^{-1} by organic mineralisation. The pH values on the plots differed lightly and fluctuated with a maximum of 0.2 pH units. The small and in the majority of cases insignificant increase on the most fields was caused by liming in the years 1993 and 1994. The fields A18 and A20 were not limed and therefore a significant decrease of pH could be observed.

In the OF system the differences between the fields were greater than in the ICC system. The fields A01, A03, A06 and A12 showed a statistical

Table 4 Some soil chemical and soil physical parameters for 1991 and 2001 on fields for the A_p horizon of the organic farming system (P/K_{CAL}, C_{org} and N_t in g kg^{-1}, bulk density in kg dm^{-3}, thickness of A_p horizon in cm).

| | pH | | P_{CAL} | | K_{CAL} | | C_{org} | |
	1991	2001	1991	2001	1991	2001	1991	2001
A01	6.31	6.04**	0.100	0.080**	0.170	0.218	14.9	14.2
A02	6.28	6.19	0.079	0.057*	0.200	0.135*	12.2	12.6
A03	6.30	5.70**	0.058	0.033**	0.201	0.131	14.8	10.5**
A04	5.66	6.19	0.046	0.054	0.146	0.188	17.0	15.6
A05	6.00	5.58	0.038	0.027	0.199	0.172	15.6	14.1
A06	6.17	6.01*	0.173	0.168	0.225	0.200	13.5	14.3
A07	5.27	5.85	0.058	0.036	0.293	0.274	16.1	13.4
A08	5.10	5.23	0.024	0.028	0.162	0.138	14.4	10.2
A09	5.85	5.86	0.075	0.072	0.223	0.113**	13.8	15.4
A10	6.00	6.20**	0.111	0.106	0.227	0.190*	17.1	13.6***
A11	6.00	6.16	0.101	0.068*	0.245	0.134**	16.6	12.9**
A12	6.06	5.75***	0.114	0.058***	0.252	0.152***	14.2	11.9**
A13	5.85	5.84	0.100	0.078*	0.218	0.202	16.8	14.9
A14	5.90	5.98	0.162	0.100*	0.236	0.151*	24.6	17.8

| | N_t | | C/N ratio | | Bulk density | | A_p thickness | |
	1991	2001	1991	2001	1991	2001	1991	2001
A01	1.46	1.53	10.3	9.1***	1.46	1.53**	23	27***
A02	1.32	1.45	9.4	8.7***	1.48	1.52	25	26
A03	1.51	1.20	9.9	8.8**	1.42	1.48	26	29
A04	1.66	1.66	10.3	9.3*	1.40	1.48	22	23
A05	1.61	1.55	9.7	9.0**	1.37	1.44	24	23
A06	1.39	1.49	9.8	9.5	1.43	1.53*	19	26***
A07	1.51	1.45	10.7	9.2*	1.38	1.46	17	22
A08	1.51	1.10	9.5	9.3	1.51	1.58	22	28
A09	1.43	1.66	9.7	9.3	1.48	1.47	19	26*
A10	2.12	1.51	9.6	9.0***	1.39	1.46*	24	24
A11	1.67	1.45	9.7	8.9	1.35	1.50**	25	26
A12	1.35	1.26**	10.6	9.4***	1.42	1.54*	21	25*
A13	1.61	1.57	10.6	9.5***	1.45	1.49	22	25*
A14	2.33	1.83	10.6	9.7**	1.46	1.48	22	27**

The levels of significance of 95%, 99% and 99.9% for changes between 1991 and 2001 are marked as *, ** and ***.

significant decrease of pH on different levels of significance, especially for A03 with a decline of 0.6 pH units (Table 4). With the exception of A01 none of these fields was limed but A01 got 30 000 kg ha^{-1} lime in 1994. Because of the high sand content and a low puffer capacity the effect of liming was only temporary. A strong but statistically not significant increase was observed on A04 and A07 as a result of liming in 1998.

The P and K concentrations decreased partially with an exception of A04. The increase of P and K correspond with the balances which are both positive on that field. For the C and N concentrations large differences occurred in comparison to the ICC system. Whereas the most fields showed a decrease of C and N in the ICC system, the C and N concentration increased in some fields of the OF system.

The decrease of C, N, P and K was more considerable on A03, A12 and A14. But this did not always correspond with the balances. For instance A03 and A12 had a surplus in P and K balance, A14 in P balance. A possible reason for the reduction of the nutrient concentrations could be the deepening of the A_p horizon. The A_p horizons on these three fields were larger in 2001 than in 1991. This was caused by deeper ploughing in the years 1993 and 1994. Thereby material of the subsoil with lower nutrient concentrations was mixed in and reduced the concentration of the newly formed A_p horizon.

On the fields A06 and A08 a deepening of A_p horizon took place as well, which caused a decrease of organic C and total N on A08 but did not decrease P and K. Both fields had a large surplus in balance for P and K which compensated the reduction of nutrient concentration by mixing of subsoil material.

Comparison with the soil inventory of 1991 – Grassland and fallow land

In the grassland and fallow land the same trends as those in the arable land could be observed. The nutrient concentrations decreased and the C/N ratio got narrower. Differences in both types of utilisation took place especially for the pH value which increased in the grassland that was caused by liming and decreased in the fallow land. The bulk density increased in the grassland too whereas it decreased in the fallow land.

A strong decrease of C and N concentration was observed in the grassland. However, this was an effect of sample strategy. In grassland it was planned to sample till a depth of 10 cm like in 1991 but this condition was not met during sampling. An increase of A_h horizon (Table 5) occurred and a reduction of nutrient concentration by meddling of subsoil material was done.

The decrease of P and K concentration in fallow land (Table 5) could only partially be explained by plant uptake so far only in a small area with plot experiments the plant material was removed to reduce the nutrient supply. Little amounts of nutrients were lost by leaching and the larger amount of reduction had to be explained by fixing of nutrient in chemical forms which are not exchangeable by the CAL extractant. These are for instance the fixing of K in intermediate layers of clay minerals and the

Table 5 Some soil chemical and soil physical parameters for 1991 and 2001 on grassland and fallow land (P/K_{CAL}, C_{org} and N_t in g kg^{-1}, bulk density in kg dm^{-3}, thickness of A_h horizon in cm).

	Grassland		Fallow land	
	1991	2001	1991	2001
pH	5.59	5.74	5.93	5.68[***]
P_{CAL}	0.066	0.050[*]	0.065	0.048[**]
K_{CAL}	0.203	0.116[**]	0.164	0.147
Organic C	27.7	20.3[***]	18.5	14.6[*]
Total N	2.74	2.18[**]	1.87	1.57
C/N ratio	10.3	9.3[***]	9.8	9.3[***]
Bulk density	1.31	1.35	1.39	1.34
A_h thickness	10	16	21	22

The levels of significance of 95%, 99% and 99.9% for changes between 1991 and 2001 are marked as [*], [**] and [***].

formation of less soluble phosphates from formerly well soluble fertiliser phosphates. The decrease of organic C and total N in the fallow land was considerable and indicated that there was still no balance between supply and oxidation of nutrients. Probably the decrease was caused by the removal of well available carbon species and an insufficient additional supply of fresh plant material. In the first years after changing of management the biomass production was low and the input of fresh organic material into the fallow land was small as well.

The single units in the grassland differed less relative to the change of parameters between 1991 and 2001. The reduction of C and N concentrations were the highest in 'pasture north' and 'pasture west' (Table 6). The comparative low C and N concentration on the 'meadow' and 'pasture west' was caused by the former use of these areas as arable land. The decrease of nutrient concentration is not only raised by the deepening of A_h horizon but also by negative balances. Unfortunately for the grassland plots no balance calculations were done but they should be negative because with the harvest (hay, silage) high amounts of nutrients were removed and fertilisation with farmyard manure was slow and no mineral fertilisers were used.

Effect of reduced tillage on nutrient distribution in the A_p horizon of the ICC system

The reduced tillage without ploughing indicated a strong differentiation within the A_p horizon (Table 7).

Table 6 Some soil chemical and soil physical parameters for 1991 and 2001 in A_h horizon on grassland plots (P/K$_{CAL}$, C$_{org}$ and N$_t$ in g kg^{-1}, bulk density in kg dm^{-3}, thickness of A_h horizon in cm).

	Meadow		Pasture north		Pasture west		Pasture east	
	1991	2001	1991	2001	1991	2001	1991	2001
pH	5.74	5.78	5.27	5.83**	5.63	5.70	5.45	5.62
P$_{CAL}$	0.047	0.042	0.079	0.065	0.077	0.052*	0.087	0.052**
K$_{CAL}$	0.134	0.087***	0.263	0.198	0.261	0.141	0.255	0.130*
C$_{org}$	24.6	19.6*	37.3	20.1**	24.2	16.9*	30.0	23.9*
N$_t$	0.25	0.21	3.53	2.15**	2.41	1.84	2.92	2.58
C/N ratio	10.2	9.4***	10.5	9.3***	10.1	9.2***	10.3	9.2***
Bulk density	1.23	1.32**	1.25	1.39**	1.35	1.41**	1.32	1.29
A$_h$ thickness	10	17	10	15	10	14	10	14

The levels of significance of 95%, 99% and 99.9% for changes between 1991 and 2001 are marked as *, ** and ***.

Table 7 Some soil chemical and soil physical parameters (P/K_{CAL}, C_{org} and N_t in g kg^{-1}) at separate sampling of the topsoil horizon in the ICC system.

	0–10 cm	Rest A_p
pH	5.86[**]	5.95
P_{CAL}	0.068	0.067
K_{CAL}	0.174[***]	0.128
C_{org}	14.4[***]	10.9
N_t	1.56[***]	1.25
C/N ratio	9.3[***]	8.6

The levels of significance of 95%, 99% and 99.9% for changes between 1991 and 2001 are marked as [*], [**] and [***].

With an exception of P_{CAL} all the investigated parameters varied significantly in both layers of the A_p horizon. Whereas the pH value was in 0–10 cm lower as in the soil beneath, the nutrient concentrations were higher in the upper layer. The C/N ratio was considerably closer in the lower layer.

In all fields the pH value was lower in the upper layer (0–10 cm) but a significant difference was observed only on the fields A18, A20 and A21 (Table 8). The P concentrations differed barely the K concentrations were significantly higher in the upper soil with an exception of A17. The C and N concentrations were in all fields significantly higher in the upper layer.

By reduced soil tillage which caused a meddling of soil only in the upper centimetre, the incorporation of nutrients from plant residues and farmyard manure took place only cursorily whereas in the lower soil beneath 10 cm a nutrient supply hardly happened. In comparison with the results from 1991 (Table 3) the K concentration decreased only lightly and the C and N concentrations increased. In the upper 10 cm an enrichment of C and N could be observed in fact. But for the whole A_p horizon this loss of nutrients in the lower layer overbalanced the enrichment in comparison to 1991. The decrease of C and N concentration in the lower layer of A_p horizon and the restriction of C/N ratio, which was especially evident in that soil (8.6 in 2001 in comparison to 9.7 in 1991), was probably caused by degradation of well available C fractions and the lack of fresh plant material from plant residues and farmyard manure.

Table 8 Some soil chemical and soil physical parameters at separate sampling of the topsoil horizon on single fields in the ICC system (P/K_{CAL}, C_{org} and N_t in g kg^{-1}).

	pH		P_{CAL}		K_{CAL}	
	0–10 cm	Rest A_p	0–10 cm	Rest A_p	0–10 cm	Rest A_p
A15	5.97	5.97	0.055	0.049	0.205	0.139***
A16	6.23	6.34	0.068	0.059	0.189	0.139***
A17	5.89	5.98	0.061	0.060	0.157	0.139
A18	5.86	6.03**	0.085	0.090	0.195	0.130***
A20	5.63	5.84**	0.073	0.072	0.163	0.118**
A21	5.39	5.71**	0.023	0.018	0.101	0.061**

	C_{org}		N_t		C/N ratio	
	0–10 cm	Rest A_p	0–10 cm	Rest A_p	0–10 cm	Rest A_p
A15	13.3	9.8***	1.40	1.11***	9.5	8.8**
A16	13.9	10.0***	1.54	1.12***	9.1	8.4***
A17	13.5	11.2***	1.52	1.33***	8.8	8.4**
A18	14.8	11.0***	1.60	1.25***	9.2	8.7***
A20	14.7	10.6***	1.54	1.12***	9.5	9.0***
A21	16.1	12.0***	1.71	1.35***	9.5	8.9**

The levels of significance of 95%, 99% and 99.9% for changes between 1991 and 2001 are marked as *, ** and ***. Plot A19 was not analysed because the sampling was conducted in two depths only at a few points.

Geostatistical analysis

The intention of geostatistical analysis was the transformation of punctual data into spatial information. Therefore, in the following figures only the results of kriging were presented.

pH value

On the arable land the pH value varied only a little and was mostly in the range of 5.5 and 6.5 in the ICC System as well as in the OF system (Figure 3). In grassland and fallow land the variation was larger but the most values were within a range of 5.5 and 6.5 too. For instance on the 'meadow' and on 'pasture east' pH values at about 5 were observed and on W51 pH values were less than 5. The lowest pH values were recorded in the fallow land close to A21.

The changes in pH are illustrated in Figure 4. In the arable land in the ICC system there was only a little change of pH values. On A17 and some

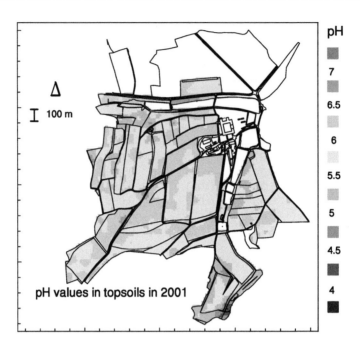

Figure 3 pH values in topsoils in 2001. (For colour version of this figure, please see page 431 in colour plate section.)

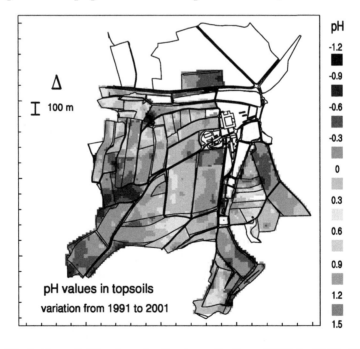

Figure 4 Variation of pH values in the topsoils from 1991 to 2001. (For colour version of this figure, please see page 431 in colour plate section.)

parts of A15 and A21 the pH values increased lightly whereas on A18 and A20 a decrease of pH was observed on the larger parts of the fields. In the arable land of the OF system a heavy decrease of pH was observed on the fields A01, A03 and A02 though at least A01 was limed between the soil inventories. An increase of pH was found on A07 and the northern part of A04. The largest change in pH was observed in 'pasture north' with an increase of about 1.5 pH units caused by liming. The highest decrease was found on the fallow land with a reduction of about 1.5 pH units.

The pH values on the experimental station were predominantly within the desired range. In Figure 5, the lower limit of pH values is presented which is required according to the management and the existing soils. In Figure 6, the pH values are given which were found in 2001. A necessity of liming was only asserted for parts of the fields A17 and A18 and especially for A12, A20 and A21 (green areas in Figure 6 on these fields). Liming was not necessary in the grassland.

Organic carbon

The organic carbon concentration on the agricultural area of the ICC system was mostly within a range from 1.0% to 1.5% (Figure 7). In small areas of A15–A17 the concentration was less than 1.0%, on A19–A21 partially higher than 1.5%. In the OF system the C concentration in the agricultural area was within the range from 1.0% to 1.5% as well but inside of

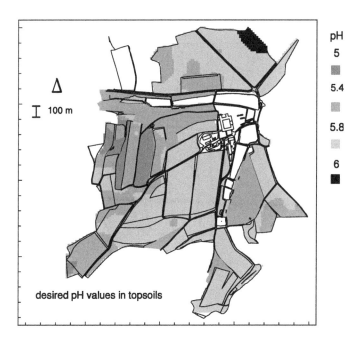

Figure 5 Lower limit of desired pH values in 2001. (For colour version of this figure, please see page 432 in colour plate section.)

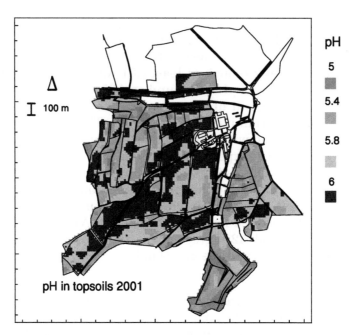

Figure 6 Lower limit of measured pH values in 2001. (For colour version of
this figure, please see page 432 in colour plate section.)

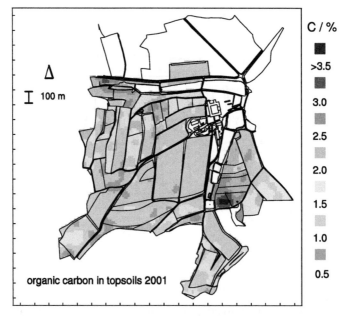

Figure 7 Organic carbon contents in the topsoils in 2001. (For colour version
of this figure, please see page 433 in colour plate section.)

single pots the variation was higher (for example A01, A12). The higher C concentration in the northern part of A09 was caused by the earlier use of this area as grassland.

The C concentration in the fallow land was within the same range as the arable land, the concentration in meadow and pasture was more than 1.5%. The 'new' grassland plots (W02–05, W19/29) which were used as arable land before 1992, showed much lower C concentrations than the 'old' grassland plots (W06–W12, W18, pasture north and east). Strong variations between 1.5% and more than 3.5% were observed in the 'pasture east'.

The reduction of organic C was especially high in the grassland (Figure 8) but this was caused by the sampling (see *grassland and fallow land*). An increase of organic C was observed only on the 'new' grassland plot which was generated by the change of utilisation from arable land to grassland. In the fallow land the concentration of organic C decreased too (for example in the south of A19 and around A19 and A20). In the agricultural plots A15–A18 and A21 a considerable decrease of organic C was observed whereas on A19 and A20 partially a slow increase could be registered. On the northern part of A18 organic C and all other nutrients decreased considerably. This reduction was limited to a former hop field. Probably the change of the management resulted in an increased turnover of organic material after the end of hop production. Investigations of Filser

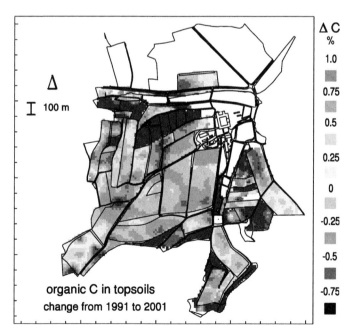

Figure 8 Changes of organic C concentrations in the topsoils from 1991 to 2001. (For colour version of this figure, please see page 433 in colour plate section.)

(1998) and Filser et al. (1997) showed an increased microbial activity and an increase or recurrence of earthworms on former hop fields.

The arable land in the OF system also showed a decrease of organic C. On A01 and A13 partially an increase could be observed. The considerable reduction in the northern part of A09 was caused by the degradation of organic substance after the former grassland use.

The change of nutrient reserves was calculated for the arable land taking under consideration the thickness of A_p horizon, bulk density and coarse material. A higher inaccuracy resulted from the calculation with several parameters and therefore no exact data could be shown. The calculations showed trends which expressed the change of nutrient reserves.

On the fields A15–A19 a reduction of reserves was observed consistently whereas on A20 and A21 partially an increase could be registered (Figure 9). In the OF system the situation was slightly different. On some fields (A02, A09–A11) organic C decreased on the whole field whereas A05 and A07/08 showed an increase on the whole field. In a few fields (A01, A04, A06) areas with decrease and increase of organic C alternated.

Total nitrogen (N_t)

The allocation of total nitrogen was very similar to the allocation of organic C (Figure 10). The change of nitrogen concentration also showed the same pattern as organic C so that no further discussion is necessary (Figure 11).

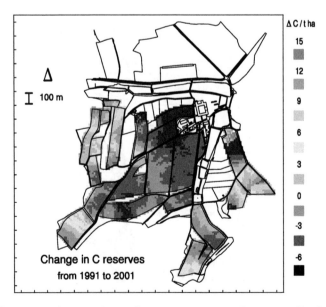

Figure 9 Changes of organic carbon reserves in the topsoils from 1991 to 2001. (For colour version of this figure, please see page 434 in colour plate section.)

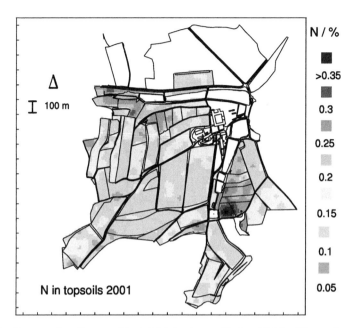

Figure 10 Concentrations of total N in topsoils in 2001. (For colour version of this figure, please see page 434 in colour plate section.)

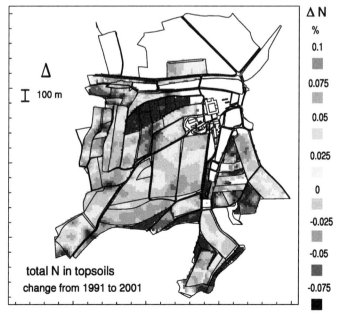

Figure 11 Changes of total N in topsoils from 1991 to 2001. (For colour version of this figure, please see page 435 in colour plate section.)

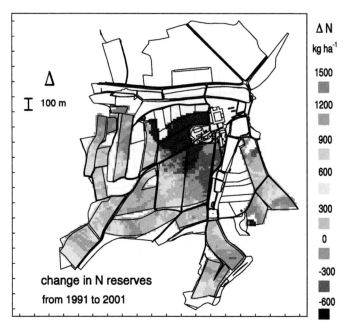

Figure 12 Changes of N reserves in topsoils from 1991 to 2001. (For colour
version of this figure, please see page 435 in colour plate section.)

Still the changes in N reserves differed in comparison to organic carbon.
In the ICC system only on A18 a decrease of supply was noticed whereas
in the other fields decrease and increase of total N alternated (Figure 12).
In the fields A15, A19 and A21 mostly an increase of N supply was
observed. Considerable differences took place in the OF system. The fields
A10 and A11 showed a decrease of total N on the whole field whereas on
A01–A08 and A13/A14 an increase of total N occurred.

C/N ratio

The C/N ratio varied between 8:1 and 9.5:1 on the most fields. Especially
on the fields A02 and A16–18 the C/N ratio was close to values from 8:1
and 9:1 (Figure 13). The arable land in the ICC system showed a slightly
narrower C/N ratio than the fields in the OF system. A differentiation
between the types of utilisation (arable land, grassland, fallow land) was
not possible. C/N ratios larger than 10:1 took place sporadically in the border area of the fallow land.

Larger differences took place when the change of C/N ratio from 1991
to 2001 was considered (Figure 14). All plots of the arable land in the ICC
system with an exception of A15 showed a considerably narrower C/N ratio
than 1991. An alteration between increase and decrease of C/N ratio was

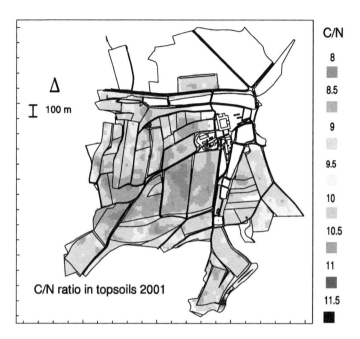

Figure 13 C/N ratio in topsoils in 2001. (For colour version of this figure, please see page 436 in colour plate section.)

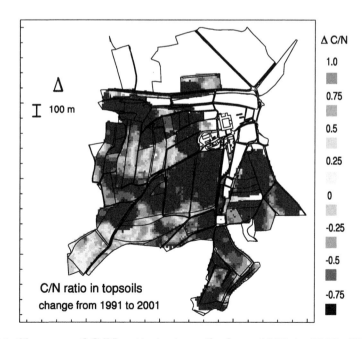

Figure 14 Changes of C/N ratio in topsoils from 1991 to 2001. (For colour version of this figure, please see page 436 in colour plate section.)

observed on A15. In the OF system the C/N ratio got narrower on the most fields as well. But on some fields (A01, parts of A02 and A09) widening of C/N ratio took place. On the grassland the C/N ratio got narrower as well with an exception of W19/20. On these fields the widening was probably caused by the change of management from arable land to grassland.

Potassium (K_{CAL})

The amount of plant available potassium was within the range of optimal to very high in the predominant part of the fields. The K concentrations (Figure 15) were transformed into the K supply classes taking into consideration the different values depending on soil texture (Figure 16). All arable fields with an exception of A21 were located within the classes C and D which represented supply from optimal to high. The field A21 was located predominately in the class B and showed therefore a necessity of fertilisation. The reason for this exception was that A21 did not belong to the experimental station but was leased after the foundation of the research association.

The arable land in the OF system was better supplied than the ICC system and all fields were located in the classes C to E (very high supply). The fallow land was located within the supply classes C and E as well. Only in the grassland the supply was slightly lower. Some areas of the

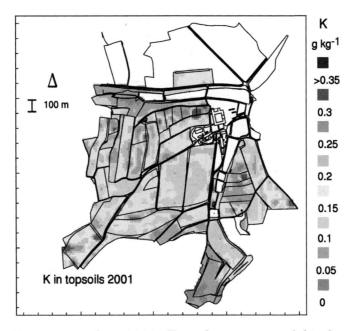

Figure 15 K_{CAL} in topsoils in 2001. (For colour version of this figure, please see page 437 in colour plate section.)

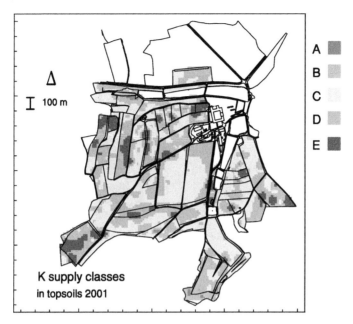

A
B
C
D
E

K supply classes
in topsoils 2001

Figure 16 K supply classes in topsoils in 2001. (For colour version of this figure, please see page 437 in colour plate section.)

pasture and the meadows were located in the supply classes A and B and had therefore K fertiliser requirements. Especially, the meadows needed fertilisers as high amounts of nutrients were removed by the hay and silage. Without fertilisation the productivity decreases.

The decrease of K concentrations on arable land of the OF system was less in comparison to 1991 than the decrease in the ICC system (Figure 17). Nethertheless, the most plots showed a considerable decrease of K concentrations. The exceptions were the fields A01 and A04 where an increase was observed on some areas. In the grassland and the fallow land it was noticed as well but the reasons were different. In the grassland the reduction of K supply was caused by negative balances and the sample strategy (see 'grassland and fallow land'). In the fallow land where no nutrients were removed by harvest the reduction could be caused by leaching of K into the subsoil and groundwater or by fixing of nutrient in chemical forms which are not exchangeable by the CAL extractant (see 'grassland and fallow land').

The change of reserves was calculated for K in the arable land (Figure 18). The reserves decreased in all fields of the ICC system with an exception of A21. The fields A02/03 and A10–12 of the OF system showed a considerable decrease of K supply whilst on the fields A01 and A04–A08 the supply increased in comparison to 1991.

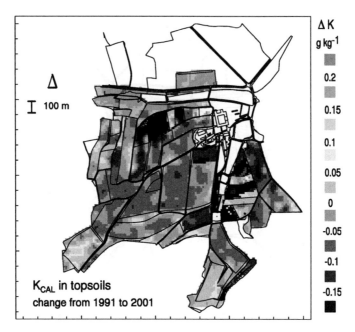

Figure 17 Changes of K_{CAL} in topsoils from 1991 to 2001. (For colour version of this figure, please see page 438 in colour plate section.)

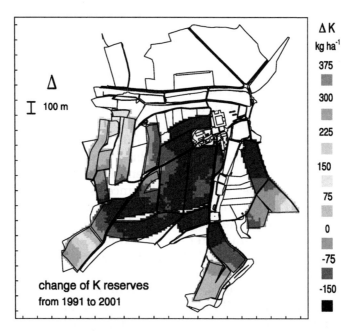

Figure 18 Changes of K reserves in topsoils from 1991 to 2001. (For colour version of this figure, please see page 438 in colour plate section.)

Phosphorus (P_{CAL})

For the concentration of plant available phosphorus (P_{CAL}) the results were similar to the results for K (Figure 19). However, the level of supply was slightly lower than the K supply. For phosphorus the concentration was transformed into supply classes as well as for K. (Figure 20).

Therefore, similar spatial distributions were observed than for K. The field A21 differed from the other fields in the ICC system as well because the plot showed a supply within the classes A and B (very low and low) and indicated a considerable necessity of fertilisation. This was asserted by a recorded yield decrease of maize. Most of the other plots of the ICC system were located in supply class C whereas A19 and the northern part of A18 were located in class E. This was caused by the former use of these plots as hop fields because hop got a very high P fertilisation.

In the northern part of A06 the supply was also very high because this was a former hop field with high P fertilisation. The other plots in the OF system were within the class C with exception of A05, A08 and parts of A03. These were located in class B and required fertilisation. In the grassland the meadows and parts of the 'pasture north' and the 'pasture east' were less supplied.

The P concentration was reduced between both inventories on almost the whole area of the farm (Figure 21). The most considerable was the

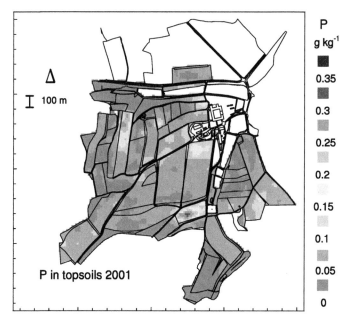

Figure 19 P_{CAL} in topsoils in 2001. (For colour version of this figure, please see page 439 in colour plate section.)

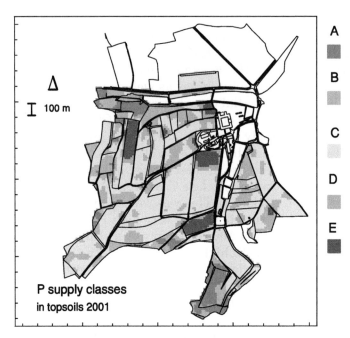

Figure 20 P supply classes in topsoils in 2001. (For colour version of this figure, please see page 439 in colour plate section.)

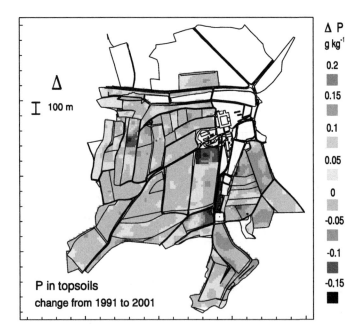

Figure 21 Changes of P_{CAL} in topsoils from 1991 to 2001. (For colour version of this figure, please see page 440 in colour plate section.)

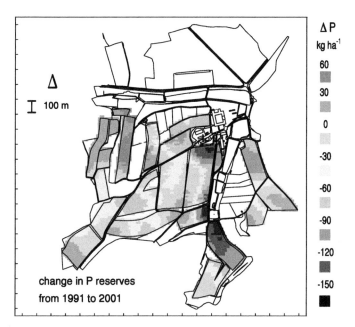

Figure 22 Changes of P reserves in topsoils from 1991 to 2001. (For colour version of this figure, please see page 440 in colour plate section.)

decrease on the former hop plots (A19 and parts of A06 and A18) and A12, A14 and A20.

The P-reserves decreased on all plots in the ICC system (Figure 22). This decrease was especially designated in the field A20 and the former hop field in the north of A18. The fields A04–A08 in the OF system showed an increase of P reserves on a low level whereas on the other fields a slight decrease of P was observed which was less expressed than in the ICC system.

4.3.4 Discussion

On the farm scale the results of the soil inventory 2001 corresponded to the results of the balances. On the level of the fields the results of balance calculations and those of the soil inventory differed at least for P in the ICC system (Table 9).

It has to be mentioned that the fields differed in their pedological characteristics. So the changes of CAL available P-reserves were not only influenced by input and output of nutrients with organic fertiliser and harvested products but also by the contribution of subsoil to plant nutrition, the supply of non-exchangeable reserves and immobilisation of easily soluble forms of P. This was especially designated for the former hop

Table 9 Balances for P on single plots in the ICC system calculated with REPRO (1993–2000) and calculated by changes of nutrient supply (1991–2001) (in kg ha^{-1}).

	P balance (REPRO)	P balance changes of nutrient supply
A15	−40	−63
A16	−120	−72
A17	−80	−50
A18	−134	−158
A19	−92	−169
A20	+8	−224
A21	−64	−18

fields where the calculated changes of nutrients could not be accounted for by the removal of the harvested products. The decision made in 1991 to refuse of mineral P and K fertilisation was approved for the arable land with exception of A21. The high reserves which existed in 1991 could be reduced. Although the balances for P and K were mostly negative in both farms, the arable land was still within optimal range for plant nutrition. For the grassland no balances existed but the analysis of the plot card index for the grassland indicated that considerable amounts of P and K were removed from the grassland with the harvest and therefore the reduction of P and K in the grassland could be accounted for.

The changes in organic carbon corresponded to the results of the balance calculations. With the model REPRO a level of humus supply of 95% was calculated in the ICC system and of more than 100% in the OF system (Gerl; personal communication). Nevertheless, the decrease of C reserves in the arable land of the ICC system was unexpected because so far it was assumed that an increase of the carbon concentration could be realised by the intensive intercropping in the ICC system and therefore the soils could serve as a carbon sink. The factors possibly responsible for the reduction of C reserves have to be investigated. It is possible that an increased activity of microbial biomass and climatic influences (mild winters during the period of investigations) played an important role.

Considerable differences for nitrogen occur between the balance calculations and the soil inventory in the ICC system. An enrichment of about 400 kg N ha^{-1} was expected by the balances but the soil inventory showed a slight reduction of N reserves. This indicated that the expected function of the soils as a sink for N was not responsible for the lack of 40 kg N ha^{-1} yr^{-1}. The research has to be done to look for other than the investigated losses because the results of the inventory showed that these N amounts did not remain within the topsoils. For example it has to be examined if a loss of N as N_2 can be derivated. As another possible way of losses the leaching of N as NO_3 via macropores has to be discussed. Investigations of Honisch and Klotz (1999) indicated an influence of macropores for the transport of soluble substances and studies of the Fraunhofer Institute for Molecular Biology and Applied Ecology (IME) showed that in soils with a high silt content preferential flow via macropores was the most important way of loss for pesticides.

In the arable land of the ICC system a differentiation started within the old A_p horizon caused by reduced tillage. In the upper 10 cm, which were more influenced by tillage, the C and N concentrations did not change or showed a slight increase. In the lower part of the old A_p horizon a considerable decrease of nutrient concentration was observed.

Acknowledgements

The scientific activities of the *FAM Munich Research Network on Agroecosystems* were financially supported by the German Federal Ministry of Education and Research (BMBF 0339370). Overhead costs of the Research Station Scheyern were funded by the Bavarian State Ministry for Science, Research and the Arts.

References

Bortz J, Lienert GA, Boehnke K, 1990. Verteilungsfreie Methoden in der Biostatistik, Springer-Verlag, Berlin, 939 pp.

Dutter R, 1985. Geostatistik, Eine Einführung mit Anwendungen. Teubner Verlag, Stuttgart, 176 pp.

Entz T, Chang C, 1991. Evaluation of soil sampling schemes for geostatistical analysis: A case study for soil bulk density. Canadian Journal of Soil Science 71, 165–176.

Filser J, 1998. Bodenorganismen. In: Filser J (Ed.), FAM Schlussbericht 1993–1997, FAM-Bericht 28, 67–76.

Filser J, Mommertz S, Angermeyr L, Jell B, Lang A, Mebes H, 1997. Steuerung von Nährstoffumsetzungen durch Bodentiere – Beitrag zur Minimierung der Belastung von Wässern und der Atmosphäre durch C- und N-Verbindungen. FAM Jahresbericht 1996, FAM-Bericht 13, 83–94.

Hartge KH, Horn R, 1989. Die physikalische Untersuchung von Böden. 2. Aufl., Enke Verlag, Stuttgart, 175 pp.

Honisch M, Klotz D, 1999. Numerische Simulation des Wasserflusses in Lysimetern. GSF-Bericht 1, 67–72.

Kaemmerer A, 2000. Raum-Zeit-Variabilität von Aggregatstabilität und Bodenrauhigkeit, PhD thesis, TU München, Shaker Verlag, Aachen, ISBN 3-8265-7331-5, FAM-Bericht 40, 207.

Matthes U, Gerl G, Kainz M, Gutser R, 2002. Stickstoffverluste durch ressourcenschonende Bewirtschaftung – dargestellt am Beispiel des Versuchsgutes Scheyern. VDLUFA Schriftenreihe 57.

Sinowski W, 1995. Die dreidimensionale Variabilität von Bodeneigenschaften – Ausmaß, Ursachen und Interpolation. PhD thesis, TU München, Shaker Verlag, Aachen, ISBN 3-8265-0994-3, FAM-Bericht 7, 158.

VDLUFA, 1991. Methodenbuch, Band 1: Die Untersuchung der Böden. VDLUFA-Verlag, Darmstadt.

Webster R, Oliver MA, 1990. Statistical Methods in Soil and Land Resource Survey. Oxford University Press, New York, 316 pp.

Weinfurtner K, Matthes U, Dreher P, Gerl G, 2002. Nährstoffbilanzierung auf der Versuchsstation Klostergut Scheyern mithilfe des Betriebsbilanzierungsmodells REPRO. VDLUFA Schriftenreihe 57.

Outlook

Needs, for a FAM, new tools and issues.

The *FAM Munich Research Network on Agroecosystems (Forschungsver-bund Agrarökosysteme München)* was started many years ago under a special and dramatic situation in view of human made changes concerning the rural environment as well as the contamination of food with pesticide residues and nitrate, food that possibly for the first time in human history had been of a highly equilibrated nutritional quality. Agricultural practice reached in the second parts of the 20th century at first time an insuring of yields by the introduction of minerals fertilisers and of synthetic pesticides together with mechanical developments. With a delay of few years only, the negative consequences of this maximisation of yields for at least our environments and some food become evident. The loading of soil water with nitrate and pesticides leads to eutrophication and contamination of limnic ecosystems, to contamination of groundwater and eutrophication of water systems. Atmosphere received high fluxes of nitrous oxide and methane as the agricultural soils were not longer limited with nitrogen and had accumulated high concentrations of organic matter, both trace gases being of high relevance for atmosphere chemistry and for global climate. Furthermore, the redesign of landscapes for the use of powerful machines and techniques with high performance and hence higher economical efficiency of human and monetary inputs, resulted in a loss of landscape issued and relevant values: loss of diversity and biodiversity, loss of recreation, esthetical and ethical value. Extended fields without hedgerows or other structures were subjected to losses by erosion, to wind erosion as well as to water erosion depending on the type of landscape and climate.

Perspectives for Agroecosystem Management
Edited by P. Schröder, J. Pfadenhauer and J.C. Munch

It was obvious, that the understanding of matter fluxes from fields into adjacent or far away ecosystems (to rivers as feeder for marine systems) is not feasible on the scale of single fields. Matter fluxes are in parts very slow at the scale of soils and landscapes. Restoration and dissemination of biodiversity in local habitats and at landscape scale also needs research on the long time scale as well as on the dimension of neighbouring habitats. Sustainable agriculture based as far as possible on internal resources of the field and the farm is not to be based on monocultures. Instead it needs the integration of multiple effects of a crop rotation, a reasonable rotation comprising five or more crops as well as intercrop with special functions in soil protection against erosion, in nitrate uptake and nitrogen conservation, in biological N_2-fixation. Therefore, the interdisciplinary research on the project FAM was conceived to have the duration of 15 years and it was realised in several project networks with intermediate evaluations over this time.

The investigations in the frame of farming were based on the typical land use management of the region in a way to possibly transfer the results into agricultural practice. Also economic investigations were established that included environmental and energy balances and the study of societal needs and transfers.

The FAM realised a unique field experiment with an unusual network of scientists. Farming systems were established and fields as well as margins, hedgerows and functional fallows designed in the landscape, based on the actual knowledge and experience in agriculture and agronomy in order to minimise environmental constraints and to support diversity. These farms had to be self supporting economically, being near to the local agricultural practice and at the same time progressive and in evolution toward possibilities of optimisation of both yields and environmental services. Agronomists investigated these farming systems from soil functioning up to farm economy. Balances of main nutrients were investigated for soils and farms, up to the consumers. Fluxes and losses of compounds including exports of nutrients out of the systems to the consumers were analysed and controlled for the production at fields and farms in order to understand yield developments and long time soil fertility for the two farming systems and both landscape sections. New methods of soil, field and landscape analysis were established, so the use of GPS and geo referencing for yield mapping, the use of non- destructive techniques, remote sensing and proximal techniques for analysis of soil units and characterisation of landscapes, characterisation by different information systems. However, the sampling grid established at the beginning of the project was still active for a repeated monitoring of changes. Site-specific farming was established, as possible now by the new quality of information on soil and plant parameters. These new promising techniques lead to a new form of agriculture often called precision agriculture which now has worldwide application. Agroecosystem models, indispensable tools to integrate for landscapes sections, were adapted and adopted as a new basis to develop decision support systems. Colleagues from the ecological sciences investigate the effects of these two newly established farming systems in a heterogeneous and varied landscape section on the vegetation variety and the dissemination as well as on the chosen parts of fauna (micro to macro fauna). The scales of investigations were the fields as well as the whole landscape section, the site Scheyern offering the possibility to study a connected agricultural unit as landscape section and as part of a small water catchment.

The land use management implemented by the *FAM Munich Research Network on Agroecosystems* lead to conservation of yield potentials and in total to higher yields on both farms during the investigation period, despite a reduction of the inputs of nutrients. Loading of environmental compartments was reduced for some aspects and better elucidated at several scales. Heterogeneity in fields and soils, first time analysed with the given precision, was recognised as one of the causes of losses of nutrients and imprecision of some model previsions. The organic farming system, a system with less yield compared to the integrated farm, but with the regional yield level of low input farming systems without application of synthetic chemicals showed lower fluxes and losses of nutrients. However, these lower fluxes and losses were similar to those of the integrated farming system when compared by harvested product amounts per field surface. It is concluded that precision agriculture should be the tool to increase nitrogen efficiency in fields and reduce environmental loading.

On the other site, FAM as a model to integrated resource protection and nature conservation together with agricultural land use showed the possibilities to increase biodiversity in fields and landscape sections in a landscape devoted primarily to production of goods for humans. Both farming systems with their margins, hedgerows and set aside sections had positive effects for species diversity, so for vegetation and for fauna (global microbial biomass and activity, microbial populations chosen for their impacts on nutrients transformation or biocontrol of pathogens or degradation of pesticides, some species of soil mesofauna and of macrofauna being chosen as models, so Collembola, earth worms and spiders, latest having a function as predator for pest organisms, and chosen as indicators). Here also, new methodological developments were achieved, especially by implementation of molecular tools to understand diversity and functioning of biota. However, limits of the richness and dissemination of species of vegetation and fauna were also identified, the dimensions and distribution of margins being the limiting dimension. General effects of field sizes were also investigated at the farm level, showing the economic impact of little fields for the farm and the feasibility of this kind of diversity in the landscape, but also the necessity of financial support for agriculture keeping more biodiversity.

This interaction and cooperation of agronomists together with ecologists was a special event in FAM and the promising way to understand the agricultural landscape in their various functions and to integrate some production services together with environmental services. FAM was contributing not only to a gain of performing knowledge on the agricultural landscape and to the development of performing farming systems, but also to new promising fields of integrated use of the landscape for production as well as for environmental services and the conservation of biodiversity.

Knowledge from FAM as a research basis for actual developments in the rural landscape.

Despite the knowledge about the necessity of soils with high yield potentials to provide the future world population with food in sufficient amount and quality and by minimising environmental constraints, agricultural soils are still object of active deterioration by humans for their infrastructural purposes. However, the most important global cause of losses of soils is still the erosion. This erosion may only be partially controlled by adequate cropping systems and efforts toward soil stability. Climate, weather events and exposition are main drivers in some landscapes.

However, a new, still actual challenge for agricultural soils is given by actual global changes in climate and atmosphere chemistry as well as in society needs.

The climate changes will lead to a different distribution of precipitation and to water stress in soils of nearly all northern European countries. Agricultural soil management has to be adapted based on the actually extended knowledge on soil system functioning and soil properties at the landscape scale.

Agriculture will provide in future, in a process that is still beginning, the society with energy and possibly with raw materials for industry by the production of plant biomass for transformation processes (into methane by biogas fermentation or into fuels so by conversion of cellulose) or extraction of compounds. Production of plants linked not to protein content but to carbohydrates and furthermore to maximum biomass amounts will cause important changes in nutrient cycling as well as in water regimes, leading to additional water stress. The return of residues from plant biomass treatment to the production areas is to be assured and the recycling of nutrients from this new quality of input and manure to be linked to plant need, to avoid enhanced transfer of nitrogenous compounds into adjacent environmental compartments and to avoid further climate change, in particular, by enhanced greenhouse gas emissions from soils.

Agroecosystems as well as the agricultural landscape are subject to new developments. To react adequately, we need to understand their drivers in the ecosystems and landscapes, the consequences of ecosystem changes at soil and water, field, farm and watershed level and rural landscape. Management systems are to be and may be adapted at various scales.

The knowledge issue from the FAM-Project together with the high level information on soils and farm scale and on the dynamics of changes induced by the farming systems at the site Scheyern are an excellent basis for the beginning investigation of the effects of the new constraints for soils and society. So, *FAM Munich Research Network on Agroecosystems* leads not only to performing knowledge and immediate transfer to agricultural practice. After the end of the project we dispose of a high informative and technical platform to counteract the actual challenges by adapted field and farm level research, accompanied by the translation of information into the practical issues.

The scientific activities of the *FAM Munich Research Network on Agroecosystems* were financially supported by the German Federal Ministry of Education and Research (BMBF 0339370). Overhead costs of the Research Station Scheyern are funded by the Bavarian State Ministry for Science, Research and the Arts.

J.C. Munch

Munich, in January 2007

Index

Colour Plate Section

Plate 1 Aerial view of Research Station Scheyern. (Please see page 8 of this volume.)

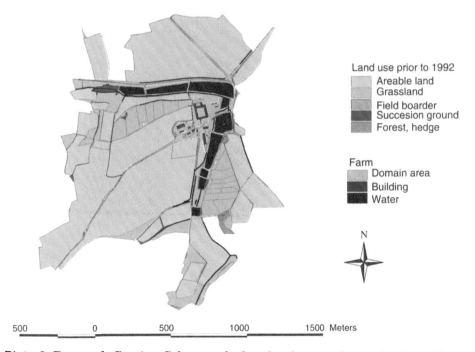

Land use prior to 1992
  Areable land
  Grassland
  Field boarder
  Succesion ground
  Forest, hedge

Farm
  Domain area
  Building
  Water

N

500   0   500   1000   1500 Meters

Plate 2 Research Station Scheyern before land use redesign in 1992 (from Schröder et al., 2002). (Please see page 11 of this volume.)

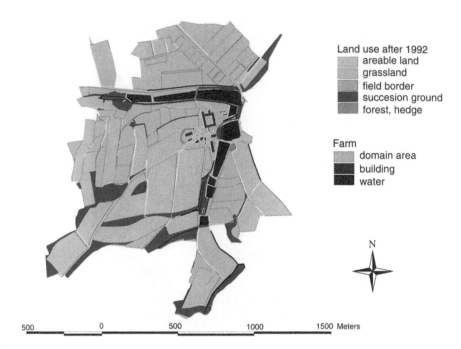

Plate 3 Research Station Scheyern after land use redesign in 1992 (from
         Schröder et al., 2002). (Please see page 12 of this volume.)

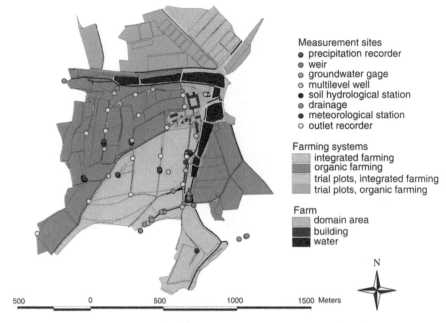

Plate 4 Research Station Scheyern after 1992. Management systems and
         permanent monitoring equipment are shown (from Schröder
         et al., 2002). (Please see page 12 of this volume.)

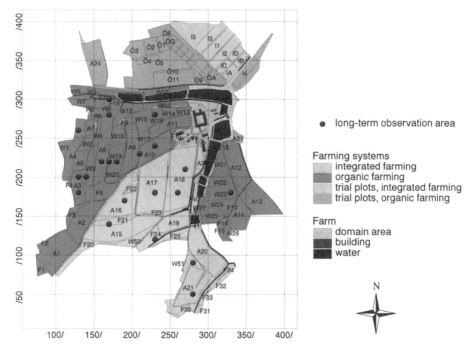

Plate 5 Research Station Scheyern after 1992. Management systems, field numbers, grid system and long-term observation areas are shown (from Schröder et al., 2002). (Please see page 13 of this volume.)

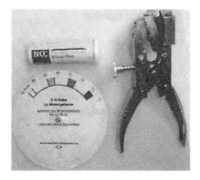

Plate 6 Nitrate-N-test: gripper and test strip (by courtesy of Sächsische Landesan-stalt für Landwirtschaft, Leipzig, Germany). (Please see page 131 of this volume.)

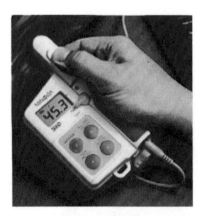

Plate 7 SPAD-Meter (by courtesy of YARA GmbH & Co. KG, Dülmen, Germany). (Please see page 132 of this volume.)

Plate 8 LAI-2000 Plant canopy Analyzer (by courtesy of LI-COR Biosciences GmbH, Bad Homburg, Germany). (Please see page 134 of this volume.)

Plate 9 Crop-Meter (Ehlert, 2004; by courtesy of agrocom. GmbH & Co. Agrar-system KG, Bielefeld, Germany. (Please see page 134 of this volume.)

Plate 10 N-Sensor on tractor roof (by courtesy of YARA GmbH & Co. KG, Dülmen, Germany). (Please see page 138 of this volume.)

Plate 11 Laser-induced chlorophyll fluorescence sensor Planto N-Sensor (Planto GmbH, Leipzig, Germany, by courtesy of Technical University of Munich). (Please see page 141 of this volume.)

Plate 12 Research farm "Klostergut Scheyern": field boundaries (arable land as A1, A2, etc.) and soil raster (for EM38: blue boundary = test area; yellow boundary = unit 1, red boundary = unit 2, green = unit 3). (Please see page 353 of this volume.)

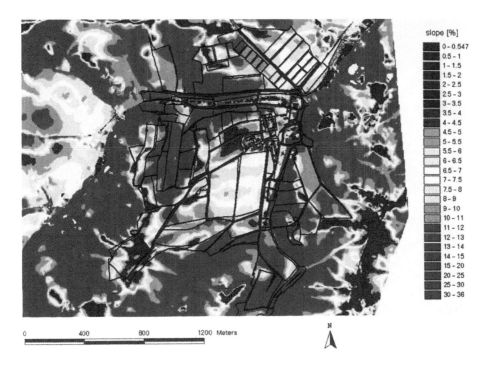

Plate 13 Terrain Analysis with SAGA on basis of the DEM 5; slope (%).
(Please see page 354 of this volume.)

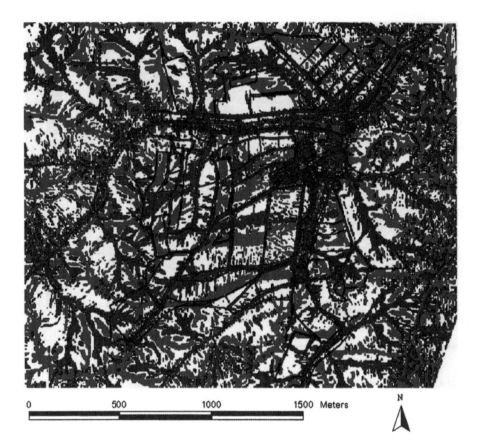

Plate 14 Terrain analysis with SAGA on basis of the DEM 5; divergence
        (red) and convergence areas (blue), linear elements (white).
        (Please see page 355 of this volume.)

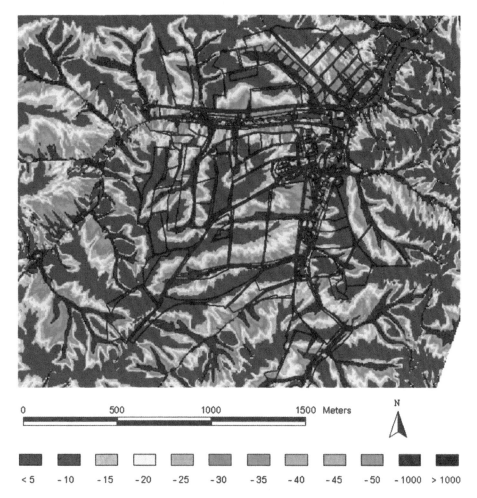

< 5	- 10	- 15	- 20	- 25	- 30	- 35	- 40	- 45	- 50	- 1000	> 1000

Plate 15 Terrain analysis with SAGA on basis of the DEM 5; local catchment area (no. grid cells of 25 m^2). (Please see page 356 of this volume.)

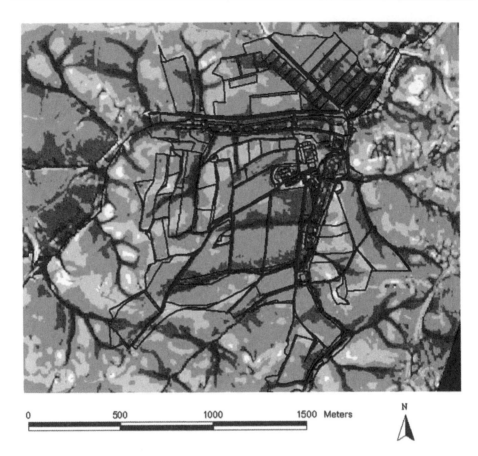

0　　　　　500　　　　　1000　　　　　1500　Meters

N

Plate 16 Terrain analysis with SAGA on basis of the DEM 5; topographic wetness index, defined as TWI = ln [local catchment area / tan (slope)], values as gradient (blue) from 4.3 to 23.5. (Please see page 357 of this volume.)

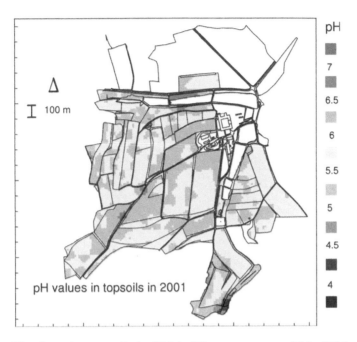

Plate 17 pH values in topsoils in 2001. (Please see page 390 of this volume.)

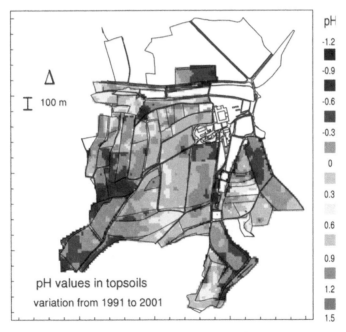

Plate 18 Variation of pH values in the topsoils from 1991 to 2001. (Please see page 390 of this volume.)

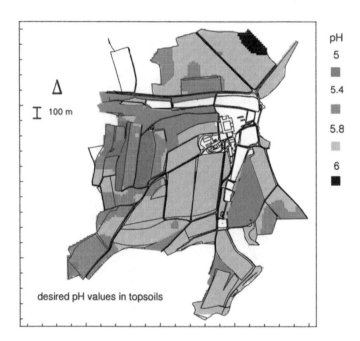

Plate 19 Lower limit of desired pH values in 2001. (Please see page 391 of this volume.)

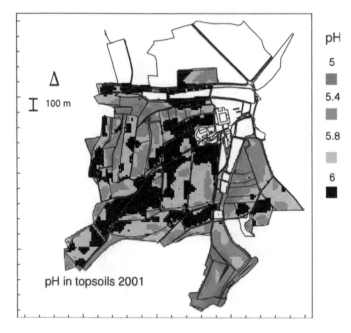

Plate 20 Lower limits of measured pH values in 2001. (Please see page 392 of this volume.)

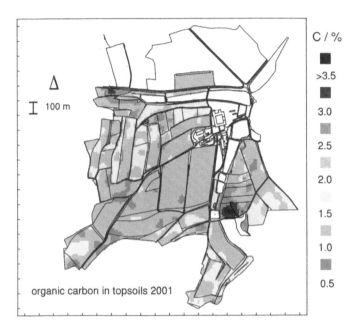

Plate 21 Organic carbon contents in the topsoils in 2001. (Please see page
392 of this volume.)

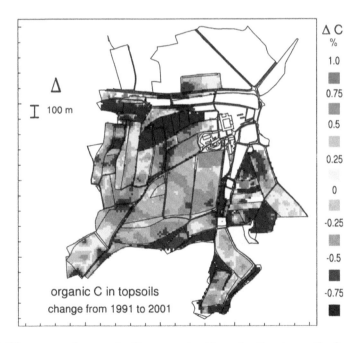

Plate 22 Changes of organic C concentrations in the topsoils from 1991 to
2001. (Please see page 393 of this volume.)

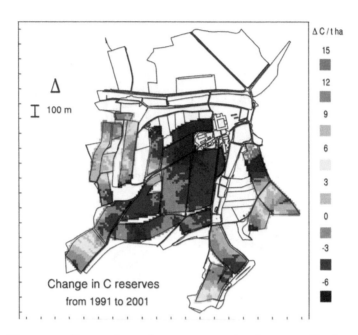

Plate 23 Changes of organic carbon reserves in the topsoils from 1991 to
2001. (Please see page 394 of this volume.)

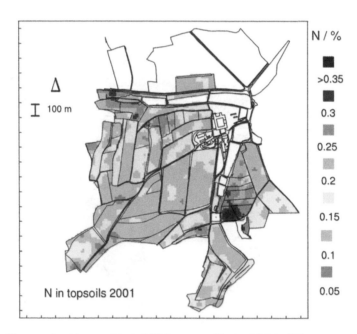

Plate 24 Concentrations of total N in topsoils in 2001. (Please see page 395
of this volume.)

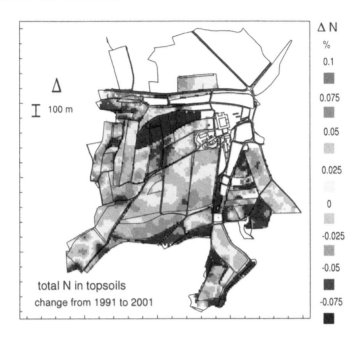

Plate 25 Changes of total N in topsoils from 1991 to 2001. (Please see page
395 of this volume.)

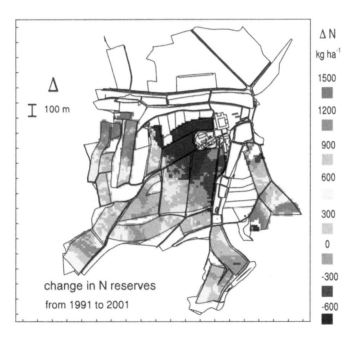

Plate 26 Changes of N reserves in topsoils from 1991 to 2001. (Please see
page 396 of this volume.)

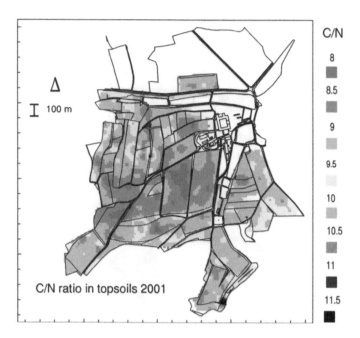

Plate 27 C/N ratio in topsoils in 2001. (Please see page 397 of this volume.)

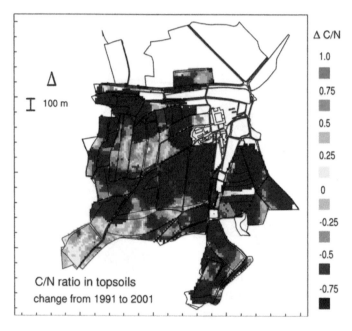

Plate 28 Changes of C/N ratio in topsoils from 1991 to 2001. (Please see
          page 397 of this volume.)

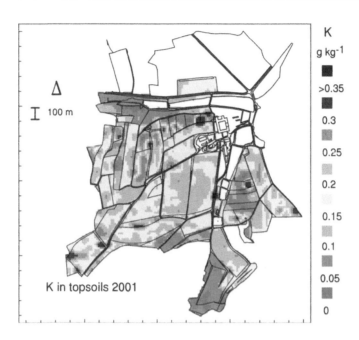

Plate 29 K_{CAL} in topsoils in 2001. (Please see page 398 of this volume.)

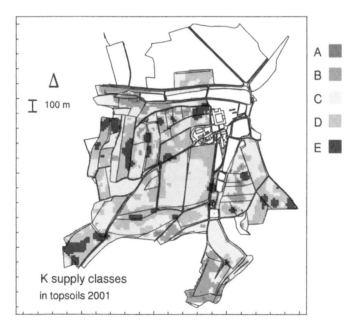

Plate 30 K supply classes in topsoils in 2001. (Please see page 399 of this volume.)

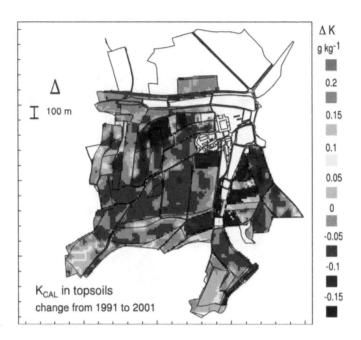

Plate 31 Changes of K_{CAL} in topsoils from 1991 to 2001. (Please see page 400 of this volume.)

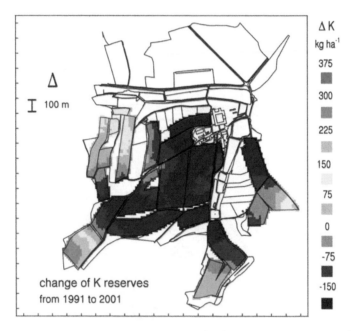

Plate 32 Changes of K reserves in topsoils from 1991 to 2001. (Please see page 400 of this volume.)

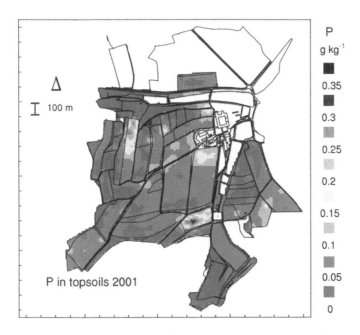

Plate 33 P_{CAL} in topsoils in 2001. (Please see page 401 of this volume.)

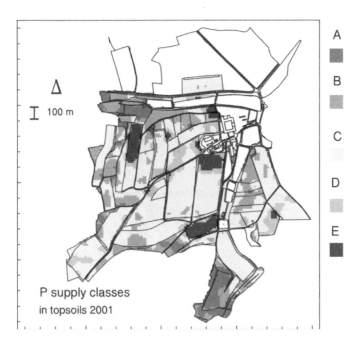

Plate 34 P supply classes in topsoils in 2001. (Please see page 402 of this volume.)

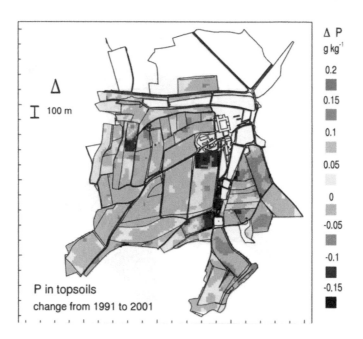

Plate 35 Changes of P_{CAL} in topsoils from 1991 to 2001. (Please see page 402 of this volume.)

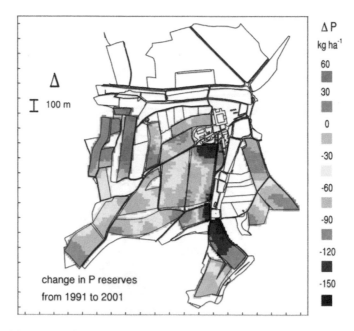

Plate 36 Changes of P reserves in topsoils from 1991 to 2001. (Please see page 403 of this volume.)

Printed and bound by CPI Group (UK) Ltd, Croydon, CR0 4YY

03/10/2024

01040332-0009